SMARTS

BY ELAINE DEWAR

Published by

DEBONAIR [e] BOOKS

SMARTS

computing slime molds,
political primates,
masterful plants,
altruistic robots,
amoeba machines,
high IQ chips, philosophers
of mind using screwdrivers,
signals, spies, the brilliant life
and mysterious death
of Alan Turing,
and the boundary-busting
story of intelligence

SMARTS

computing slime molds, political primates, masterful plants, altruistic robots, amoeba machines, high IQ chips, philosophers of mind using screwdrivers, signals, spies, the brilliant life and mysterious death of Alan Turing, and the boundary-busting story of intelligence

Published by
Debonair[e] Books
an imprint of Debonaire Productions Inc.
www.debonairebooks.com.
Toronto, ON, Canada

Cover and Book Design by Carmen Dunjko,
www.pod10.com.

First Debonair[e] Books trade paperback edition 2015

Library of Congress Cataloging-in-Publication Data is available.

ISBN-10: 0994051204
ISBN-13: 9780994051202

Printed in the United States of America

For
Lilah and Grace

Table of Contents

The Back Story

Every book has a back story, a how-we-got-here narrative. Usually it's found in the last few pages, stuck like an afterthought among the thank yous as if it doesn't matter. Most science books leave the back story out altogether, along with the personal pronoun and thoughts that struck the writer as she was doing her research. Science books are supposed to be succinct and impersonal arrangements of facts (as if any facts are uncontested). Like a long letter to *Nature.*

So this is not a science book. It's an exploratory narrative of the ideas, experiments, and lives of some very unorthodox scientists who are shaping our future in more ways than you know. Some of them think they're making revolutions. Some of them actually are. I tell their stories, along with my own, because I am a journalist who has long understood that stories can convey as much information as essays, and that no story can be well told without a point of view. You need to know mine and how I came to it in order to judge whether my reporting is biased and unfair, or honest and true. If you don't know why I think what I think, you won't know what to make of what I tell you and what I leave out.

To begin at the beginning: this book got its start when I was about nine years old. That's when I first started to ask myself: what *is* intelligence? And who displays it?

These questions kept coming back at me: they made a habit of presenting themselves at critical times in my life. The context was always personal and always pressing. I usually wanted to know: am I smart? Am I smart enough? What about my children?

When I first asked them, I was pretty sure that I was stupid. Everybody who mattered to me seemed to be *so* smart. I wanted that special something that they had, but what was it exactly? Many years went by before I learned that there is a whole literature devoted to these questions which bedeviled and beguiled smart folk such as Aristotle and David Hume, and notably Charles Darwin, his son Francis, and their cousin Francis Galton. Since the nineteenth century, generations of anthropologists, animal behaviorists, ethologists, psychologists, educators, neurologists, and computationalists have spent the best years of their lives trying to figure intelligence out. (Did I mention the microbiologists and the plant people?)

Through most of their ruminations, the focus was on us—*Homo sapiens.* We got our Linnaean name from our capacity to think about and solve problems: intelligence was supposed to be our defining characteristic, the attribute that raises us above all the rest of nature. Even the father of modern computing, Alan Turing, set the mimicry of human intelligence as the goal for any artificial one: it must be able to fool us into thinking it's one of us, he said.

I'd heard of none of the above when these questions first occurred to me. (Well, maybe I'd heard of Aristotle. The Harvard Classics were arrayed like the blue ribbon for Best in Western Culture above the medical texts in my Dad's den). Computers had not yet become our defining technology and no one I knew spoke about Turing or his fantastic story, though he should have been known to every school aged child as the Hero who helped save Britain. When I was in school, only his former colleagues sworn to secrecy forever knew that Turing had broken the Nazi Enigma ciphers by inventing a framework of logic that could be used by machines. His ideas were embodied in the first digital programmable electronic computer (called Colossus) and eventually in the search engines that many years later have made Google a great fortune.

The questions popped up again when I was in my early thirties. I was a first-time mother with a newborn daughter. As I hunched protectively over her miraculous development, watching it unfold with fascination and fear, abject adoration and dread, I was fixated on how smart she would be. Could be. Or might not be. I thought—then—that intelligence is the one quality that makes human lives better, or worse. If stupid, you could be born into heaps of money and lose it all. If smart, you could be born with nothing and surmount all barriers. Both sets of my grandparents had been smart enough to see what was coming and get out of Europe long before the Nazis came hunting Jews like them. As desperately poor homesteaders, they had come to the bitterly cold Canadian prairies at the turn of the last century and prospered. It seemed to me that their intelligence saved them and gave me my life. Smart was good while lack of same meant trouble, even annihilation.

Before I knew it, I had two daughters mysterious and surprising and I really wanted to know for certain if intelligence is innate or learned. By then I had almost convinced myself it is entirely dependent on circumstance, so a suitably doting and hardworking mom could plump it up high in her children like so many feather pillows. But in the long nights of feedings and wet diapers and ear infections and fever dreams, doubts crept in. The experts always nattered on about the importance of inheritance. If it really *is* just a given, I thought, what then? Would my daughters think like me or like my husband? Would one or the other be a math whiz (like him) or math fool (like me), a think-out-of-the-box rhythm freak (him) or a pattern, melody and word nerd (me)? And really, what is intelligence anyway? Is it a single trait or many jammed under a single rubric? Would their minds be made from combinations of the best of us? And/or the worst?

Questions about who besides people might display intelligence also arose after I gave birth to my daughters. I had been seduced by my husband Stephen into writing and story editing for his television nature series. Back then, in the early 1980s, I was a magazine writer and editor. I was also a feminist (still am) and I wanted to keep a sharp divide between his ventures and mine. TV was *his* business. And though script writing paid much better than magazine journalism, the words mattered so much less than the flow of images that I thought it would be very dull. But then he said he couldn't find enough writers who could weave new science into film melodramas. (Was this manipulation? Definitely.) So kicking and complaining, (and yet so gratified), I had jumped in.

The series was called *Lorne Greene's New Wilderness.* Lorne Greene, the Canadian-born actor/announcer who terrified the nation with his WWII news reports (nickname: the Voice of Doom)[1] was then one of the best known stars in the world, especially for his portrayal of Pa Cartwright on the TV western *Bonanza.* He was the lead partner as well as the narrator of the show. His son Charles was Stephen's business partner.

New Wilderness was perfectly timed. Science had been redefining the human role in nature from the biblical (made in God's image with dominion over all) to the post Darwinian (top predator run amok). We had begun to learn that we are much less important than we thought, and much more dangerous than we had ever imagined.

James Lovelock, the renowned British chemist, along with the American biologist Lynn Margulis, had pointed out that land, water, and atmosphere endlessly recycle their chemical products through living things. They suggested that life is the planet's circulatory system, that this coupled system, this biosphere as they called it, also behaves like a living entity in that it is self-regulating, maintaining conditions suitable for itself and for everything alive. Lovelock (upon advice from *Lord of the Flies* author William Golding), called the whole system Gaia[2].

But Lovelock's invention of the Electron Capture Device which discerned industrial contaminants in air and water in concentrations as low as parts per trillion showed that humans were destabilizing Gaia even as we discovered her. As well, ecologists were demonstrating that human development was devouring nature, pushing many species to the brink and beyond.

That's why the promo line for *New Wilderness* was: "on the cutting edge wherever man and nature meet."

It was the great anthropologist and paleontologist Louis Leakey who had kicked off this first major revision of the human story. Interested in the natural history of humanity, he wanted to know when complex social behavior started, and which of our predecessors displayed it first. But old bones and ancient hearths only reveal so much about behavior. Leakey thought we could learn about our own prehistory by studying the great ape cousins from whom we'd diverged only a few million years earlier. He suspected the great apes are much smarter and closer to humans than we like to think.

In 1959, Leakey gathered enough money to send the first of his "angels"—Jane Goodall—into the wild to study the natural behavior of chimpanzees. He thought a woman would succeed where men had failed because women are more observant and patient (women who worked with Leakey *had* to be patient)[3].

Goodall rocked behavioral scientists everywhere when she reported that wild chimpanzees make tools. Until then, toolmaking had been considered to be the objective product of humanity's defining characteristic, our intelligence. She went on to describe chimpanzees as social animals living in complex societies. She documented their power hierarchies, their inter-group warfare, their loves, jealousies, capacity for sharing, even their adoptions, and how they hunted smaller mammals than themselves. Goodall made the chimpanzees of Gombe famous, naming each one (to the horror of conservative academics), and writing their life stories as if they were human.

Dian Fossey came next, studying the mountain gorillas in the Virunga National Park in the Congo, near the border with Rwanda, moving later to the Rwandan side. She also gave the gorillas names and made them famous. Just as we were making the *New Wilderness* series, Fossey was murdered while fighting to keep those gorillas safe from poachers and so-called conservation tourism. Even as she died, Leakey's third angel, Canadian Birute Galdikas, was beginning to report on the extremely clever orangutans of Borneo. Their existence was threatened by the cutting and burning of the rainforests they relied upon.

By that time too, other scientists were publishing on intelligent behavior displayed by other species that they were studying in the field. They often shot films to document their work. *New Wilderness* bought the rights to show them and we turned their pictures into stories.

The first episode was about the orcas (killer whales) of Argentina who teach their young to hunt. We had footage showing an orca pod hunting off the coast of the province of Chubut. One of the mothers was teaching her pup how to kill young sea lions. Yet teaching the young was supposed to be a humans-only trick.

Language-like communication was also displayed by the animal heroes of the *New Wilderness* shows. Individual orca, gray and humpback whales had been identified and followed by scientists who recorded their whistles and songs which were clearly meant to communicate something, and which varied between individuals and groups just as languages and dialects vary among human groups. Similarly, it had been known for many decades that bees use a dance language to indicate locations for food and places for nests, and that ants use a chemical signaling system to pass information along. We did several shows about both. Goodall's former husband, Baron Hugo van Lawick, had footage showing wild African dogs hunting cooperatively. One pack he'd followed for years even knew how to kill zebra. The dogs would chase around after a herd until one of them got close enough to leap and grab a zebra's upper lip, a ferrier's trick to make horses stand still. That zebra would stand immobile long enough for the pack to gut it. Van Lawick and J.R. Malcolm reported that the group passed the skill on from one generation to the next, which amounted to cultural transmission, also supposed to be a humans-only intelligent behavior.

The idea that animals might be smart was not new: but it had been shoved off the table of science many years earlier. George Romanes (born in Canada but educated in Britain) had published a book[4] in 1882 in which he tried to demonstrate that intelligence is displayed by a wide range of species. Romanes' methods were similar to those of his mentor, Charles Darwin. Both collected reports from various correspondents that they deemed to be reliable. Romanes' book was stuffed with anecdotes sent to him by people bowled over by the amazingly intelligent behavior of their dogs, parrots, monkeys, etc.

But psychologists and animal behaviorists after Romanes argued that these field reports could not be trusted. It was far too easy for untrained observers to be fooled into seeing what they wanted to see. Only controlled experiments in labs could actually tell us something reliable. Observations in the field made good "anecdotes," but not good science.

I thought the films we made added up to a lot more than mere anecdotes and that observation is the beginning of any good science, but when the series wrapped, I moved on to other work, specifically a magazine article about a very smart and powerful human family who weren't too keen on having their intelligent behavior described in print. There would be tears and woe when the story was published.[5] My questions about intelligence soon narrowed down to whether I was smart enough to withstand the incessant barrages of due process dreamed up by a fleet of lawyers determined to shut me up.

I didn't think computers had anything to do with intelligence when I bought my first, a beige IBM clone that required a desk all its own. My husband Stephen (always an early adopter) had already blown through two generations of desktops, starting with a portable that had a screen so small it could barely display twenty-five words. Everything about computers was Greek to me yet the magazine editors I worked for insisted that I must hand in my stories on floppy discs to save them time and money. Whenever I moaned and complained, Stephen would mutter about the twentieth century I'd missed.

I thought my computer made my work suffer. It certainly slowed me down. I had to learn all the operations it could do that were supposed to save me time, but which actually took forever. All the to-ing and fro-ing between the manual and the keyboard as I moved text around made me forget the direction of any argument I was hatching, never mind the over-all story line. My so-called spelling errors— putting a u in behaviour because I'm a Canadian—would be red lined because my software was American. My so-called grammatical errors—style choices, dammit—got underlined in green. And then there were the total disasters when I forgot to click save and a day's work vanished into the digital Wherever.

I was accused of being a Luddite, but that was not true. I did not think my job was being taken over by machines. (That will come soon enough, but back then narrative software wasn't even on the horizon.) I was just unnerved by the creepy liveliness of these clever devices that were worming their way into my life. First came the desktop that made comments on my work; then came the car that warned in a neutered voice about the door being ajar whenever we opened one. ("No it's *not,*" my kids would chorus from the back seat, "a door is a *door.*") And worst of all: the annoying quasi-musical bleet bleet issuing from the bowels of my purse which was the home of the new cell phone we had to have so as to stay connected (even when we were trying to escape).

Yet instead of probing the meaning of this new kind of smartness, my questions about intelligence remained fixed on the potential of the two small humans who depended on me. Should I bribe them with cookies to get them to practice their scales? How much exercise fuels a better brain? How many languages? What about a dog? (Yes, lots, three, and yes.)

Slide the timeline cursor forward—but before Facebook, and before YouTube, long before allegedly sane people began to upload videos of themselves doing things they will forever regret and can never escape. It was the early 1990s. The Internet was still new and it was dial-up slow. I had begun to use it as a way to get around the phone company's absurd long distance rates and the incredible sloth of snail mail. There were databases out there, and chat rooms and websites, but nothing so useful as Wikipedia. Google wasn't even a gleam in the eyes of Larry and Sergey.

I acquired the teensiest glimpse of the age-of-intelligent-machines-to-come when I was asked by a big communications company to write a pamphlet to be handed around to certain "elites." (By elites they meant the politicians and regulators who might—if persuaded—protect them from their competitors.) I had to explain all about the fibre optic information superhighway that was about to materialize in every home. Huge data capacity at tiny cost, digital everything from soup to nuts, vast two-way traffic, convergence between all media, yada yada, that's what the male executives crowed in my ear. For one shining moment I was breathless, transfixed by the possibilities. We could all be publishers, we could all be producers, the means of distribution would be cheap as dirt, almost free.

You mean, I asked, we'll all be able to make movies on our computers and send them out into the world?

"Don't be absurd," sneered one of the VPs. His drumming fingers said 'who *hired* this woman?' "Who in hell would want to watch amateur crapola?"

The Millennium came and went. My kids were in university or soon to be. They had become smart, talented women who made music without the lure of cookies. But my mother and father were shrinking into retirement. Their behavior was subtly different than it had been when they were younger, when they could do anything at all if they put their minds to it. Their minds were, well, different. And I began to notice a change in mine: words had begun to slip from working memory (coincidental with the arrival there of computer catchphrases). There were moments, hard moments, when I would sit at my computer searching for a simple word. Like *hose.* Like *kettle.* I was perfectly able to see the image of the thing I wanted to name yet I was unable to say or write it. And then the word would just emerge from my moving fingers before I could form it in my mind. Was I becoming stupider?

At about that time—surprise!—I became interested in the new biology. The idea of living well, and forever, had a certain appeal. I wanted to know about genomes, stem cells, chimaeras, and especially clones. I was amazed to find a revival of interest in the ideas of Jean-Baptiste Lamarck who had argued that the environment induces changes in organisms which can be inherited. There was evidence that he'd been right. It was all part of a new understanding that living things are shaped in dynamic interactions with the world, we're not just automata mindlessly following chemical orders issued by genes. In Vancouver I was introduced to Don Riddle, a geneticist who studied a certain worm. The use of *Caenorhabditis elegans* (*C. elegans*) for parsing how genes guide development had been pioneered by Sydney Brenner, who would soon earn a Nobel for this work. Riddle studied the way *C. elegans'* genes interact with the environment at critical stages of its early life.

Over a pho soup lunch, Riddle explained that this worm, which has a grand total of 959 cells when fully developed, makes a decision as a *larva* (in the first eleven hours after hatching) about whether to continue to develop or to hibernate. These larvae sense the environment around them, particularly the availability of food (bacteria) versus the numbers of their competitors (other larvae). Too little food? They hibernate and wait for a better day. Too many competitors? Same judgement. Everything just right? They develop.

I sat there calmly spooning in the soup and writing his words in my notebook until they drifted in to that part of my brain where meaning erupts. I almost choked. I remember thinking: I've heard him wrong, how can a wee beastie with only a few neurons (humans have about 85 billion) make decisions? He seemed to be saying that these tiny larvae intelligently solve a life-and-death problem.[6]

How does that work, I demanded? Can they calculate? Think?

His answer was disturbing: it was all about chemistry and thresholds—a pheromone rising to a certain level and triggering a molecular switch that turns a gene on, or off. And there was something about another pathway involving molecules that interlock. It was all automatic, pre-programmed, a chemical Rube Goldberg machine. And yet the larvae adapted their behavior to a changing reality, a signpost of intelligence. But also, no, as in no arithmetic.

So it made me wonder: how do *I* think? Is it also just a matter of hormones and pathways? Am I just a chemical machine that senses and reacts? And if the chemistry changes, what then?

I remembered the awful struggle to write a simple book review article just two months after giving birth to my eldest daughter. I remembered being absurdly happy, yet finding it to be the hardest work I had ever done. My brain seemed to belong to someone else. Writing one coherent sentence that led to the next and the next was almost impossible. It was as if the lactating hormones had changed the very nature of what could go on inside my head. At a certain point, I panicked: what if this turned into a permanent state? Yet as breastfeeding wore down, thinking returned to normal. Or what seemed to be normal. Was I the same as before? Or different? What is normal? I really couldn't say. Then I tried not to think about any of it ever again (though I gave myself six months off after my second daughter's birth before tackling something serious).

Don Riddle's clever worm raised those questions again: what is intelligence? Is it just a product of reactive biochemistry? Is that all? And if it is, can just about anything be smart?

Slide the cursor forward to 2004. A news story caught my eye. It was about the discovery of ancient, human-ish bones, including a fairly complete female skeleton, found deep in a cave on the Indonesian island of Flores. These 'people' had extremely long arms, tiny heads and their remains were found surrounded by a litter of stone tools. They had been dubbed hobbits by the *Nature* editor who ran the report as a cover story, getting ahead of Fleet Street editors who would try to tag the remains their way, or so he later explained it. I thought the nickname was meant to attract Fleet Street. If so, it definitely worked. The story and the follow-ups bounced around the world's press for years.

The scientists who found the remains called them *Homo floresiensis*, meaning they are of our genus (*Homo*—still defined as tool makers in spite of the fact that others make tools) but of a different species. They were like us, yet not us, and weirdly they had been living among us until—here was the breaking news—around 18,000 years ago. Anthropological dogma had it that we modern humans replaced the last remaining other species of *Homo*, the Neanderthals of Europe, 30,000 years ago. We were supposed to have been alone ever since, the last of the *Homo* lineage surmounting all of nature because we are so smart.

I watched the news stories proliferate via Google (which had become a habit and also a public company) as a vicious tug of war erupted over the hobbit bones. They had been snatched from the control of Australian archaeologist Michael Morwood (who'd lead the dig) by one Teuku Jacob, the leading figure in physical anthropology in Indonesia. He had the bones taken to his own lab and denied access to the Australians (though he called other experts in to look at them), until he was finished examining them himself. He soon announced that Peter Brown, the Australian physical anthropologist who'd claimed they represent a previously unknown *Homo* species, had got it all wrong. These were just the bones of a modern pygmy human distorted by defects commonly found on Flores, specifically a form of microcephaly (tiny head syndrome). Jacob was supported in his opinion by an American and two other Australians, one a fierce enemy of Peter Brown.

Out flowed a river of accusations. People told me that Jacob was jealous because he didn't get to measure the remains first; that Jacob cracked a mandible and the pelvis and made an improper mould of the impossibly tiny skull. (The size of a small grapefruit! Bones the consistency of sponge!! Damaged it in the process!!!).

It was the usual paleo *sturm und drang*.

Paleoanthropologists have been known to squabble like hyenas over the tiniest bits of bone upon which large theoretical empires are then built. It is a case of making much more out of much less. But I followed the story anyway. I was drawn by the mystery of how these creatures had gotten to Flores in the first place. Flores is on the far side of the Wallace Line[7] which physically divides Asia from Australia and New Zealand. The Line is a water barrier comprising a treacherous sea with dangerous currents which is why many animal species native to the Australian region are not found in Asia. The first modern humans got to Australia from Asia by boat or raft about 50,000 years ago. But the hobbits had probably been on Flores before that, perhaps descended from *Homo erectus* long before modern humans appeared in Africa, let alone Asia.

And yet it seemed unlikely that the hobbit female was a member of the *erectus* species. The top of her tiny head had barely reached three feet, while *erectus* had been much taller. Her skull was less than 420 cubic centimeters in volume, versus *erectus'* which ranged from 800 to 1100 ccs. Adult modern human brains are 1500 cc's and up. Could a creature with a brain the size of a chimpanzee's build a boat or a raft sturdy enough to cross the Wallace Line? Could a human with a brain so very small be that intelligent?

The tools found near the bones said the hobbit had been smart. There were two kinds on the site. Large, crude tools were found in a lower layer. They were like those found on *erectus* digs. But there were also micro tools, tiny blades that looked to me like some found on sites in the Arctic. I showed the *Nature* article to an expert on Asian archaeology at Toronto's Royal Ontario Museum. He thought they were similar to the micro tools found on ancient sites in China—made by modern people with modern brains.

There is a story in this, I thought. If you get it, you might find out what intelligence is made of and who displays it.

I got on a plane and flew to London.

On a dark January afternoon, I sat in a jammed lecture hall in the basement of University College London's Cruciform Building waiting for a talk on the hobbits to begin. I had noticed Joseph Lister's profile on a memorial plaque as I made my way down the stairs. Lister invented ways to clean wounds and sterilize instruments for surgery which saved the lives of millions. It was good to be reminded that science can be balm as well as sword. I'd spent the earlier part of the day staring at the measurement devices and eugenics posters from Francis Galton's former lab, and reading some of his letters in a University College London archive. Why? I thought the story should really start with him and his study of human intelligence.

But I wanted to wash out my brain with soap to get rid of the things he'd written to his mother while on a Royal Geographic Society mapping expedition of southern Africa (now Namibia). Galton, a Victorian-era polymath of distinction, was in his late twenties, a young British gentleman on the loose, when he made that trip. His grandfather was Erasmus Darwin: like his older cousin Charles, he had tried medical school, disliked it, and then gone to Cambridge. After studying mathematics, he became a traveler and had already ventured up the Nile to Sudan before he went to Namibia. His letters from there were rife with vile, racist language and opinions. He mainly despised the societies he encountered and the inter-ethnic slaughters he witnessed. He casually used the N word, but that was by no means the worst of it.[8] He assumed (along with physical anthropologists), that humans are divided into races arranged according to a hierarchy of capacities—from the smartest and most moral on top down to those beneath contempt at the bottom. He assumed that the best white Anglo-Saxons defined the pinnacle. The blacks of Africa in his view were mentally inferior and incapable of building any sort of civilization. And these opinions were not sloughed off with his youth. Twenty years after his Namibian sojourn, he famously suggested in a letter to *The Times* that the Chinese, though downtrodden, were highly capable. He argued they should be encouraged to colonize empty East Africa and he hoped they would eventually replace (did he mean kill off or just starve out?) the blacks of the entire continent.[9]

After his cousin Charles published *On the Origin of Species by Means of Natural Selection* in 1859, Galton, who was thrilled by Darwin's ideas, plunged into the study of intelligence, especially genius. Darwin was interested in intelligence too and ran many experiments to study it in many different species, often with the help of his son, Francis. But Galton focused on the intelligence of humans. What is genius and who has it, he asked. Is it inherited or the product of experience? He coined the phrase nature versus nurture. He believed human genius is inherited.

He thought he could prove it with twin studies. He asked: would the intelligence of twins separated at birth differ if they were raised differently? Galton assumed intelligence would show itself equally in twins, regardless of their circumstances. He was also interested in developing proxy tests for intelligence. He thought the speed of reaction times mattered, convinced that the fastest reactor would also be the smartest person. He had theories about how hair colors and limb measurements presage certain behaviors, illnesses, even criminality. The artifacts from his study included some measurement devices and hair color samples he'd collected. He amassed reams of data about the human form, and, as a good mathematician, invented new ways to extract meaning from it. He introduced the concepts of correlation, standard deviation, and regression to the mean which are fundamental to modern statistics.

While Galton loved Darwin's theory, he didn't like what he thought it implied for British society. Instead of leaving the results of human breeding to random nature and the invisible hand of selection, Galton wanted to take control of it. (It must be mentioned that Galton, though married, remained childless.) He invented a new science—eugenics—to save the nation from itself. The purpose of eugenics was to get more children from the brightest and best who, Galton found, didn't have as many children as the dullest and the worst. He assumed that if this situation remained unchecked, the worst would outbreed the best, lowering the overall quality of the race (though Darwinism predicts the opposite). Galton argued that the resultant misery of these lesser breeds was both cruel and unnecessary. Science could show the way to something better. It would be so much smarter to prevent the births of idiots instead of bringing them into the world to suffer. A sufficient number of appropriate measurements correlated with life histories would allow for predictions. Measurements could sort wheat from chaff, intelligent prospective parents from the weak-minded. He wanted everyone to be measured, but had to settle for volunteers. (He got about 9,000 people, mainly from the middle classes, who paid for the privilege.) His program aimed at persuading those who didn't measure up to refrain from having kids.

But the eugenicists who followed Galton were not content with persuasion. They lobbied politicians in the UK, Canada, the United States, and Germany to pass laws to enable the forced sterilization of the mentally deficient. The word "hygiene" appeared frequently in their writings as if intelligence is clean and moral, while stupidity is dirty and evil. Many American states, and some Canadian provinces, passed such laws and sterilized people against their will. German eugenic scientists took these ideas to a whole new level. Hitler's program to cleanse the German world of undesirables (including Jews, Roma, Slavs, the mentally ill, the mentally deficient, the developmentally imperfect), so as to protect Aryans from pollution, was firmly rooted in Galton's eugenics. In other words, Galton's science of human intelligence had paved the road to Auschwitz.

Physical anthropologist Peter Brown began his hobbit talk in the lecture hall.

I'd already interviewed Brown at London's Museum of Natural History where he was studying ancient remains from southern Asia. I wondered if he would tell this audience what he had whispered to me. He'd said the arms and shoulder bones of *Homo floresiensis* reminded him more of the *australopithecines* than *homo.* That the hobbit might be a tiny-brained *australopithecine* was such a radical notion he was afraid to even hint at it until he'd studied many more old bones from south Asia, bones stashed in places like the Museum of Natural History. It had long been assumed that *australopithecines* (such as Lucy, unearthed by Donald Johanson in Ethiopia) last walked the African earth more than a million years ago. They had very small brains (about 380 to 600 cc's). But if the hobbits had descended from *australopithecines,* how could they have found their way to Flores? How could they have made either type of tool? Early anthropologists had been convinced that intelligence is a function of brain size. (That's why some nineteenth century American anthropologists measured the circumferences of native peoples' skulls—to try to prove they were smaller and therefore inferior.) And the idea that size matters still resonated.

To his audience, Brown mentioned the tiny size of the skeleton, the length and shape of its limbs, but then darted back to safer territory, the idea that the hobbits might be an early form of *erectus* that had become dwarf size on Flores. He did not deal with the issue of intelligence other than to say that endocasts had been made of the skull which would tell us about the organization of the brain. That work had not yet been published.

So I got on another plane, this time to Milwaukee. At a conference on physical anthropology, I had lunch with a woman named Dean Falk who had studied the hobbit skull endocasts. At nearby tables, the warring proponents of the two competing theories about the rise of modern humans ignored each other while holding forth to their devotees.[10]

Falk studies the markings imprinted by brains on the ancient skulls that once cradled them. She infers from these marks what the shape and organization of the brain must have been. After doing various computer analyses of the hobbit endocasts and comparing them to endocasts of a pygmy, a microcephalic, a modern human, and primates' skulls, she had come to a few conclusions. First, contrary to Teuku Jacob, this hobbit skull did not belong to a modern pgymy deformed by microcephaly. It might have been a dwarf version of something, but not a modern human. [11] *Floresiensis'* brain was small, yes, but it was also organized differently from the brains of other primates or modern humans. Despite its size, it showed clear signs of capacity for higher cognitive functions.[12]

Organization matters when it comes to intelligence, she said, not just size.

I remember thinking: so maybe one can be intelligent with very few neurons so long as they're packed the right way. It's not just about the chemistry, it's about the arrangements.

Yet I put the hobbit story aside: I didn't think it would help me learn what I wanted to know. There was too much contention, too little light.

Move the time cursor forward to early summer 2007. By then, the word smart had begun to attach itself like a trademark to cars that sense danger and bleep warnings, household systems that turn on or off at the sound of a hand-clap. There were rumors that Google had a secret project to build a car so smart that it would be able to drive itself. Google already kept track of my searches and was starting to offer smart corrections, as in, 'did you mean?...' as if the Google servers knew better than I did just what I wanted to know. What they were learning about me and my interests made me nervous, but not nervous enough to stop googling. The meaning of smartness, of intelligence, seemed to be morphing to include this dynamic give and take between a person and a machine.

I was puttering in my office doing something I can no longer recall when my desktop made a gulping noise heralding mail. A query via my publisher: did I want to be contacted by two women who claimed to know me?

To be available, or not to be, was still a valid question in 2007, at least for most adults. It had never been a question for journalists. It is our duty to take every call and more than that, it's useful. How could whistleblowers send us brown envelopes stuffed with scandal that we would share with the world if our phone numbers weren't in the White Pages? Yet by the summer of 2007, smart devices and things like Facebook were beginning to change even that basic principle of my trade. Soon, scandal in the form of screen shots of politicians' penises and videos of police beatings and the outrages of war would zip around the net without journalistic intervention thanks to new kinds of phones that did everything except make lunch. Most of the reporting skills I had acquired in my youth were about to become obsolete.

I had prided myself on my capacity to find out about people. Finding out was fun but it was also hard work requiring knowledge and guile, not just basic cognition (the current term for thought). But by 2007, rivers of data had begun to pour off all of us even as we emailed, Facebooked, phoned, searched, read, bought, sold, even as we walked down the street with nothing in our pockets but a locked cellphone. We were each becoming well known to everyone, but especially to intelligent machines that can predict what we want and serve it up on a digital spoon. To the young, this all seems both eternal and good. But when I was young, even science fiction writers hadn't imagined it.

The two women trying to contact me turned out to be Kathy and Marion, old friends that I had not seen or heard from in many years. Both of them had first appeared in my life on the first day of grade five, in a certain class, in a certain public school, in a certain prairie city. Kathy had trailed a fine aroma of horse around the classroom because she was a rider. I had seen her last many years earlier in Calgary when I was there on a story. She had appeared in my hotel lobby wearing a perfectly cut canary yellow coat, tall, dark and so chic I was astonished. Marion was the feistiest, sweetest, and most athletic girl in the class. I had seen her last when we were barely out of our teens and I still went by my born-with last name. Marion had drifted up to the door of my fifth floor walk-up in Toronto, her bare feet black with street grime, her thick brown hair swinging at her waist. A hippie is what she was then, pure but not simple.

There followed a back-and-forth of phone calls and emails. Marion was teaching ESL at a prairie university. She had five kids. She'd once hauled them by car all the way from her former husband's First Nation reserve down to Belize and back again. Kathy, still in Calgary, was a long distance traveler too, but for fun not life. Though she had started in journalism, she'd become a go-getter in the oil and gas business and had a son.

We're all going to have a big birthday in 2008, one of them said. What if we get the whole gang together for a reunion?

There was no mistaking which gang they meant. They were talking about our public school class, though the three of us had gone to the same high school, too. Our public school class was special. We were special. Or so we were told. That class was the reason why these questions—what is intelligence and who displays it— had grabbed hold of me when I was young.

Our phone calls and emails unleashed a flood of memories. The questions came roaring back. They led me by the nose from Montreal to Florence, to Lausanne, London, Lewes, Edinburgh, Florida, Gulfport, Chicago, Vancouver, and to a little city down the road from Toronto called Waterloo. I delved into the brilliance of the octopus, how squids speak with their skins, how slime molds run mazes. I met those who teach plants to distinguish the annoying from the dangerous and remember it well, who are building robots to mimic living things and who are turning living things into robots. I learned that 'evolved' brainless machines are capable of altruism and communication, not to mention that it is possible to do philosophy of mind with a screwdriver. I was even introduced to an intelligent machine that scores right up there with university students on an IQ test.

What I heard and saw has changed how I think about the world and how I move through it. But never mind about me. What matters more is this: some very smart human beings have learned how to machine intelligence.

And that will change everything for everyone.

Part 1
The Second Copernican Revolution

1

The Special Class

I am standing on the stone terrace of the university faculty club in the prairie city where I grew up. It is late afternoon in late spring, and the sun filters through the pines and spruce, painting zebra stripes on everything. Marion, Kathy and I, along with some of our former classmates, have gathered here for part one of our reunion. We are not at full strength. Some of us couldn't be found, some had other plans, some won't show until tomorrow, one said he wouldn't be caught dead at an event like this. Two, who would have come if they were alive, died eight years earlier at exactly this time of year. These women—my friends Kathy and Sandy—both came to Toronto to study a few years after I did, and, like me, stayed on. They became lawyers, acquired husbands, had kids, did work of value to the community and then, suddenly, they were gone.

The mosquitoes buzz bomb my ears as we talk plans. There will be a dinner, a group hike to a classmate's house for lunch tomorrow, another dinner downtown. Drinks are circulating, and one of us leans against a wall smoking something interesting, his eyes squinched and red. I think I can smell the river swirling its way under the bridges and over the dam.

Every one of us at this reunion has been certified smart. This is like being certified sane, and about as meaningful. In grades three and four, in the different schools we then attended, all of us were subjected to IQ tests which purported to measure our intelligence. This eugenical method to differentiate between very smart and very dull human children is still used in schools and psychologists'offices in most countries of the world. In this town, doing very well on the tests got you (still gets you) labeled as special and an invitation to join a segregated class with an enriched curriculum commensurate with your bright prospects. Doing very badly got you (still gets you) labeled special and an invitation to attend a segregated school where prospects are bleak, the downside of up. For many years in Alberta and in the US, doing extremely badly could also result in incarceration and involuntary sterilization. In Ontario, any found wanting were sent to isolated residential 'training schools' where they were barely educated at all. Instead they were set to work like slaves with little payment or none at all, abused, their diseases left untreated. Many ended up in unmarked graves. These 'schools' weren't finally shut down until 2009. Class action lawsuits had begun to wind their way through the Ontario courts. I knew the lawyers handling them.

We spent grades five through eight together in the same room in a very nice public school because our IQ tests said we were intellectually *gifted.* I use italics because I want to convey my skepticism of IQ tests (there are several types) as measures of inherited, innate, un-teachable traits associated with intellectual merit, which is what they purport to do. They do measure abilities valued in school, which is a form of intelligence, but are these abilities innate and inherited?

Back then, I didn't know enough to be skeptical. Being classified as smart mattered to me. Until that point, I had been certain that I was stupid.

I can't believe how weird this reunion has turned out to be. Public school faces peer out at me from under wrinkles and sags. It's like seeing double—as we were, as we are this day. Marion has short hair now and is as plump as I am. Yet her freckles and bright eyes are as I remembered them. Kathy's formerly dark hair is *The Devil Wore Prada* white. There is also this group thing going on, as it once did. Someone says something interesting, and everyone moves in tight around that person. We used to do that on the playground, our feet synchronizing as we headed in a direction chosen by some unspoken process, our individual selves merging into something more than the sum of us, something almost intelligent in its own right.

I had been reading about this phenomenon: it is called swarming when bees do it but apparently the same sort of thing can be seen across the kingdoms of life (even among the roots of plants). Swarms operate under simple rules. Individuals stay the same distance from those on the left and right, in front and behind, above and below. An entity assembles itself from all these individuals working together, and this persists until whatever it came together for is finished and the swarm breaks up. There are looser forms—schools and shoals. Human swarm behavior can be very stupid (consider the mob) or very smart. Groups can exhibit a collective intelligence that is better at solving certain problems than any expert individual. (Francis Galton was the first to demonstrate this.)[13] We make art from our ability to synchronize: in symphonies, marching bands, ballets, chorus lines, choirs. At the microscopic scale, each of us is also a swarm. Every living multi-celled entity is made of different kinds of cells that must coordinate and cooperate. And we also cooperate with other swarming species: each of us carries a vast array of bacteria within us and on our skin that help us stay alive.

I keep my journalist's wits about me and ask everybody the usual questions. What do you do? What did you study? What about your family? Are you happy? Almost everybody has a degree, sometimes three, yet I am struck by how very ordinary we are. No one has achieved outrageous fame and fortune. Many of us are communicators: one makes TV documentaries, another writes novels, another writes nonfiction, another does PR and there are several former journalists. There's also a teacher, a medical administrator, one man who runs a recycling operation manned by the disabled, and a musician/professor. One guy brags about getting into the domain name game early and shows pictures of his boy toys. Another is a property developer who mentions former wives. I had expected more doctors and lawyers. It appears that none of us ever ran for public office, though one married a politician and another, who calls himself a hack, has advised the Liberals in Alberta, which makes him more like Sancho Panza. Then there's Alexander, who does genealogies for First Nations and let's not forget Jack, a high ranking civil servant. Jack is the biggest surprise. We'd all expected him to be a nerdy scientist in a white coat with pens in his shirt pocket because he had pens in his shirt pocket when he appeared on day one of class and kept them there throughout. Yet here he is, an historian by training, and a wiry, aging double for Hugh Grant in *Notting Hill.* Every female turns to gawk, synchronously.

I had been terrified/ thrilled about starting grade one where reading was the center of the action. I was sure I wouldn't be able to learn how to do it. I don't know why I thought I couldn't, but maybe it was because everyone around me was a reader, my mother, my sister, our housekeeper (a deeply committed Christian who read from the Bible every morning, but was so scrupulous that when I, a nice Jewish girl, crawled in bed with her she only read to me from the Old Testament lest she seem to be proselytizing me.) My Dad read comic books (especially Donald Duck) with his late night snack, but he also spent hours reviewing dense medical texts with mesmerizingly hideous illustrations and photos of surgical techniques. Somewhere along the line he'd also memorized many literary passages which he spat out on (in) appropriate occasions. "A horse," he'd shout on his way to the car. "My kingdom for a horse!"

Though I quickly learned to read (thanks to phonics) I was sure my stupidity had just gone underground and would reappear. I was right about that because my grade two teacher saw it plain as day. Q.E.D., stupid as charged.

Grade three. My new teacher was kind, beautiful, pregnant and gave me a long line of As. I went to her desk and told her she'd given me someone else's report card. No mistake she said, looking at me strangely. But I knew she was wrong. By then I was taking piano lessons and music theory made no sense to me at all. My music teacher would thumb through my theory homework like the queen reviewing obscene correspondence. She'd say: what's the matter with you, it's just fractions, you learn them in school, are you stupid?

Yes, but not *always*, I wanted to shout.

The grade three teacher sent us down the hall for some tests. There were things to read and questions to answer with a clock ticking. No one said why. Yet I remembered those tests the next fall when my grade four teacher sent me alone to an unused classroom where a stranger, a man, presided over quizzes involving blocks, patterns, mazes. What was it all about? He wouldn't say.

So we come to that special day in 1958.

The world hovered between the dangerous brightness of winter's end and the sweet softness of a prairie spring. Birds were everywhere, frantic and raucous. The sun was hot on my face as I ran through gutters where slush stopped the drains and made lakes.

My parents were in the kitchen when I got to the house. It must have been a weekend because my Dad was rarely home before we went to bed during the week. He was toying with a letter. They both had this look, very grave, as if I had done something very bad, punishment to follow. I'd seen that look on my father after he'd surgically investigated a patient only to find a cancer so widespread it was inoperable. The death sentence face.

I sat down.

There's a new class being formed, said my mother. It's for kids like you.

I was sure she meant stupid kids like me. I knew my grade four teacher had doubts about me regarding long division. My Hebrew school teacher had threatened to hold me back a year. I often took refuge in asthma supposedly triggered by allergies, thus Mittens the cat had been dispatched to the gas chamber. My tonsils had been sliced out in grade two as a preventive, but their removal hadn't prevented anything, least of all wheezing. I had learned how to make myself wheeze so I could stay home and finish any story I was reading, as opposed to going to school and doing long division.

It's an experiment, said my Dad. You don't have to go if you don't want to.

The word experiment was clever bait. I knew science dealt in experiments and I was a devotee of science, like everybody in the West after the wicked Soviets launched Sputnik. That's why, though I was sure I was stupid, I was still determined to be a doctor like my Dad. I could already rattle off the names of the tests he usually ordered when the hospital emergency department called for instructions about a patient. I had a small microscope for looking at tiny things. I had a chemistry set with test tubes and a *faux* Bunsen burner and a supply of chemicals in crystal form which I burned to brown smudges. I didn't want to be sent to some awful place with the other stupid kids. However, I knew better than to shout a flat no. Subtlety was called for. My Dad liked sports. He liked me to like sports.

Go where, I asked, and what about the school teams?

Same school, said my Mom. I could see she was eager for me to say yes. So you can keep your teams and your friends too, she said. Some of them will be in it anyway. We've been to the meeting. We know who's involved.

You don't have to go, said my Dad, which meant they'd been fighting about it. It's an experimental enrichment program. No skipping. I don't believe in kids being skipped, he said.

What was he talking about? Skipping was for smart kids. Three friends of mine had already skipped. They were smart, not me.

But I'm not smart, I said, thinking ah ha, gotcha.

The IQ test says you are, said my Dad, to my shock.

A test had to be right, I thought. Didn't it? So I said I would try.

I have our class photo from grade six pinned to my office corkboard. I am the tallest. I am almost as tall as the teacher, Mr. N. Kathy and Marion are there along with Marian and the other Elaine, plus the shy Helen (not shy now) and the even shyer Sandy who soon developed diabetes and had to shoot up so we called her The Crank. The other Kathy is the second tallest in the picture: she had been a friend since our sandbox days. Her cousin, who lived down the street from me, was my best friend. There were more boys than girls. Michael and Alan I knew from Hebrew School. A boy I will call Cameron and tiny David had been friends since grade one.

When we pooled the bits of information we'd picked up from our parents, they generated more questions than answers. Nobody knew what an IQ was, only what it stood for: intelligence quotient. Most of us believed that this class was some kind of experiment (and still believed that five decades later). A couple of the boys thought it had to do with a brand new lab that we were supposed to get, but no lab materialized. Our enrichment entailed a conversational French teacher, a music teacher, and English grammar lessons from the dry lips of the Principal himself. Professors from the university occasionally came to demonstrate their specialties. More to the point, we had three male teachers out of four.

Sexism was so ubiquitous in those days: it was like air, like water, it was everywhere and unquestionable so it had not yet acquired its name. Public school teachers were almost always women, poorly paid. After all, why would anyone pay real money to *women*? Any male who became a public school teacher was quickly hauled up the administrative ladder where he could earn more, because he was a man. It followed that since *men* taught us in the first two years, we were important. In fact they said so, as in, much has been given to you, so much will be expected: maybe one of you will be Prime Minister one day. But they were really talking to the boys. Ellen Fairclough had just been appointed to the federal cabinet, the very first woman to be so honored. The notion of a female Prime Minister was laughable.

We knew everyone had an IQ number—just a single number—which was said to represent how smart we were. We believed that our number wouldn't change, and would define our futures. They were secrets never to be revealed to us, but someone heard that you had to have an IQ over 130 just to get into the class. (This at least was accurate: gifted is still defined as a test score that is two standard deviations above the norm—100—when one standard deviation equals fifteen IQ points. Some school systems use 125 as the cutoff, ours used 130. In my hometown now, 120 is the real cut off as the special classes held in one undesirable neighborhood attract too few children.[14])

Since everybody had a number, there was a hierarchy of smartness, so one of us was the smartest of all. We heard that someone had an IQ over 180. Everybody knew that famous geniuses such as Albert Einstein behaved oddly, such as going around in public without socks, so the smartest in our class had to be Alexander, the kid from Germany. Like Einstein, he spoke with an accent. He also ate cake for lunch and sat at his desk drawing landscapes full of rounded hills rendered in colors of fire as if his imagination declared it forever autumn. Sometimes these hills were peopled with Roman soldiers in full regalia. All he wanted to do was draw and that's all he did do and nobody could make him do anything he didn't want to do, not even come to school on time.

I was sure I had gotten in with the lowest number possible: how else to explain that I continued to be stupid more often than smart? My Hebrew School teacher had to come to our house at the end of the summer to tutor me. My former piano teacher who had entered me in the Kiwanis Music Festival and planned for me to play before her fellow music teachers threw up her hands in fury when I said I was going to play softball instead. Then she fired me. My new piano teacher, a concert pianist during the 'season,' ranted on about the brilliance of another student named Martine as he slipped his gloved hands under his armpits to keep them safe from all dings and scrapes. He was shocked when I did well.

No surprise, then, that questions about the nature of intelligence came alive for me in that classroom.

Kind reader, you may be perplexed that I have not yet defined intelligence. That is because its meaning has been morphing since the days of Aristotle. You would think that those who use IQ tests to make judgments about the potential of our children would at least be able to define what it is that they measure, but as you will see, that is not the case. If you must have a definition to work with, here are a few plucked from *Webster's New World Dictionary*—because I have the book to hand, not because they shed any light:

"**intelligence**

1. a) the ability to learn or understand from experience; ability to acquire and retain knowledge; mental ability
   b) the ability to respond quickly and successfully to a new situation; use of the faculty of reason in solving problems, directing conduct etc., effectively
   c) *Pschol,* measured success in using these abilities to do certain tasks
   d) generally, any degree of keenness of mind, cleverness, shrewdness, etc.,

2. news or information
3. a) the gathering of secret information as for military or police purposes
   b) the persons or agency employed at this
4. intelligence personified; an intelligent spirit or being"

And further:

> **"intelligence quotient**
> A number indicating a person's level of intelligence; it is the mental age (as shown by intelligence tests) multiplied by 100 and divided by the chronological age"

And also:

> **"intelligence test**
> standardized series of problems progressively graded in difficulty, intended to test the relative intelligence of an individual"[15]

The last two definitions are actually circular arguments. IQ tests claim to measure intelligence which is defined by how well or poorly people do on IQ tests. Note that the word consciousness doesn't come up.

Another thing not spelled out—though it's relevant—is who displays intelligence. The dictionary definition seems neutral on that, though other animals besides humans are not mentioned. In 1958 only humans were thought to be smart.

The main war crimes trial before the International Military Tribunal at Nuremberg had finished its work twelve years before our class was formed. The defendants, leaders of the Reich who had been hapless enough to be captured, had had their IQs tested before the trial began. Nine out of the twenty-one in the dock would have made it into our class.[16] Their evil-doing had been justified in their minds by Galton's eugenical science, the same science that led to the tests that certified me and my classmates as smart.

You would think that after the trials at Nuremberg, anyone who wanted to rank the intelligence of individuals, and races, would at least not carry on as if nothing had happened to bring these ideas into disrepute. Yet in Canada, the same year my classmates and I scored well on IQ tests, an anthropologist examining the IQ tests of First Nations children noticed that they scored poorly. These children (taken from their parents to residential schools by the RCMP when they were five, forbidden to speak their own languages, forbidden to refer to ancestral beliefs, often beaten for infractions, sometimes sexually abused, and even starved as part of a science experiment) didn't speak English very well. But they were tested in English. The anthropologist was concerned that those who placed faith in IQ tests ("many," as he put it) would conclude they are an "inferior race."[17]

His concern was well placed. The people who first created IQ tests were, by and large, racists who did believe that some races are more intelligent than others. These early leaders in the field of educational psychology mainly accepted Galton's basic ideas and passed his torch on to their students.

Professor Charles Spearman, a leader in the field in the early part of the twentieth century, studied psychology in Leipzig, Germany. While working on his doctorate there he became fascinated by Galton's work on genius. Spearman eventually headed the psychology department at University College London where he invented and applied statistical methods to measure variations in human intelligence and taught generations of graduate students. He became a member of the Eugenics Society of London (later called the Galton Institute). Spearman became well known for his belief in a fundamental, stable *something* which does not vary when measured by repeated IQ tests, and is discerned through mathematical means called factor analysis. Spearman believed that this special, fluid factor is applicable to all forms of human problem-solving and artistic ability and is inherited. He called it "g." [18]As Arthur Jensen put it in his biography of Spearman, which is posted on the Galton Institute website, Spearman was a hardboiled eugenicist. He thought people should be forced to be tested and not permitted to vote or "to have offspring" if their scores were low. He killed himself at age eighty-two by jumping off the top story of the London University Hospital (now called the Cruciform Building).

Lewis Terman, born a generation after Spearman, accepted Spearman's argument for the existence of a general quality of intelligence known as g. He too thought IQ tests measure it. Raised and educated in the US, Terman spent most of his academic career teaching at Stanford University. Terman helped turn the first Binet-Simon IQ test (used to identify mentally handicapped children) into the Stanford-Binet which is still widely used. He also helped to develop IQ style tests (one for the literate, one for illiterates) used by the US Army in WW I to keep out unsuitable recruits and to determine who should be officers. As a student, Terman, like Spearman, had become fascinated by Galton's book on genius, and by Galton himself. It was known that Galton had experienced something like a nervous breakdown and had said that genius can be close to insanity. Terman delved into the facts regarding the belief that geniuses are difficult, unhappy, odd, sickly, and close to insane. He found the opposite to be the case. The geniuses he studied were generally happy, sociable, healthy, stable, etc.

Early on, Terman published a paper in which he estimated Francis Galton's IQ. He based his conclusions on Galton's biographer's report that Galton could read, write, spell, do basic arithmetic, and had learned a little Latin and French before he was five. Terman decided this meant Galton functioned with the capacity of someone twice his age, so he calculated Galton's IQ at about 200. He had by then tested thousands of children in California and had not found a single child who achieved a score over 170.[19]

Like Galton and Spearman, Terman argued that intelligence—that vague and mysterious quality known as g—is mainly inherited and is measured by his test. He did note that intelligence seemed to be comprised of two capacities, one having to do with learning and adaptability, the other related to abstract reasoning. He argued in a book published in 1916 that everybody displays the first kind of intelligence, but that blacks and "Spanish-Indians and Mexicans of the southwest" were not capable of abstract reasoning. He considered this to be a eugenically "grave problem because of their unusually prolific breeding." According to his wiki[20] he arrived at these conclusions after giving English IQ tests to Spanish speakers, written tests to black illiterates.

Terman, in other words, was also a fire-breathing eugenicist. He supported US legislation to enable forcible sterilizations of those considered dangerously inferior.

He also developed California's gifted program. He started a famed longitudinal study to keep track of the State's gifted students— nicknamed Termites. He believed that the gifted should be taught differently than other children, but that gifted children were often not identified as such by their teachers. He thought IQ tests could distinguish the smart from the norm in an objective manner. He wanted gifted children to be saved from the boredom of the ordinary classroom so they could realize their full potential. This was important for them as individuals, but also for their nations. These ideas were shared by educational psychologists to the north.

Samuel Laycock, a leading Canadian educational psychologist for three quarters of the twentieth century, was born in Ontario, educated at the University of Toronto, the University of Alberta, and then at University College London by Spearman. Just as he was granted his doctorate, he was asked by John Diefenbaker, then a renowned Saskatchewan lawyer who would ultimately become Prime Minister, to measure the IQ of a seventeen-year old boy charged with murder who was in danger of being hanged. Laycock determined the boy had the IQ of a high grade moron and therefore was not criminally responsible. Shortly after this courtroom drama, Laycock was hired by the University of Saskatchewan to teach educational psychology; he eventually served as dean of the department and taught generations of teachers.

Like Terman, Laycock was keen on gifted programs. He oversaw their introduction to the City of Saskatoon's public schools in the early 1930s. He also believed in educating the public about public education and he was involved at the end of his career in advising home and school groups across the country. He co-wrote a booklet for teachers on "mental hygiene." He was a well known public speaker on education and psychology, appeared frequently on the CBC and wrote columns that appeared in various magazines and newspapers.[21]

Like his teachers and their teachers he was an elitist. He was not interested in erasing the differences between the bright and not so bright, the lucky and not so lucky. He was perturbed that Canadian high school matriculation requirements were so low that merely average students could get a diploma. He pointed to the Soviet Union as the model to follow in the encouragement of gifted children because "in the struggle between East and West, the race to educate the genius is on..." The Soviets, he argued, cleverly channeled gifted children into engineering and science. These children were rewarded, stipended, paid more than their fellows to the end of their careers, whereas in the West, the gifts of the gifted were often wasted when they did not acquire university degrees. He insisted that among the most important things to teach the gifted is the scientific method.

"The teacher," he wrote to teachers, "must help the gifted to understand and use the scientific method as a general tool for the solution of man's problems."[22]

After Laycock retired as a distinguished educator of educators, he wrote a handbook on teaching gifted children. It was published in 1957, one year before our class was formed.[23] He relied on Professor Spearman's definition of intelligence—g, for general—as the indefinable essential factor in all creative achievement.[24] Like Spearman, Terman, and Galton, Laycock was convinced that intelligence is inherited, unalterable, and that science had shown this to be true. He said so plainly in his handbook:

> "...Yet biologists tell us that a child is born into the world with the top limits of his brain power fairly well determined by internal regulators called *genes*, which children receive from their parents. With our present knowledge, there is little that can be done to improve these genes. Given two children of differing capabilities, good teaching will undoubtedly increase rather than decrease the difference between them."[25]

In the same volume, Laycock descried the sort of scientist who would be scrupulous within his own field yet "credulous or even superstitious" about everything else. He quoted an economist who had declared to him: "'I know what happens to gifted children. They all go haywire,'" though as Laycock pointed out, the evidence said otherwise.[26] But he seemed to have no insight that he had committed the same error. When Laycock wrote his handbook, almost nothing was known about genes. Watson and Crick had just published their 1953 paper on how the DNA molecule is structured. In it they had suggested that DNA's double helix format might be useful in the transfer of genetic information during cell division, but that's about all they knew. No biologist in 1957 had any idea where genes reside on a molecule of DNA, or what kind of code genes employ to direct protein assembly or cell development. And no one had identified any genes responsible for intelligence. Six decades later they still haven't.

Laycock should have stuck to his own field and read Don Hebb. In 1949, Hebb, a neuro-psychologist at McGill University, published a seminal work called *The Organization of Behavior.* Struggling to interpret what he'd observed in countless experiments designed to show how the brain thinks, Hebb explained that mammalian brains are generally alike in their capacity to adapt to insults, though 'higher' primates are slower to learn when young than baby rats. He thought it was of vital importance that brains can adapt to damage that happens after a certain point in development. He worked with rats, blinding some at birth, some as adults, to compare their capacities. He had found that rats temporarily blinded when young were slower to learn visual patterns than their siblings blinded at maturity. He also found that rats taken home and treated as pets, provided with an "enriched" environment, did much better on what he called his rat IQ tests than rats kept in cages. More to the point, their speed of uptake improved as time went by, while the caged rats did not improve their capacity to learn.

While Laycock may not have been interested in enriched rats, he should have paid close attention to Hebb's discussion of human adults who performed as they always had (or better) on IQ tests after a prefrontal lobe or half their cortex was removed. The then mainstream belief was that human abstract thinking takes place in the frontal lobes. Yet Hebb, who worked for a time with the renowned neuro- surgeon Wilder Penfield, believed that could not be true if a lobectomy had no effect on IQ. Hebb also showed that if a baby's brain was damaged in the same places, mental function never developed properly. At minimum, these studies showed that the mature brain had the capacity to reorganize itself and that intelligence is shaped by experience. So how could intelligence be described as innate if it could be damaged or derailed early in development, but sustained and even improved if damage happened later?

Hebb saw intelligence at work in all kinds of animals. Like consciousness, he found no specific location for it in any special group of cells. He saw it as a plastic process distributed across brain regions, dancing with whatever challenges the environment presented. Hebb reminded readers that others had already shown that the twin studies (suggested by Galton), which purported to demonstrate that intelligence is strictly inherited, and not driven by environment, had been improperly done.[27] Hebb concluded that while there must be some innate form of intelligence that is inherited, it cannot be measured directly. Babies can't do IQ tests. IQ tests done at an early age (five) are only able to predict adult performance at a rate 20% better than chance. Whatever is measured by IQ tests after so many years of development and immersion in a culture has been shaped by that environment. Since no two people have exactly the same environments, no one's IQ can be directly compared to anyone else's. Racial rankings are also without merit since IQ tests must reflect the impact of experience. Experiences, especially between whites and blacks in North America (or whites and First Nations peoples) were obviously not the same.

Hebb's book should have shut down eugenicists like Laycock for all time. But of course it didn't. Ideas, good or bad, are feisty things that cling to life for no good reason other than that someone important repeats them. Once they have been accepted and passed on, it's almost impossible to kill them. The ranking of races by IQ, the belief that intelligence is inherited and innate, the comparative ranking of children by

IQ, all continued on long after Hebb had flayed these notions with surgical precision. And they are still with us.

When I look at our grade six class picture I can see something dark in motion, though all of us children are white as snow, not an Asian, black, or First Nation child among us. Alan has a glowering, unhappy intensity, as if he's trying to tear himself away from himself. Michael almost smiles, yet there is something aggressive about the way he focuses on the camera. Cameron, squeezed between them, seems to be scrunching his shoulders tight as if to make himself smaller. Ken, who is tall, but not yet as tall as me, looks a little hangdog, a little cynical. I'm looking at the camera sideways, as if I don't want to face it head-on.

Our teacher, Mr. N., looks confident and cheerful, as if he knows he is going to be a principal soon, and then get himself a doctorate. He was only twenty-three when he began teaching us: he had a freshly printed BA/B.Ed in his pocket, a wife, and a child on the way.

We came back at the start of grade seven to find that Mr. N. had been replaced by a woman, a middle aged woman with odd habits. There was the ubiquitous cloth handkerchief she draped over her belt, or dragged out of her sleeve, dabbing, dabbing at her flaring nostrils with what she seemed to imagine was lady-like gentility. We had had two years of being treated as intellectual royalty—taught by men! Why had they sent us this woman who made us listen to stories we could have read ourselves? The boys, their heads down on their desks, practically wept with boredom. The little gang that formed around Alan squawked and moaned, muttered and swore.

The other teachers loathed us, treated us as smart Alecs. Some of us were competing to see who could exceed their worst expectations by reading the most outrageous stuff. I read Ayn Rand's *Atlas Shrugged*: she had written such a fat book it didn't matter how bad it was, I was impressed with myself for plowing through it. I liked the idea of society crumbling after the withdrawal of services by the clever useful few. Like us? But Alan got the blue ribbon. He was reading *Mein Kampf*.

My friend Cameron, who had been the most popular boy for most of grade six, suddenly became the butt of vicious jokes. A small gang of boys lay in wait for him, mocked and hassled him day after day after day right through grade seven. I told myself that they'd get tired of it, they'd stop on their own.

One day, I came out of school late after a game, and there they all were, behind the candy store near the corner. The gang was beating the crap out of Cameron. They really, really hurt him. And I did what?

I kept walking.

If I'd stood up for him it might have changed things.

I kept walking.

This is when it became perfectly clear to me that our so-called intelligence had no connection with moral behavior, the moral hygiene psychologists muttered about.

Cameron's Dad demanded that something be done. Eventually, there was a meeting between the Principal and the Vice-Principal and the miscreants. One of the boys blustered that there'd soon be a big fat lawsuit on the Principal's desk if he was punished, so the Principal told Cameron's father that Cameron should move to another school. Cameron's Dad— have I mentioned that he was a leading journalist?—made it clear he would not stand for that. He went to the Superintendent. He got nowhere there too until he said he was going to go see the Minister of Education.

It was resolved that our special class would be split.

This broke the structure of power as Marion put it. Kathy left the school entirely: she couldn't endure the idea of another year with that woman teacher.

My new teacher, a man, had been teaching in a high school. He was told there had been relationship issues, that some kids had been sent to another class and a few new kids brought in who were not gifted but had good skills. (He told me later that he could spot who was who right away.) He had no knowledge of gifted programs, so he read something that sounded suspiciously like Laycock's handbook, which said there was a high dropout rate among gifted students. He was determined to stem this tide of waste. He read up on the ancient Greeks and Romans and put together a program on ancient history that had nothing to do with the curriculum, along with a big required reading list that we had to report on. He taught different number systems, assigned us beginner's algebra problems. Nobody mentioned anything to him about the class being an experiment. (In fact, there was no experiment.)

At his first parent-teacher meeting he was totally intimidated—because of who the parents were. Many were professors; the father of one student was an education professor. His ancient history program was both lauded and hated, but it certainly inspired some of us—I am fascinated by archaeology to this day. He also inspired me in another way.

About halfway through grade eight, after a flush of excitement about what he had to offer, I realized it didn't matter if I attended class or not, I could do all the reading he required at home. Besides, my asthma had gotten really bad: I wheezed whether I wanted to or not. One rare afternoon when I was in school he asked me to stay for a moment after class. Was I in trouble? Couldn't be. My marks were high, my assignments were in on time. High grades meant anything would be forgiven. Well almost anything.

Elaine, he said, leaning a bit too close. Elaine you miss a lot of school. Did you know that if you miss too much your IQ will drop? It's probably started to go down already.

My heart went thump, thump. Ohmigod, I thought. Could he be right? Would I be able to tell if it started to slip? You're losing it, I said to myself. You're sliding back down to stupid.

But then questions popped up, as they always did, and still do. We'd studied Gregor Mendel's probability system. IQ was supposed to be an inherited trait and therefore unchangeable. Eye color is also inherited. Does it change because you stop going to school? It does not. Maybe if

I developed a brain tumor, or squashed my head in a giant grape press, a lot of brain cells would die and then I might get stupider but how could my IQ go down just from missing school? Yet what if he was right? That would also mean IQ could improve because what can go down could also rise up. If IQ could go up, why had no one tested our IQs again to see how the enrichment program was working?

I decided either he had invented it, or this whole business of declining IQ was a crock. I bet on the former. Because I liked being certified smart.

You made that up, I said, beginning a long career of *j'accuse.*

Certainly not, he said. There are lots of studies that show it. If I were you, I wouldn't skip school. Think about it.

He was certain. I could feel it. So the studies had to be there if I wanted to go find them. But what was the point? I thought I knew the score: IQ was a sham.

So, when my high school guidance counselor decided to reveal our IQs to us, four years later, I didn't do what he required, which was to make an appointment to see him privately. The man taught eugenics but called it genetics and didn't appear to know the difference, so why would anything he had to say be of interest? He sought me out in the hall one day to remind me that I had to make an appointment for the big reveal. I told him I didn't need one, I didn't want to know my IQ.

What do you mean you don't want to know, he said. You're mature enough to handle it. Why not?

Because it doesn't mean anything, I said.

The look on his face made me want to dance through the school.

I had decided if I was smart, it was because I had been lucky enough to be born into a family that gave me everything I needed, who taught me how to study and acquire new skills by carefully working at things. Since I didn't believe in IQ, it didn't matter to me that others did. Besides, I was sure I would always be smart. Why not? I would always work at things. I also thought maybe the guilt I carried for failing to stand up for Cameron would vanish now, just because I'd stood up for myself.

I hereby excuse myself. I was a teenager: I thought it was all about me.

Most of us make the time to go to dinner together in a Greek joint downtown. We arrange ourselves around one long table. Michael sits across from me. Cameron, who seems to have forgiven me, is to my left. He has brought along a beautiful black and white snapshot that his father took of him, me, and my best friend when we were eight, in 1956, before the class was formed. We are in waist high grass on a plateau overlooking the river. Our heads are raised as if we've just heard a warning signal, like timorous dik diks on the Serengeti. We are the same age as orangutan youngsters when they first head into the forest without their mothers—but with the help of their friends.

If you really want to find out what intelligence is made of, the photo whispers to me, go back to your *New Wilderness* days. Pick up where you left off.

2

Hunting the Wild Galdikas

In the first year of its operations, *New Wilderness* borrowed some dingy offices and editing suites in a building that belonged to one of the show's investors on Toronto's King Street. Stephen, Charles (known as Chuck), Lorne, and their other partners had managed to get a year's worth of production financed after CTV bought the rights to air twenty-six half hour episodes. By then, the pilot had also attracted a major sponsor (Alpo, makers of pet food) and attention at a major television trade show (NATPE) in the US. It would later win three Emmys.

Chuck wandered around looking worried all the time, trying to keep track of chaos. Stephen's office was so awash in paper it looked as if a hurricane had stalled overhead. When he wasn't locked in one of the editing rooms trying to make a show work, or out in the field shooting Lorne's stand-ups, he was being berated by *New Wilderness's* so-called entertainment accountants ("You're a million bucks short, you gotta shut down!" "But you forgot to count the CTV license fee…" "Oh…"), or by the panicky line producer ("Stephen for fuck's sake if you say just one more time all-you-gotta-do-is I'm gonna fucking kill you…."). Production staff lined up outside his door waiting for decisions, which often changed though delivery deadlines loomed. This gave rise to a sweat-shirt emblazoned with the words:

> Dewar's Law:
> for every vision,
> there is an equal
> and opposite
> revision.

That year, he barely made it home to sleep for a couple of hours a night, which meant I had to figure things out for myself.

The first show I had to story edit and write was on orangutans. I screened Rod Brindamour's footage in an airless editing room. It featured Brindamour's former wife, Birute Galdikas, and her work at what she called Camp Leakey. The outpost was just a few rickety wooden huts in a clearing in tropical swamp/peat forest beside a river in the Tanjung Puting Reserve (now a national park) in Kalimantan, (Borneo). The wild orangutans lived in that forest. In addition to studying their behavior, Galdikas tried to re-introduce to the wild many young orangutans whose mothers had been killed by hunters or foresters and who had been sold as pets (until they became old enough to be really dangerous). These rehabs lived with Galdikas in the camp.

Galdikas, Leakey's last angel, got to Borneo in 1971, a decade after Goodall made her first discoveries about chimpanzees. Like Goodall, Galdikas was an outsider to primatology, which may well be why Leakey decided to back her. She was an outsider in many other respects too. Her parents were Lithuanian refugees who trekked to Germany to get away from the Russians during WWII. They met and married in a refugee camp, and Galdikas was born in another one in the American zone.[28] An aunt who traveled with them had been born in the US, but came to Lithuania to live before the War. After it was over, she moved to Los Angeles. Galdikas' parents got papers to come to Canada.

She grew up in Elliott Lake, Ontario, and in Toronto. She studied biology at the University of British Columbia, then moved with her family to Los Angeles. In 1965 she earned a degree in psychology from UCLA and UBC. She was a graduate student in anthropology at UCLA, specializing in archaeology, when she saw a picture of a male orangutan and felt herself drawn by the deep intelligence of his gaze.[29] She asked her professors how she might get to Borneo to study orangutans in the wild. A few said that kind of field study would be impossible. Despite the discouragement, she applied to the Malaysian government for permission to study them. When Louis Leakey came to Los Angeles to give a lecture, she stood in the crush of his admirers until she finally caught his eye, and then told him that she wanted to study orangutans, as Jane Goodall had studied chimpanzees.

It took three more years before he was convinced she could do the job and found the money to send her into the field. In the meantime, she'd met Brindamour and married him. Though she had her MA (Leakey had sent out Goodall and Fossey without a BA between them),[30] no science granting agency would have supported this early work. After all, Galdikas had never studied apes in the wild and almost no one else had either. As with the other angels, the task Leakey set for her lay outside mainstream science.

When Brindamour sold his footage to *New Wilderness* ten years later, Brindamour and Galdikas had already split up. No doubt the orangutans came between them, in as much as she carried the young ones on her body much of the time. She also accused him of being in love with a young Indonesian woman she had brought to the camp to help care for their son. Later, after they separated, Galdikas fell in love herself with a man named Bohap, a Dayak (an animist Bornean forest people) from an important family who worked on the reserve as a tracker. After she and Brindamour divorced, she and Bohap married. This gave her entrée to the world of the Dayaks and to the higher echelons of Indonesian politics.

I marveled that she managed any of it, never mind all of it. By day, she followed the wild orangutans on the ground as they swayed their way from fruiting tree to fruiting tree twenty meters above the swampy forest floor. From dawn to dusk, when they finally made their complex sleeping nests and settled in, she stuck with each particular wild one she was trying to document, wading through the swamp or standing stock still, head craned, covered in sweat from the heat and humidity, attacked by bugs and leeches, and, often enough dripping with orangutan feces thrown down at her because they got annoyed about being watched. She watched everything they did, from feeding, to parenting, to fighting, from rapes to consensual sex, and the shameless attempts by young females to lure older males in to their clutches. Back at Camp Leakey, she cared for her own small son Binti (and eventually also Fred and Jane), while still carrying the youngest rehab orangutans on her body. After she got her own doctorate, she supervised grad students.

Orangutans are constantly carried by their mothers for several years. She was trying to coax these orphans back to a forest life they knew next to nothing of, while also trying to be their stand-in mother. She put food out for older rehabs who liked to stay in the trees by day. But some preferred camp all the time and were devious and even dangerous.

Even a novice like me could see there was barely enough of this complex life of hers in Brindamour's footage to get a half hour out of it. But we managed to capture some of it. Galdikas's furred charges didn't swim (at least they didn't swim then), but we had footage of them stealing the camp's boat to get across a stream. Apparently there were no doors that could withstand their sneaky determination, no researchers' clothes they didn't want to wear, no makeup they didn't want to try, no hidden foods they could not sniff out. In the tradition established by Jane Goodall and followed by Dian Fossey, all had been given names. Names of members of the same family had the same first letter, the better to keep track of relationships. One of the rehabs was called Princess. Our story featured her.

As Galdikas would later explain in her faintly mystical autobiography, *Reflections of Eden,*[31] she soon saw that it was in the area of language and sociability where the sharpest contrasts emerged between orangutan and human intelligence, specifically between her two-year-old son Binti and his orangutan playmates. Binti and the young orangutan rehabs played together like siblings. Binti tried to move as they did. Though he couldn't follow them high into the trees, he quickly picked up their social signals, their facial expressions and the sounds they used to communicate their feelings. However, he also picked up words and sentences with no effort while Princess, about three years older, struggled to learn American Sign Language (ASL) from researcher Gary Shapiro who gave her formal lessons. Binti learned signs without being instructed in them at all.

But at the time we made our show, the *New Wilderness* researchers were able to find very little in print by her about these animals. She'd only published a few articles in peer reviewed journals. (A few more would appear the following year.) The researchers did find a paper by anthropologist Peter Rodman and another by John Mackinnon (who described the orangutans he observed in his study area as migratory). Both had published on what orangutans eat, where and when. When we put together everything we had, we thought we could say that orangutans are not social: or at least that the males are solitary after adolescence. Adult males spent their days searching for food and resting, and sometimes searching for older females for sex, warding off competitors with their booming long calls and the terrifying crash of the dead trees they toss aside like so many matchsticks.

Galdikas had learned that adult females are more social than males, and that at least in Kalimantan, were not migratory. Females spent their days carrying a child, searching for food, and resting, and sometimes consorting for a few days with males. Orangutan females don't show physical signs that they are interested in mating: males have to parse their social signals and woo them. The children stay with their mothers until around the age of eight, and then wander and explore with other adolescents. The females don't have children until at least fifteen, and most have fewer than five babies in their whole lives though they may live into their sixties.

Young orangutans stayed with their mothers for so many years because they had so much to learn, specifically the life cycles of hundreds of edible fruiting trees, how to open and eat the exceptionally unyielding fruits of the forest while maintaining intricate mental maps of where good things grow—in a forest that is always changing. They also had to learn how to forage for lesser foods in periods when no fruit was available, how to eat ants and termites, how to eat bark if necessary. Their movements through the forest shaped it: the seeds they scattered as they ate, or that passed through their digestive tracts, generated the trees their children would use as they moved along. The dead branches the males tossed around opened the forest canopy for new growth.

Galdikas could see that sisters and daughters cooperated, that their paths crisscrossed at favorite trees. But adolescent sons were banished. Outsiders were not accepted either, which made the whole business of rehabilitation more than tricky.

That was what we had. And as I would later learn, most of it overturned most of the orthodoxies about whether or not animals learn, whether or not their intelligence is anything like our own. The limited information from Galdikas made me nervous about what we could say with authority about her work. In those days, it was a given that nothing a scientist says should be treated as fact until it had been published in a peer-reviewed journal or as a thesis under the guidance of a properly constituted committee.[32] While we had Brindamour's film, it needed to be interpreted by someone who knew what the images meant or who could explain what had been left out. We couldn't phone Galdikas in the rainforest to fact-check what I wanted Lorne Greene to say. Yet Brindamour had let it be known that she was extremely difficult. So I had to bob and weave.

After the program went to air, I waited anxiously for screams of outrage from Borneo. But we never heard from Galdikas.

Even if we had got things wrong, it is unlikely Galdikas would have objected. To save the orangutans and their forest at Tanjung Puting she needed to get charitable donations. To do that, she needed to drive home to the Western public through shows like *New Wilderness* that orangutans are interesting, and almost as intelligent as humans, yet are under severe threat due to human activity. This is advocacy, not science, yet doing an end run around science was how Leakey had got the science going in the first place.

It had been accepted by most anatomists since the 1880s that humans and other primates are closely related, but when it came to intelligence, a line had been drawn. Why? Partly it was due to the Judeo-Christian idea about the singularity of humankind— made in God's image, granted dominion over all— that still persisted in scientists' minds. Before that, Aristotle had divided the world into categories according to complexity of motion, starting with the planets, moving down to humans, animals, plants, and finally to inert minerals. Aristotle's hierarchy, his *scala naturae* at it was called in Latin (ladder of nature), had been modified to suit Christian doctrine. Christians put God at the top of the ladder, then the angels, then the fallen angels, and so on down through human kings, nobles, peasants, and finally animals far below. Humans were divided by race with the last race to be discovered, native peoples of the Americas, on the bottom. Everyone had a proper and immutable place. Long after Darwin generated his theory of the origin of species by means of natural selection, and replaced Aristotle as the patron saint of natural science, biologists still taught that humans occupy the tallest branch of Darwin's evolutionary tree, utterly alone on the pinnacle.

In my gifted class in 1958, the proof of human exceptionalism was presented in the form of a list of things only humans can do. We were told only humans can:

> invent and make and use tools;
> invent and use languages and other symbols to pass information to each other and down through the generations;
> make art to express the ineffable;
> have a moral sense of right versus wrong;
> have a sense of number and time;
> live in ordered societies thanks to a capacity for empathy and reason;
> go to war;
> venerate the dead because we know we will die (and meet God or the fallen one below.)
> etc.

Our intelligence, which defined us, was demonstrated by these capacities. So perhaps it is not *so* surprising that almost no long-term scientific observations of our closest cousins' behavior in natural settings had been made until Goodall went to Gombe.

Her main practical training for the job of making a revolution in science involved the animal tracking lessons she got from Louis Leakey for whom she worked as a secretary. But her heart and mind were wide open. And what she saw—when she took her eyes in her hands and looked, as my mother would say—were many of the allegedly definitive intelligent behaviors supposedly restricted to humans (as above) being displayed by chimps. Dian Fossey saw equally complex intelligent behaviors displayed by gorillas. And Galdikas said orangutans were the smartest of all. Galdikas said they didn't just learn by trial and error, as animals are supposed to do, but often through contemplation and insight, through mental manipulations involving memory and foresight and judgment. They learned like us, in other words, a notion that was anathema in Galdikas' original field, psychology.

For most of the twentieth century, most North American psychologists had confined their curiosity about thinking and intelligence to reformulating IQ tests, or to watching what people and animals do in labs. Up to the early 1970s, the ideology known as behaviorism was utterly dominant in psychology departments where questions about intelligence were studied. Behaviorists argued that no psychologist could ever puzzle out how humans think because our brains are too complex to be studied. Neuroscience was in its infancy and could not show how brain activity gives rise to either mental illness or normal operations or how to locate the pilot light called consciousness (though lots of drugs were available to shut it off).

So psychologists treated the brain as an impenetrable black box. They restricted themselves to studying human and animal behavior by delivering a stimulus and observing the response. Some went so far as to say that stimulus/response is all there is, that even the most intelligent human behavior—the use of language for example—is just the byproduct of stimuli producing inevitable responses. Only in labs could stimulus and response be studied in a controlled fashion.

In Europe, things were different. Animal behavior was mainly studied by ethologists—usually zoologists who worked in zoos or in the field. But ethologists were not asking direct questions about the nature of intelligence either. Ethologists wanted to know about behaviors—natural ones, not lab test induced responses. They focused on the evolutionary history of behaviors over time, on the way certain behaviors persist across species which share a recent or even a distant common ancestor.

The ethologists argued that animals are born with instincts—innate behaviors ready-made to cope with the world. They argued that the same successful behaviors have been hit upon over and over again by many species over the four billion years of evolution, in the same way as comparable structures (such as similar eyes in humans and the octopus) appear in animals far apart on the evolutionary tree. They insisted behaviors are not passed on by teaching but by the physical organization of tissue and nerve as ordered by genes. As the new language of computers began to creep into their writings, they referred to these behaviors as "hardwired."

Ethologists such as Karl Von Frisch, Konrad Lorenz, and Nikolaas Tinbergen tried to show that many animal behaviors are stereotypical. Von Frisch had described the complex, information rich bee communication dance and explained it in terms of the bees' extraordinary sensory equipment, particularly the ability to see polarized light unseen by humans. Tinbergen and Lorenz explored the way the Greylag goose automatically rolls eggs, and even objects that look like eggs, back into the nest. But Lorenz also showed that emotions too are behaviors: he said geese, like humans, exhibit jealousy when in love. He argued that when behaviors are shared by distantly related species it must be because they aid fitness so therefore they are necessary to the life of that species and only alterable so far.

Behaviorist psychologists objected to Lorenz's comparisons between the emotions of geese and humans and derided his work as anthropomorphic. The epithet "anthropomorphic" had long since become a slur, something all students of animal behavior sought to avoid even when studying the evolutionary relationships expressed through shared behaviors. There still there had to be a firm line drawn between humans and everything else.

Lorenz replied [brackets mine]:

> "Psychologists have protested that it is misleading to use terms like falling in love, marrying or being jealous when speaking of animals....Since we know that the behavior patterns of geese and men cannot possibly be homologous [meaning similar because produced by similar life pressures]—the last common ancestors of birds and mammals were lowest reptiles with minute brains and certainly incapable of any complicated social behavior—and since we know that

> the improbability of coincidental similarity can only be expressed in astronomical numbers, we *know for certain* that it was more or less identical survival value which caused jealousy behavior to evolve in birds as well as in man."[33]

By the late 1960s, just before Galdikas went to Borneo, Robert Ardrey and Desmond Morris had published best-sellers describing paleobiologists and ethologists' findings. Ardrey argued that humans' very special brains and social complexities are direct results of the evolution of a human behavior called hunting, a complex specialty rooted in an ancient primate ancestor's aggressive behavior. We, he said, are killer apes. Morris saw many more similarities between humans and apes, including the capacity of chimps to make art, but still that barrier remained. His book *The Naked Ape: A Zoologist's Study of the Human as a Revolutionized Animal,* was a big hit in 1967. But all the ethologists' work—with its insistence that some behaviors are innate and unchangeable having been shaped by evolution and necessary for survival—triggered ferocious counter arguments.

Egalitarians, especially the new second wave of feminists, argued that all human behaviors, including gender role behavior, are learned, not inherited, and that sexist behavior is specifically taught in order to maintain male dominance. The arguments between feminists and ethologists, between psychologists and ethologists (some called sociobiologists in the US) represented a new version of Galton's nature versus nurture dichotomy.

After Goodall reported on tool use by chimps, she was sent by Leakey to Cambridge where she earned her doctorate.

As she prepared to write about the things she'd witnessed among the chimps of Gombe, including war, rape, murder, the hunting and killing of bush-babies with sharpened sticks, adoption, love, and devotion, some colleagues begged her to suppress all the information she had about their aggression. Why? It would support the ethologists' arguments that aggression is so deeply embedded in evolutionary history that war will always be inevitable among humans, no matter how much smarter we are than any other animals. In other words, ethology and psychology were knee-deep in politics.

In 1973, Konrad Lorenz, Nikolaas Tinbergen and Karl Von Frisch won the Nobel Prize in Physiology or Medicine. Lorenz's lecture carried hints of radical new brooms just beginning to sweep through biology including a new understanding of dynamic interactions (described by general systems theory and observed by ecologists); the unpredictable nature of natural systems (what later came to be known as chaos theory); information theory, and the rapid advance of computation. Lorenz grabbed at ideas from these other disciplines as if they were life preservers. He desperately needed them to explain the problem at the core of his work, how something as ephemeral as a behavior can be stored in physical structures and passed on:

> "There are other functions that are equally indispensable to the survival of all living systems, ranging from bacteria to cultures. In any of these systems, adaptation has been achieved by the process, already mentioned, which hinges on the *gaining of information* by means of genetic change and natural selection, as well as on the storing of knowledge in the code of the chain molecules in the genome.

> This storing, like any retention of information, of knowledge, is achieved by the formation of *structure.* Not only in the little double helix, but also in the programming of the human brain, in writing, or any other form of "memory bank" knowledge is laid down in structures.
>
> The indispensable supporting and retaining function of structure always has to be paid for by a "stiffening," in other words, by the sacrifice of certain *degrees of freedom*...All the adapted-ness of living systems is based on knowledge laid down in structure; structure means *static* adaptedness, as opposed to the dynamic process of adaptation."[34]

Galdikas arrived in Borneo amid this great ferment. I wanted to know what she had learned in the thirty years since *New Wilderness* went to air. I thought she was the right person to go to since she'd been drawn to the study of orangutans as a psychologist, but went to the field instead of the lab. It was almost as if she'd made herself a bridge between the behaviorists and ethologists.

I knew she taught a semester every year at Simon Fraser University in BC where she has long had an academic appointment. Most academics list on their websites everything they have published, often making PDFs available. I found a paragraph on hers that said at Tanjung Puting people were studying primate behavior, ecology and evolution, with a focus on orangutans' sociality and cognitive potentials. That seemed promising. But her publication list showed only two books published by her years before, and two books about her.

Too busy out there in the field, I thought, no time to upload her papers. So I searched for her work on Google Scholar. There wasn't much recent work there either, almost nothing where her name was first or last. Leading scientists usually have large publication records. On the other hand, she was not like most scientists. She'd helped open a whole field.

Okay, then, I thought, I'll just go see her. She is a Canadian public figure, an Officer of the Order of Canada, the subject of various films including an IMAX production shown in 2011. She won't be shy.

But there was no answer at her office at Simon Fraser when I called, and no one in the Archaeology Department where she has an appointment knew when she'd arrive back in the country. So I called the Orangutan Foundation (OFI) which she'd set up years earlier in Los Angeles. I called numerous times until someone finally answered the phone. I asked where I could catch up with Dr. Galdikas, thinking she probably crisscrosses the continent to drum up donations when she's not teaching. (Goodall is on the road 300 days a year.) But the Foundation had only two smallish events planned for Galdikas on its fall calendar, neither one a sure thing. The woman I spoke to asked for an email setting out what I wanted to do and promised to forward it to Dr. Galdikas. She could not guarantee when she'd read it.

I sent it. Time went by. No response. I phoned the department again. Yes, she'd be teaching there in the fall, I was told, but as to whether I should leave a message on her office phone, probably that wouldn't help much.

Was that sarcasm?

More weeks went by. I tried OFI again. I was asked for another email, which I sent. I waited until the week before classes started and called the Foundation once more. Had Galdikas arrived in Canada? Yes, said the woman on the phone. Well, I said, I emailed and I left messages at her office but have had no response. Do you have phone numbers for her in Vancouver? She gave me Galdikas' cell and her unlisted home phone number.

I tried the cell. Galdikas answered. She said she was on her way out, please call her at home tomorrow.

I called her land line at the appointed time. No answer. I tried again later. No answer. I tried the cell. She answered and began to shout: I've never heard of you. I haven't seen any emails, and how did you get my phone numbers? When I explained, and told her I'd like to come out there and sit in on her class, and then spend a day with her to talk to her about her work, she yelled that what I wanted was more than even her *friends* would ask of her.

I found that an odd thing to say since she leads guided tours of Camp Leakey for those who will pay for the privilege. I was more disturbed that she seemed frightened by my call.

I called back to OFI. Dr. Galdikas doesn't know who I am or why you've given me her phone numbers, I said. Please explain to her that I am a reporter. She's very upset.

Why don't you send an email to send on to her, the woman said to me.

At which point I finally got it. There would be no field update from Birute Galdikas. I'd have to think of something else.

3

Discovering the Clever Russon

Thirty years is a long time in science. Many people had studied orangutans in the field since Galdikas established Camp Leakey. Linda Spalding, a novelist who wrote a terrific nonfiction book called *A Dark Place in the Jungle,* in which she describes trying to 'follow'Galdikas, suggested I speak to a professor of psychology named Anne Russon. One Internet search and phone call later and I was walking across the grass of the Glendon campus of Toronto's York University to meet her.

It was so humid it might as well have been raining. Though it was early afternoon, the building lights glimmered behind a dark veil of pines and cedars. If I ignored the big-shouldered 1980s condo that towered by the entry gate, the campus seemed soothing and calm, which is not how I felt the first time I saw it.

My father had honored his promise made under duress to send me away to university if I won a scholarship. I got one: to York University. My parents put me on a plane and I flew from the prairies east to Toronto in the care of Auntie Ida who was on her semi-annual buying trip to Toronto and Montreal to stock her ladieswear stores. I had two suitcases (trunk to follow), a small amount of cash, and a big hope that York University would turn me into the new Simone de Beauvoir, the only feminist female philosopher I had heard of.

My aunt put me in a taxi telling the driver to take me to York University. Which campus, he asked. One of my acceptance letters said Glendon, another said Vanier College. I told him Glendon. It was advertised on the brochures as the bilingual Harvard of the North. It was bilingual, but it was no Harvard.

And there was no *there* there either, nothing but spindly trees and ugly brick buildings set around a grassy sward. I had pictured turrets, stone walls, things both American and French and therefore revolutionary in the beating heart of the big city. You had to take a bus to the subway to get anywhere interesting but a worse problem was that while my residence room was at Glendon, I was registered for classes in Vanier College which was on the other campus on the other side of town. Every day I had classes I had to take a forty-minute ride on a little red school bus to a windswept prairie studded with a few brutalist concrete towers. The place felt as anonymous as a car factory. How could I become Simone de Beauvoir in a factory?

I stood in Glendon's dining hall staring up at the flags that still hung from its ceiling forty years after I last ate a wretched meal there. I thought I spotted Anne Russon at the coffee stand. There was a picture of her on her web page, but this woman seemed much smaller and older, my age in fact, with a wild cloud of dark hair threaded with silver. Did she look like a person who spends half her year in Borneo watching orangutans? What would such a person look like?

I followed her to her office. She was small, and moved very fast. As she unlocked her door, I introduced myself. She thumped her sensibly-shod feet across one of her pulled out desk drawers and began to explain how she had got herself to Borneo. The hows and whys turned into a mini-history of how psychology and ethology merged through the study of intelligence.

It turned out we shared many points of connection. She's also a prairie girl, from Regina, and her Dad was also a doctor though he specialized in psychiatry whereas mine, a family practitioner as well as a general surgeon, used to say psychiatrists are witch doctors. First her Dad worked in prisons. Then he switched to dealing with troubled juveniles. Meanwhile, she fell in love with nature because of the Disney shows she and her siblings watched on TV. (She still remembered one sequence of a small bear cub tumbling down a snowy hill. She was horrified to learn later, from a documentary about nature shows, that the cub had been tossed downhill by the director to get the shot.) Her dad built a trailer. Every summer he hauled the whole family, including her grandmother, off to Lake of the Woods east of Winnipeg for three weeks. There she learned to camp, which would prove to be very useful many years later.

She talked with the rat-a-tat speed of the very smart kid. So I had to ask: gifted class? There were none in Regina, she said, but she was skipped in grade seven mainly because she was extremely good at math. She applied to the only eastern university she'd ever heard of, McGill, where she took her BSc and MSc in mathematics (and a course in psychology given by Don Hebb). In the late 1960s, AECL, the crown corporation that developed the CANDU nuclear reactor, recruited math and physics graduates to its experimental station in Chalk

River, Ontario. They hired Russon, but she hated Chalk River and eighteen months later she was back in Montreal with a job at Control Data Corporation.

Later, she traveled to Australia with a boyfriend, and got another job with Control Data, at the University of Melbourne. Five years after she left school, she decided a life serving computers wasn't for her, that she really wanted to be more like her father and to work with kids. So she was accepted to do graduate work in psychology at York, later winning a fellowship from the Canada Council. But her boyfriend was offered a post elsewhere. When he told her that in their life together one person's career was always going to be more important than the other's...

"Well," she said, giving me a sardonic look. "Long pause...."

She dumped the boyfriend and arrived at York in the fall of 1973, just as the ethologists Von Frisch, Tinbergen, and Lorenz won their Nobel. She knew nothing of ethology: she was studying the psychology of motivation but she soon realized she had picked the wrong subject. She did her experiment anyway, "a stupid experiment," she said, and got her masters in 1975. In the meantime, she found the track she did want via a course on primates during which she read Robert Hinde, Jane Goodall's mentor at Cambridge. He had published a book called *Biological Bases of Human Social Behavior*. This was not psychology as she knew it.

"North American psychology was based on flatworms, rats, cats," she said. And, she said, in order to bolster their image as *real* scientists such as physicists or chemists, some psychologists wore white lab coats, which was "ridiculous," and tried to discover universal laws of behavior. At McGill, in the psychology courses she took, "humans were not mentioned. No studies were done on humans. It was frowned on." Evolution was not mentioned either. "Psychology dumped Darwin in the 1920s," she said. There were different branches, "the mainstream and the others," as she put it. "Yerkes and Kohler, they did cool stuff, showing what chimps and gorillas and orangutans in labs or zoos could do. But they were swamped by the behaviorists. The bulk of psychology on intelligence was on learning and memory, low level stimulus and response kinds of processes, trial and error, reward and punishment, the Pavlov thing, getting dogs to salivate to the sound of a bell, the associations of two stimuli. That was the basic focus until the 1960s. Psychology had to be experimental and so-called objective."

Hinde's work flew in the face of all that, but it flew in the face of feminists' arguments, too, and she was a feminist. Feminists in the 1970s argued that "it was all learned…nothing could be biological," Russon said.

Even sexual orientation? I asked. As these words emerged from my mouth, I remembered some of my radical feminist friends determinedly having sex with each other because they believed their attraction to men had been learned and could be jettisoned. Sexual attraction was considered to be just another part of the patriarchal social construct, another way to control women.

Yet Russon found her interest piqued by Hinde's book. She had to pick a primate species to study for the course and she chose chimpanzees "for some not well informed reason." The man who taught the class, "a contract guy," then got a job in Quebec. Through a combination of what she called "conniving and academic climbing" he became a collaborator with a research group at Université de Montréal working on Piaget's theories of development. Their plan was to get four baby chimpanzees and raise them in a nursery so they could follow their cognitive development, measuring it against Piaget's theory. "He asked me if I wanted to do a PhD with him on chimps. So I thought, I'll do that."

Money was no problem. Her fellowship was good for three years and she could apply for an extension for a fourth. She had no husband to worry about. "I never got married. *Nobody* got married. I went to almost no weddings."

I got married, I wanted to say, but kept my mouth shut.

She moved back to Montreal. She reported to the chimp lab which was just a trailer on the campus. It had a bathroom, sleeping rooms, a kitchen, and a video room with a one-way mirror. The first two chimps arrived when they were only two months old. They had been bought from a colony in Oklahoma— a notorious place that bred chimps for research labs—run by a man who was described to her as being like a cowboy in a safari suit. She worked with the two babies: they were called Spock and Sophie. "Their 'mothers' were two university students who worked eight to five. The rest of us rotated evening and night stuff. I'd do two nights a week."

And she loved it. There was no time to sit down when the chimps were up and moving, "they could shred their dirty diapers in five seconds," she laughed. She studied the social interaction between the two babies. And that's when she got interested in intelligence.

Why?

"Because they were so *smart.*"

Examples?

"I took a friend into the play room, there was a kid-proof hook lock way up on the door. Sophie was weirdly nice. She draws us in to the farthest corner from the door, plays, then runs to the door, unlocks it, and she's at the fridge in the kitchen with the door open before I caught up with her... I didn't even *think* about their intelligence until then....They're not like dogs or cats. This is something different. They look you in the eye and you know major wheels are going on in there and it's calculating. And this is a one year old. That got me."

It took her a long time to get her doctoral thesis done, which she finished after taking a teaching job back at York: Glendon needed psychology taught in French. Then, she decided she ought to study something "serious," which at that point still meant intelligence in humans. She started a project on learning in infants nine to eighteen months old.

"Ankle biters. So cute. Like fifteen of them. Studying imitation with them. It was okay, but so boring after chimps. Changing a chimp's diaper took four hands plus teeth...I liked apes better."

So, in 1985, she started attending primate conferences to see who was doing what, to figure out where she might fit. She wanted out of the lab, but could she live in the field? Would she like it? She went to Africa with a friend so she could get a close look at Fossey's mountain gorillas. She thought of the trip as a kind of test. If she liked the hiking and roughing it, then she thought she could take on studies of her own anywhere.

Fossey had been murdered not long before she got there. Gorilla tourism was going strong (just one of the things Fossey had tried to stop before she was killed). Russon joined a tour. She and sixteen other people bounced over bad roads in a converted army vehicle to get to gorilla territory where they did a whole day's walk up into the mountains, slept overnight at high elevation, all to get one hour of watching the gorillas.

She made up her mind she would do it. Now she had to find a place to do it. She knew she wanted to work with rehabs, not follow the behavior of wild primates from a distance. "I'm a psychologist. For what I wanted to study, I needed to see up close," she explained. She knew there had been studies done with rehab chimps and tool use in Liberia. But a friend, a keeper at the Toronto Zoo, had been to Camp Leakey in Borneo. Russon found out whom to contact and went to Borneo with another friend.

"Would have chickened out without that friend....I wrote to Galdikas to see if I could come and she said yeah."

When she got to Camp Leakey in 1988, there was relatively little literature on orangutan behavior, and even less on their intelligence. The Dutch had been studying them at a site in northeast Borneo. She'd read a thesis by Herman Rijksen. Peter Rodman worked from a site about one kilometer from where she would later be based herself. But their work didn't answer her questions.

"People weren't out there looking for stuff on intelligence," she explained. And even if they had been looking, they probably would have missed important things. "As a psychologist, I see stuff that biological/anthropological researchers don't see," she said. "Their eyes aren't tuned for it."

After four days of looking around at Camp Leakey, she decided it was feasible to do her study there. They put out food for the orangutans on the boat dock twice a day. Sometimes as many as thirty or forty orangutans would come to feed there, a mix of wild ones and rehabs. There was also a nursery for orphans, ten little babies under the age of four to five years. Russon was interested in studying imitation in orangutans—probably possible from age three and up. Some of the rehabs had kids of about the right age, enough for a study. Her major concern was that the orangutans wouldn't hold her interest,that she might get bored. They weren't fast like chimps.

"The orangs are slow as sludge," she said. "One took two hours to walk 200 meters down a bridge. They are more reflective. They *think*. They are just mulling it over."

In 1989 she went back to Camp Leakey to do her first field study. She was interested in learning through imitation, one of those supposedly human-only traits. She ended up spending two and a half months focusing on the rehab mothers who turned out to be much better imitators than the three or four year olds. Galdikas "was helpful for a while." (Galdikas is listed as a coauthor on several of Russon's early papers on this subject.)

Russon had a look on her face. So I told her of my experience trying to interview Galdikas.

"She's paranoid," she said, flatly. But, she said, she had good reason to be that way. "She built up this camp for twenty years and others want to parasite on that. It could make you a little strange."

Russon's framework for her study was basically Piaget's development theory, which she had already applied to young chimps. "His theory is that intelligence is epigenetic," she said, by which she meant it develops alongside the growing brain according to what that brain encounters. "There is a biological and experiential part of building the brain you have as an adult. It grows differently according to what you do. The question at that time, the *claim*, was that *only humans* could learn by imitation, learning a new behavior by seeing it done rather than by practicing it, that other species couldn't. The idea was you could copy, but you could only copy things you already know. Neonates less than one month old and infants up to eighteen months old can copy others' behavior, but only behaviors they already know."

As it turned out, Russon saw imitation going on everywhere among Camp Leakey's orangutans. As she eventually learned, orangutans tend to be choosy about who they imitate, there has to be some kind of emotional attachment. Young adolescent males are not interested in mimicking just anybody, they are interested in each other or older males. Older females copy their important peers, unless it's a rehab who identifies with humans. Russon had lots of stories to illustrate what she saw. She believes that stories are what everybody remembers from her work, no matter how committed they are to reducing the orangutans' behavior to graphs and numbers.

"My favourite is Supinah makes a fire," she said.

She pulled a picture book out from under a pile and showed me a picture of Supinah.

One day, she said, Supinah decided to copy the camp cook's method of making a cooking fire.

At Camp Leakey, rice was generally cooked slowly over an open fire in the midmorning. The orangutans were not allowed to get close enough to be taught how to make a fire, and certainly had no chance to practice. But one day, Russon and a student watched as Supinah ambled over to the abandoned smouldering remains of the morning's cooking fire. She began to go through the same motions the cook would normally use to get a fire going.

"She comes to the cooking area, goes to get a metal can with a steamer lid on top. Gets a plastic cup. Using the cup, she dips liquid out of the square can, which contains kerosene that the cooks used to start their fires. She picks up and blows on the end of a burning stick. Gets the cup of liquid, plunges the stick into the liquid. I said she's trying to put the fire out. My student said 'No, start one.' She looks at the stick. Dead. She took the cup, dumped the old liquid back in the can and then dipped the cup back into the big can and put new kerosene in. Tries the stick again. Gets another burning stick and blows..."

This went on for twenty minutes. Supinah even tried to fan her non-existent fire with the steamer lid. "It was the same fanning motion the cook used," said Russon. "She fanned sideways like cooks do...I saw nothing as fancy with chimps."

Then, she told a story of imitation by Princess, the same female who had starred in our *New Wilderness* show years earlier. Princess begged for hot coffee.

"I put it in a bottle," said Russon. "I said careful it's hot. But she drank a big mouthful anyway and scalded her mouth. Got another bottle, a glass bottle of soy sauce that was cold. She tried to pour from it into the bottle of hot coffee. But it had a lid on so she bit the top of the bottle off. Got the hot coffee bottle. She poured the hot coffee into the cold soy sauce bottle. You can hear it going in. Then drank the cooled down coffee from the soy sauce bottle. She did it. That is *not* shabby. She *clearly* understood how to pour between the two and cool her coffee."

And how about the Supinah-brings-the-hammock episode? Hanging a hammock properly is a tricky business to copy, but Supinah did it. "There was a wraparound vertical tie. The hammock has to be high off the ground. She gets the rope, wraps it around a tree trunk, two times around the tree. Puts a hand on the sling of the hammock to test. She pulls to make it tighter. That's smart thinking. Archimedes Orangutan."

In other words, Russon documented forms of learning that orangutans share with humans. "All learn by conditioning, but humans learn at higher levels," she said. "There's a boundary...what I knew was that the boundary we had was at the wrong place."

When she published the paper on imitation in 1991, the experimental psychologists pooh-poohed it, though it came out in *The Journal of Experimental Psychology*. "Some said it's random," she said, "or, it doesn't mean anything. That is the way this area works: you get all kinds of nonsense, but it was a very good journal and getting it published there gives credibility."

In 1996 she co-edited a book called *Reaching Into Thought: The Mind of the Great Apes.* She and her co-editors solicited chapters from all the major great ape researchers who studied intelligence, including the acquisition of language, an area that had been all but rendered off limits after projects of the 1970s and 1980s. The difficulties began when Herb Terrace and Tom Bever and their colleagues at Columbia University tried to teach American Sign Language to a baby chimp. They named him Nim Chimsky as a joke on Noam Chomsky, the MIT linguist. Chomsky had a theory that language is too complex to be learned from environmental inputs only—which was the behaviourists' theory—and insisted that language is a product of the peculiar wiring of the human brain *alone.* Duane R. Rumbaugh, and, later, his wife Sue Savage-Rumbaugh, tried to teach chimps a language too, but initially they used neutral symbols instead of ASL. Terrace later announced that Nim Chimsky never learned to sign anything beyond demands for food or tickles and never developed syntax. He said his earlier reports of important results were the product of self-delusion.

Yet certain boundaries had nevertheless begun to fall.

So far as Russon could see, the placement of those boundaries, with humans on one side and everything else on the other, came and went in cycles, one generation forgetting the work of their forebears, then rediscovering it, or flat-out stealing it. There was always something that animals weren't supposed to be able to do that was being re-examined by someone.

"Imitation—there was lots of work early in the century. But it gets dropped and thirty years later it shows up again...I saw my niche as bringing new information to the table."

So, in addition to her work on imitation in orangutans, she studied innovation and how new inventions spread within groups. She and her students hunted through databases of other scholars' observations of primate behavioral innovations. She found that many new tools or new practices appear first among juvenile or adolescent primates who tend to be highly curious. However, for a new thing or practice to be taken up by their larger social group, it has to be adopted first by a high ranking individual.

Russon described part of this pattern as displayed by a group of Japanese macaques:

> "Fish was first scavenged by a peripheral, low ranking adult male, alone; no others were interested. It was later scavenged by high ranking adult females and then spread top-down through the group. Approximately twenty years after the first fish scavenging, almost all fish eaters were from the original fish-eating matrilines."[35]

Among orangutans, Russon noted a clear difference in rates of innovation between the rehabs and wild ones. Rehabs tried more new things. Successful innovations spread among them more quickly. Example: rehab orangutans had been set free to wander on two river islands in northeastern Borneo. Fish frequently washed up dead along the shore, and human fishermen could be seen pulling fish off their lines. Her colleagues documented the rapid acquisition of fishing skills by the rehabs on one of the islands. First, one of them tried to eat a dead fish, then a few progressed to picking up fish trapped in shallow, drying ponds or snagged on fishing lines, and then they started catching them by using tools.

While paleontologists had long surmised that fish might have been eaten by our big-brained ancestors, it had always been assumed that this complex skill would have taken thousands if not million of years to refine and spread widely. But she observed that immature rehab orangutans with smallish brains of only 300 to 400 cubic centimeters had come up with the idea that fish are good to eat and mastered the art of fishing in just six years.

The rehabs were also observed swimming, though it had long been believed that orangutans avoid water because they are unable to float. Some were seen to submerge themselves in pools, including their heads, so they could see what was going on under the surface. They don't float: those who make their way across deep water lunge forward and grab vegetation on the other side, or flail their way across. In Russon's view, they may build new swimming skills on the physical motions they already used to move through the tree canopy. To get across gaps between trees, they use their weight to bend the tree they are on until they can push it far enough to grab hold of the next tree and haul themselves into it.

Russon's observations of innovation among orangutans brought to mind the way it works among humans, especially in science. The young, often outsiders to the power structure, are also often the innovators. Einstein did his most important work when he was a very young man perched on a stool in the Swiss patent office, not while he occupied a prestigious chair at a prestigious university. Darwin removed himself early from the cut and thrust of London's science salons so he could hear himself think. Turing came up with the idea of computers in his early twenties. But they all needed to convince prominent, established people to accept their radical new ideas in order to get them widely accepted. Once the field leaders took them up, they turned into dogmas taught to the young.

Russon still does not trust what emerges from psychology labs when it comes to the study of animal intelligence.

"In labs," she said, "humans can make apes look better or worse depending on the experiment." People make up criteria for intelligence, such as "the false belief test" which purports to show whether or not the animal studied has a "theory of mind." Theory of mind means the animal understands that the experimenter can have a view of the world that might differ from his own.

She described one typical theory-of-mind experiment done with children. A kid goes into a room with an experimenter. There are two boxes set out on the table. The experimenter puts a doll in one box and covers it and leaves the room. A stooge comes into the room. The stooge moves the doll from one box to the other box and leaves the room. The experimenter comes back. The test: does the kid understand that the experimenter now has the wrong information about where the doll is?

This kind of experiment is used to show, or disprove, that the child, or animal, has knowledge of another's invisible mental state.

One book chapter Russon wrote provides a long list of the orangutan accomplishments observed in the field. She described objects used by orangutans as tools, including: leaves, sticks, and grass for a variety of needs such as stabbing, prying open, hammering, probing, wiping, making gloves to avoid stinging ants, poultice-making for wounds, and sex toys.[36] Yes, orangutans have sex toys.

And then there was their story telling—another boundary that is supposed to separate humans from the rest of nature. A couple of Russon's more recent papers deal with pantomime in great apes. She and her colleagues had observed orangutans miming. So they combed through the great ape literature to see if other apes do it too. They found pantomimes described in several papers. Some language theorists argue that meaningful gestures were the first step taken by humans toward the development of spoken language. To be a proto language, gestures have to be consistent enough, yet flexible enough, to carry novel messages. And they have to have an order, which leads to syntax. With syntax, you can build a story.

Critics claimed the meaning of gestures made by other primates only exists in the minds of the beholders, which in a way Russon acknowledges because, as she put it, that *must* be true for any system of communication. Pantomime requires a shared context in order for meaning to be conveyed.

Among the examples of pantomime Russon cited was the young female chimp in the Tai forest having trouble using a stone hammer to crack open a nut. Her mother spotted the problem, came over and took the stone hammer from her daughter, rotated it slowly while her daughter watched until she held it properly, and then several times in a row, slowly and deliberately, showed the young one how to crack nuts open with it. Then she handed it back, at which point her daughter picked it up again and held it as her mother had demonstrated, practiced hammering that way, and eventually succeeded in cracking nuts open herself.

Russon explained that this incident and similar examples of primate pantomime display the requirements necessary for language: consistency, flexibility, and order:

> "Even this limited data set on great ape pantomime shows these properties....They communicated messages as complicated as what tool to use, what action to perform with it, and on what target and who should perform it....These complex pantomimes suggest understanding the semantic relationship expressed, so they imply corresponding cognitive abilities: this is consistent with other evidence on great ape language and cognition."[37]

The orangutans' capacity to communicate through gesture included the creation of rudimentary narrative. She described one orangutan called Kikan who got a splinter in her foot. Her caregiver extracted it, then swabbed Kikan's foot with latex squeezed from a fig leaf to close the wound.

> "A week later, after gaining this caregiver's attention, Kikan picked a leaf and poked its stem at the sole of her (now healed) foot in similar fashion.... This case counters the common view that narrative is a uniquely human capacity. It also shows some of the

> components of episodic memory or reconstructing one's own past experiences as situated in time in that Kikan reconstructed key elements of a personally important experience. She was only an infant (three years old) so older orangutans may be expected to show greater sophistication."[38]

And also, Russon explained, orangutans had been observed pantomiming to *themselves*, something human children do as part of a kind of general "explanatory drive," something also previously considered to be on the wrong side of the fence between humans and apes. [39]

Finally I asked Russon my question: what is intelligence? I thought she'd be able to tell me. She'd been on the trail of it for more than twenty years, ever since Sophie and Spock's obvious calculations caught her imagination. The hows and whys of *C. elegans'* calculations nagged at me. And I had begun to read about an amoeba-like, single-celled organism formed from the temporary merger of a swarm of tailed, swimmer cells. During one phase of its complicated life cycle, *Physarum polycephalum,* usually called slime mold, is basically a jelly bag filled with thousands of nuclei that divide continually.

You can find slime mold in the damp under-story of any temperate forest and see it without a microscope. Its fruiting bodies cast spores everywhere when exposed to light, just like fungi, but it isn't one. Years ago, slime molds were used in biology courses as demonstration organisms. Lately, slime mold had been taken up as a model organism by people interested in new forms of computation.

Toshiyuki Nakagaki, a professor at Hokkaido University's Research Institute for Electronic Science, and its Centre for Creative Research, began to work with slime mold in the late 1990s. He was interested in the way living things solve network problems and whether their methods, honed by evolution, might be applicable to electrical engineering. He and his colleagues, Hiroyasu Yamada and Agota Toth found that in its jelly-like plasmodium phase, the slime mold will always ooze its way along the shortest path to reach two food rewards placed in a maze. Somehow, these brainless cells are able to solve complicated labyrinth problems efficiently. As they wrote in their *Nature* report, published in 2000, "This remarkable process of cellular computation implies that cellular materials can show a primitive intelligence."[40]

Calling a slime mold intelligent sounds almost crazy. Yet rats and mice are also tested in mazes and the faster they learn the maze, the smarter they are considered to be. Nakagaki had been awarded a 2008 Ig Nobel Prize (the anti-Nobel handed out for work that may at first glance appear odd or even silly) in cognitive science.

Inspired by Nakagaki's group, another group at the Unconventional Computing Centre at University of the West in Bristol, in the United Kingdom, directed by Andrew Adamatzky, also tested these molds in mazes. They made mazes that mimicked the networks of major highways in Britain. They found the molds chose routes very similar to those designed by the human planners'. The slime molds oozed their way through the best path to link major features, always finding "the shortest path between two selected points in a labyrinth," as Nakagaki describes it.[41]

If a slime mold efficiently routes a highway system, is it intelligent, I asked Russon?

"*I* don't know," she said, exasperated. But the more she considered the question, the more it became apparent to me she didn't think so, or at least that this sort of intelligence was not the sort that interested her.

"Intelligence has to do with the speed, the level, the complexity of your learning abilities. It has to do with modifiable, flexible behavior...If it's hardwired," she said, "it's not a form of intelligence I'm interested in."

The mold generates proteinaceous tubes that expand and contract, connect and disconnect, so it can draw in food where it finds it. It moves its jelly-like self with its thousands of cooperating nuclei by means of oscillating contractions that create a rocking motion. But how do its constituents decide to merge in the first place? How does it sense direction or the presence of food? And how does it decide to cooperate with others? It will merge with another plasmodium in order to do a better job of foraging, then disengage. And what about this: without a neuron to its name, it *anticipates* electrical shocks if you shock it *periodically*. So it has some form of memory, some sense of time, and some form of foresight, which makes it "a useful model organism for studying behavioral intelligence."[42] According to Nakagaki, its calculation of the right network geometry means it:

> "...met the requirements of a smart network: short total length of tubes, close connections among all the branches...and tolerance of accidental disconnection of the tubes. These findings indicate

> that the plasmodium can achieve a better solution to the problem of network configuration than is provided by the shortest connection of Steiner's minimum tree."[43]

In Euclidean geometry, Steiner's trees represent the shortest lines that will connect points on a plane. Steiner's minimum trees are used in electrical engineering to find the optimal layout of a network of wires. That made the slime mold's solutions better than an electronic engineer's normal method for solving a layout problem, which makes it useful for computing difficult network problems and probably useful for robotics too. The slime mold changes its behavior according to the location of the food in the maze. It somehow computes the shortest distance to grow its tubes right to it, saving the energy it would lose if it was inefficient.

Orangutans also try to minimize the expenditure of energy in their search for food. A question about how orangutans do that had induced Russon to give a graduate student the task of mapping individual orangutan's routes through the forest while they forage. It turned out they weren't just following traditional routes, doing what their mothers or other community members had done before them, although they do that. Nor did they always follow habitual routes; sometimes they'd travel and forage nearby. Nor were they forced to take particular routes because there was no other way to get where they wanted to go, though sometimes that occurred as well. When their GPS positions were plotted against the locations of specific fruiting trees in the forest, it was clear that, like foraging slime molds, they follow routes which lead most efficiently to available resources.

"Surprise!" Russon said.

And, like a shocked slime mold, orangutans in the forest are exquisitely tuned to time.

"The notion they don't understand time? How stupid can you get," Russon laughed. "I know here (in Toronto) that you can go around the 25th of September to pick apples of a particular variety." But in the rainforest in Borneo you can't rely on changing seasons to predict when a fruit will be ripe. "Many tree species don't synchronize their fruiting. So time prediction? You can't build it seasonally. Orangutans are fruit specialists so if they couldn't use such time cues, they would have a hard time surviving."

In short, orangutans are intelligent because they adapt their behavior and learn from experience.

"They're *dead* smart," she said.

But slime molds adapt and learn too.

4

Boundary Issues

I read in *Macleans*.ca that the orangutans of the Toronto Zoo would soon be using iPads to communicate with orangutans elsewhere. Very quickly thereafter I drove out to talk to one of the academics involved in the Apps for Apes project at the main York University campus, once my daily destination on the Little Red School Bus. The campus's architecture had evolved in the interim. Now it offers examples of every architectural fad of the last half century. Trees—all the same height—had colonized the former wide open spaces.

I found my way to the office of Suzanne E. MacDonald, then head of York's psychology department with a cross appointment to the department of biology. She is the anti-Russon, conducting comparative cognition experiments in labs and at the Toronto Zoo. Cognition is the science term for perceiving, learning, knowing, judging—in other words, the ways and means of intelligent behavior. MacDonald tests different species to compare how they do it. She began her career mainly studying pigeons, but now spends time with gorillas and orangutans. She tests them in the small behind-the-scenes rooms at the Toronto Zoo where the gorillas and orangutans spend the cold winter months locked up. She can do her experiments there only when the zoo keepers find it convenient, the animals are willing, and nobody is sick. This means that her studies take months when they could be done in a proper lab in hours—if you could bring an adult orangutan safely into a proper lab. According to her, the gorillas and the orangutans like doing the tests she devises. It eases their boredom.

Easing zoo animals' boredom is a side expertise of hers, one she developed during her postdoctoral days at University of British Columbia in the late 1980s. After she visited Vancouver's Stanley Park Zoo (now closed) and saw the soul-destroying barrenness of its animal enclosures, she voiced her outrage to the director who asked her to help make the place more interesting for its inmates. She has been volunteering to enrich zoos ever since.

As MacDonald explained, with the right reward (M&M`s), it takes about four seconds for orangutans to learn how computer touch screens work. The orangutans won't touch screens directly with their fingers, only with sticks (and they are picky about the kind of stick too. They are concerned with the shape of it, the smell of it.) Most orangutans understand that the things that appear on computer screens are not in the room with them but are representations. Some gorillas get it, but not all: some gorillas try to eat a magazine page when it shows a picture of a banana. Bonobos (close relatives of chimps) seem to understand representation, or at least most do, as do some chimps, but again, not all.

The orangutans' capacity to distinguish real from image is what gave rise to the idea that the orangutans in North American zoos should all get iPads to play with. That way they will be able to see each other, which might be stimulating, and who knows, might even lead to love (and more successful zoo breeding programs).

It had certainly led to good PR. The story about Apps for Apes seemed to be in every paper with an online presence. Unfortunately, no one at the Toronto Zoo had returned my phone calls, which is why I'd chased down MacDonald.

She sat at her table, a tallish, red-haired, attractive woman, very energetic and full of ideas. Along with a graduate student, she had just asked a very interesting question about orangutans' taste in music. They had decided to create a music app for the orangutans. Why? She is a pianist and thinks music is intrinsic to human societies. As she put it, music's deep emotional resonance is why we remember the words of songs we heard more than twenty-five years ago. Her question was, what music would the orangutans like to hear?

I had spent an evening that very week in a neighbor's living room, listening to her son's string quartet rehearse for an upcoming concert. All of us—friends, family, the cat—fell into a reverie at the first strike of string on string. We went somewhere with Mozart, Beethoven, Handel, transported to a wordless yet structured story/world woven from the voices of cello, violins, and viola. The physical became mental, the emotional merged with the intellectual. We anticipated where a phrase was going, and were delighted when we were surprised. Players tuned to each other. Fingers tapped, toes moved, heads bobbed, as sounds became meaning. The meaning could not be reduced to words but it was there and it was shared.

Is the ability to make music part of what we mean by intelligence? Music lets us sync with each other *without* words, but more than that, organized sounds convey sadness, joy, amusement, and tell a story. Music is built from invented relationships between tone, timber, and time, but these are all conventions passed on or reconfigured like motifs in literature or painting, so music is an always morphing cultural construct. Does it increase our overall fitness, as the Darwinians would ask, or is it just something humans do that doesn't matter? What about other animals?

MacDonald explained that music has always been piped in everywhere at the Toronto Zoo, mainly to keep the keepers happy. She thought a music app could make the orangutans happy, too, though she believed they shouldn't have to listen to stuff they didn't like. She and her student had rigged up a test that allowed the orangutans to select their preferences from a variety of musical genres (kids' music, country, pop, classical) by touching different colored areas on a computer screen. Kids' music was red, classical was blue, etc. She stuck in one grey area to signify silence, as a control, and to relieve them of the burden of sound, if they preferred that.

To her astonishment, the first female orangutan she tested always chose gray, the color of silence. She even made a shoving gesture as if to say 'make the music go away.' In fact, all six orangutans she tested chose silence. It had never occurred to MacDonald that they might actually dislike all music and prefer quiet, but they did. What is so different between their world and ours, she wondered, that for us music is vital, while for them it could be unwelcome or annoying?

These differences between species are what really interest her. Humans (and birds) make music and sing. But no other primates sing, although some vocalize, and some make drumming sounds. And yet "you can train a pigeon to distinguish Beethoven from Mozart," MacDonald said. While no one had yet tested chimps or gorillas on their musical affinities, she didn't think they would have any, either. Why? "You don't see music produced by great apes. A (chimp's) hoot is not a note, it's not enjoyment. It is a communicative gesture," she said.

We are biased about everything we do, she continued, we are always thinking of ourselves as inhabiting the top rung of the complexity ladder. That's why she always asks her undergraduate students if there are some things only humans can do, and someone in the class always says yes and offers an example; recently it was only humans are moral which MacDonald had been quick to point out is not the case, at least according to the work of Frans B.M. de Waal (and others). These answers always take her to her real point. The real point is to find out what other species do that humans don't even notice.

"Dolphins can echolocate, they can hear right *through* an object in complex ways I can't even imagine," she said.

I asked if I could go see the orangutans working with the iPads at the Toronto Zoo.

No, she said. Not possible.

Why not?

Because they live in very close quarters back there, and we have to be careful about exposing them to human diseases, so the answer is no. (Months later, someone at the Zoo let in a science journalist doing a report for the CBC, though no one there had returned my phone calls. Her answer could have been for *you*, no.)

Where else should I go then to see primates up close? Yerkes? I asked.

The Yerkes National Primate Research Center just outside Atlanta, Georgia, (with a field station in another town nearby) is a famously infamous place where non human primates have been studied for generations—not just to learn about their intelligence, (although Robert Yerkes was certainly interested in that when he set up its predecessor operation in Florida before WWII). At Yerkes (and a few other research centers) they are also used as stand-ins for human beings (known as models) in biomedical studies which can be invasive and painful. Some of these research subjects have been sprung lately and taken to 'sanctuaries' such as The Fauna Foundation, near Chambly, Quebec,[44] where Russon's friends, Sophie and Spock, can also be found. (Another sanctuary in Ontario would be in the news soon as the new home of Darwin the monkey, found wandering in a parking lot in Toronto, dressed in a custom sheepskin jacket.) The US National Institutes of Health, on the advice of the Institute of Medicine, had announced in December, 2011, the end of funding for new biomedical or behavioral studies on chimpanzees ("our closest living relatives") *unless* necessary to advance human health or of sufficient scientific value to offset "moral costs." In addition, it ordered that all chimps used for such research must be kept in ethologically correct conditions (whatever that means) or in natural settings.[45] The US was far behind other counties: in 2010, the EU banned the use of apes in experiments, and placed strict restrictions on the use of all other primates. The philosopher Peter Singer, like Jane Goodall, has long argued that great apes should be recognized as persons and should have rights. The more I talked with MacDonald, the more it seemed to me that Singer et al had it right.

Yerkes is a sad place, she said, shaking her head. She said I should go instead to Des Moines, Iowa, to the Great Ape Trust.[46] She wanted to do her sabbatical there, if she could. It had specially designed rooms with glass observation walls so primates could be observed but not endangered by our germs. Not that there were many primates left there to observe. Four orangutans that used to be there had been removed by a man named Rob Shumaker when he resigned from the Trust and moved with them to the Indianapolis Zoo, after some sort of trouble there. It was mainly bonobos there now, such as Kanzi and his half-sister Panbanisha, and a baby called Teco.

Her eyes flicked to me, and then away. Without words she asked: are you wise to what I'm really saying?

But I wasn't wise, so at first I failed to pick up her signal. I was thinking instead how weird it is that a zoo development person could take four intelligent orangutans with him from one job to another like so much personalized luggage. How had he got effective ownership of animals so smart they can be taught to use computer touch screens in four seconds flat? Why should any animals that smart be locked up and displayed in zoos at all? She had said, when talking about her work with zoos, that elephants shouldn't live where there's winter but better they should live in a zoo than that there be no elephants left alive just because humans are changing everything to suit themselves. That way, when the humans are finally gone, the elephants could take over. "I feel we're a bad species," she told me.

When I asked how one guy could schlep four orangutans to Indianopolis, she said that "may be part of the story at the Great Ape Trust. Get to the Kanzi stuff. Life will be happy then." Followed by another eye flick.

Hadn't she said something about trouble? Trouble is code for a good story.

What trouble? I finally asked.

Well, you know about Kanzi, the Einstein of the primate world, she asked.

I did not, I said. I had heard about Koko the gorilla with an IQ of eighty. Or was it thirty or forty?

Oh, she said, that's Penny Patterson. Patterson, who has a doctorate, is mainly a passionate advocate for gorilla conservation. I should understand that IQ only has meaning when it's measured against a population of others of the same age. You can't have an IQ that is a stand-alone proposition. No, Kanzi is the ticket, she repeated. Then she pointed at a colorful triptych with weird symbols arrayed in lines which hung over her couch. Kanzi's lexigram, she said. And there was a book about Kanzi on her desk.

Kanzi is a male bonobo. Bonobos (*Pan paniscus*) behave very differently from chimpanzees (*Pan troglodytes),* but are such close relatives that they can interbreed. Instead of fighting for dominance like chimps, bonobos resolve their differences by making love constantly—without regard to age or gender. Kanzi, according to MacDonald, had also learned how to communicate using human-designed symbols called lexigrams without having been directly taught them. Kanzi lives at the Great Ape Trust in Des Moines in the care of Dr. Sue Savage-Rumbaugh.

Anne Russon had mentioned that Duane Rumbaugh and his wife Sue Savage-Rumbaugh had tried to teach chimps to 'talk' with computer symbols instead of ASL. They had wanted to use what they learned about language acquisition by apes to help "severely mentally retarded" children develop language skills. Sue Savage-Rumbaugh apparently came to believe that humans acquire language so easily because we are immersed in human language cultures from conception. She thought if she immersed bonobos in human language culture, they could learn to use language too.

What an idea, I thought. If language has driven the development of our intelligence, would the same thing happen with bonobos? If they could learn to read symbols and use them to communicate ideas, why wouldn't they end up just like us?

With Kanzi, said MacDonald, Savage-Rumbaugh had succeeded, and had come to startling conclusions about Kanzi's capabilities.

In 2005, Savage-Rumbuagh had left her husband/partner Duane Rumbaugh and Georgia State University where she'd worked for twenty-three years and gone to work at the Great Ape Trust. It was created by Ted Townsend, a Des Moines-based heir to a meatpacking equipment design company called Townsend Engineering. Savage-Rumbaugh had taken the bonobos she previously studied at Georgia State to Townsend's purpose-built, state-of-the-art facility, surrounded by many acres of open fields and idyllic woods, arrayed around a pond. However, things had taken a wrong turn.

MacDonald printed out a story for me that had been published recently by the *Des Moines Register.* It said that after a series of complaints by former staff, Sue Savage-Rumbaugh had been escorted off the premises while the Board investigated the allegations against her. There were hints about animals being in danger and Savage-Rumbaugh being over the edge.

Is she? I asked.

"There have been rumors about her and *boundaries,*" MacDonald said.

Galdikas had also been accused of having issues with boundaries. And we heard rumours at *New Wilderness* that Dian Fossey had boundary issues too. The word was that Fossey drank too much and slept with particular gorillas. Male gorillas.

What sort of rumors? I asked MacDonald.

"Like she slept in the cage with one of her bonobos for three months when she (the bonobo) was pregnant," MacDonald said. (Savage-Rumbaugh reported herself that she briefly slept with Kanzi when his adoptive mother Matata was removed for breeding.)

Unsaid words humped in the air.

Are there any rumors like this about *male* primatologists? I asked.

She shook her head, no, but pointed out that hardly any primatologists are male. "The primate males don't like human men. Too aggressive. Women know how to smile, to be submissive." She would never forget how frightened she was the first time she saw a male chimp do an aggressive display.

"I hit the deck," she said. "…absolutely terrifying."

While MacDonald thought that Frans de Waal was not wrong in his beliefs about chimpanzees' capacities to reconcile, she thought Sue Savage-Rumbaugh had not yet scientifically proven that she remodeled the brains of her bonobos by introducing them as babies to human language and symbols.

MacDonald made noises about having to leave to teach a class. I asked again about getting to meet the orangutans in the back rooms of the Toronto Zoo.

Too small and dark, you won't be able to shoot there.

But I don't need to shoot, I just want to look, I said.

Des Moines is best, she said. Go there. And tell me what you find out.

Then she offered me candy from a bowl as my reward.

I ate that candy and went right to my computer when I got home.

There had been several stories about the Trust, sometimes also referred to as the Iowa Primate Learning Sanctuary or IPLS, in the *Des Moines Register*: there were the usual feel good kind that newspapers like to run about animals, but there were some that reported on troubles within the kingdom. Sue Savage-Rumbaugh had been forced out of the Trust once before, in 2011. This time—2012— almost all the Trust's former staff who had resigned after the last kerfuffle, (except a relative and a long-time animal keeper) had signed and sent a letter of protest to the board. Their complaints boiled down to accusations that Sue Savage-Rumbaugh was a danger to the animals and to herself.

The board of the Trust had placed Savage-Rumbaugh on administrative leave, then issued a statement within a matter of weeks saying that she had been reinstated as Resident Scientist because the former staff members' charges could not be substantiated. The well-being of the animals in the facility, the board said, was being overseen by a veterinarian (who knew a lot about keeping cats and dogs healthy). Shortly after that, two respected academics, including Dr. Ed Wasserman, a well known expert in comparative cognition, resigned from the board.

The website carried a message from the founder Ted Townsend about what had been envisioned when he set up the Trust in 2002. It was to be a world center for the scientific investigation of the "origins and future of language, tool-making, culture, and intelligence." But there was not much noted on the website about the science done during the Trust's operations. I found only a few works (three) authored or coauthored by Sue Savage-Rumbaugh between 2008 and 2011, the same year *Time* named her as one of the 100 most influential people in the world (citation written by Frans B.M. de Waal.) I found a few recent papers with her name on them, mainly concerning language and tool use. Most had been coauthored—with a UCLA psychologist, Patricia Greenfield, with William Field, Itai Roffman, W. Hopkins, and a woman named Heidi Lyn.

No one answered the phone at the Trust or responded to emails.

Then I came across a blog called *Chimp Trainer's Daughter.* It is written by a woman named Dawn Forsythe, an Amercian civil servant who grew up under the thumb of a father who trained chimps to entertain the crowds at the Detroit Zoo. On someone else's site—the Intimate Ape blog—she described her father as a drunk who had abused his wife, his kids, and the chimps under his care until the Zoo fired him. Forsythe's blog is a fascinating mix of personal stories, investigations, and advocacy, with a sharp eye fixed on the way humans use apes for their own profit.

The former staff members' protest letter to the Great Ape Trust's board of directors had been posted on her blog, along with various anonymous contributions. The former staff claimed that Savage-Rumbuagh had insisted that the puppies of the adult dogs roaming the Trust campus talked to her in English. They also claimed she had permitted incestuous ape relations resulting in "unauthorized" pregnancy; that she had accused the staff of deliberately cutting the feet of baby Teco; that she had taken Teco for a spin in her car and to public events exposing him to human germs, and worse; that Kanzi had become morbidly obese and wore a leash like a dog. They warned there was not enough staff left at the Trust to take care of the animals and some would die.

Forsythe, who had taken a lesson from Woodward and Bernstein, tracked the Trust's spending through its reports to the IRS. Any American 501(C)-3 charity is required to file annual reports of donations received and dollars spent. Forsythe had found no such reports posted by the Trust after 2010, the year Townsend gave notice that he was unable to continue supporting the place. By then, the Trust had run through at least $15 million, most of it Townsend's money. In the interim, the Trust had failed to get significant new support. Yet, among the board's reasons for reinstating Sue Savage-Rumbaugh in 2011 was her fundraising capacity.

Allegedly, researchers at the Trust had applied for federal research grants but none were awarded after 2009. So the Trust had applied to the US Department of Agriculture for the right to exhibit the animals to the paying public. To Forsythe's outrage, that right had just been granted.

Then, in November, 2012, Panbanisha, Kanzi's half- sister, died of a pneumonia-like infection. Panbanisha had also been adept at using lexigrams to communicate. The eulogies posted by some scientists who had interacted with Panbanisha were what one would expect upon the death of a dear friend, not a study subject. One even had a tinge of the mystical about it in that it was addressed to the dead

Panbanisha, as follows: "I know you will look after us now as you did in life...we will do everything to get you the personhood you deserve..."[47]

This left me wondering: if Kanzi is so smart that he taught himself to communicate with lexigrams, he's as smart as we are. So why does he wear a leash? Is he not deserving of the same rights as the rest of us? Don't rights start with liberty? How could anyone who had proven how smart bonobos are think it is right to exhibit them?

5

Enter Politics

Move the time cursor back to 1983/1984 and the third production season of *New Wilderness*. Stephen, Chuck, Lorne, and their investors were embroiled in the kind of petty wars that can and do erupt during the production of any television series. People start to feel very important. Then the brown stuff hits the fan. The first skirmish was over who got to go to Cannes. This led to a spousal shopping battle concerning a watch.

The 1980s were like that: sunny and dark, greedy and generous, brave and stupid. President Ronald Reagan, the Great Communicator, went about offering cheery waves, toothy smiles, and long pauses between meaningless phrases. There was an explosion of cheap crack on the streets for the riff raff, and high grade cocaine for those on top of the heap, plus gated communities to keep winners and losers apart. Reagan's intelligence people couldn't get money from Congress to fund the Contra insurgency in Nicaragua so they sold missiles to the Iranian Revolutionary Guard, supposedly in exchange for hostages, and used the cash to buy guns for the Contras. Some middlepeople made a lot of money. The movie character Gordon Gecko mocked/exposed the Wall Street investment bankers who toyed with the lives of millions as they bought and broke up industrial conglomerates. We had special words and phrases to characterize their machinations: greenmail, corporate raiders, hostile takeovers, insider trading. Magazine writers like me were expected to cover business as epic drama.

We had celebrated *New Wilderness's* financing at a Santa Monica restaurant so chic that it had no name on its door. If you knew about it, you were important enough to eat there. But soon enough, things had turned real. The *New Wilderness* production offices had to move to a new location, a semi-derelict house-turned-office-building on the corner of Toronto's Queen and Simcoe streets. The paint peeled, the radiators leaked, the stairs creaked, the bathroom was past vile. The big room beside the main entrance had window walls on the street corner. There were a lot of accidents. Desks there were called the death seats.

CTV was happy with the series because it drew up to two million viewers a week to its pre-prime time slot. It was selling in the US in what was called the syndication market. Figuring out who was owed what was all very complicated. Wham, along came a law suit in California, launched by a fired distributor, a very expensive place to get sued. The Los Angeles defense attorney said things like: "don't try to make more than ten points, judges get confused..."

So, by the end of the second year, there wasn't enough money to finish the promised shows unless the masters became the slaves. Stephen, Chuck and Lorne deferred their fees. I had to do lots of story editing and script writing or everyone in the Dewar household would go blub, blub as we went under. I would have preferred doing magazine pieces and being paid next to nothing for the privilege, so I wasn't happy. Anyone interested in irrationality in economic decision-making should have studied us.

At a certain point Stephen said, hey, babe, there's a chimp show we want you to do.

A fucking chimp show about fucking what? I bellowed. When you're not happy, you talk like that.

He rattled on about a group of chimps who lived in the Burgers Zoo in the town of Arnhem, in the Netherlands, in what he called a colony. They roamed together in a big outdoor enclosure during the summer months. In the winter, they were kept inside in a big hall by day, separated by gender in cages at night. They had interesting politics, he said.

I was willing to believe that chimps are very intelligent. I had read enough of Goodall's work to know chimps in natural settings organize themselves in hierarchies with everybody showing submission to an alpha male, the females making way for a dominant female. But politics?

Politics feature finesse, duplicity, aspirations, fears, coalitions, plots, common causes, big tents, betrayals, schisms, not to mention monarchs, Parliaments, dictators, and of course, baby kissing. In other words, politics require great intelligence. Stephen had specialized in political theory. I had studied it too. We both took the political philosophers more seriously than all the rest. Libertarians and followers of Ayn Rand who argued that the State should be slashed to release the pent-up creativity of individuals had begun to pop up in public life like so many prairie dogs. But humans without societies don't last long, and societies need functioning institutions, such as governments, to mediate between conflicting interests. When a libertarian gibbered within range, either one of us could be counted on to raise Aristotle's banner and shout: a man without politics is a beast or a God!

Here is a snippet of what Aristotle said about the origin of politics and states in The Polis section of *The Politics* (available on line thanks to Fordham University):

> "...he who by nature and not by mere accident is without a state, is either a bad man or above humanity; he is like the 'tribeless, lawless, hearthless one,' whom Homer denounces, the natural outcast is forthwith a lover of war: he may be compared to an isolated piece of draughts...."

Aristotle acknowledged that some animals are social but maintained that there is a difference between the sociability of humans and beasts. He located the boundary line precisely at politics, specifically at the power of speech, which he assumed that only humans possess, and which has a political purpose. For Aristotle, everything was defined by its purpose, meaning its developmental end point, known as *telos.* He said:

> "...language serves to declare what is advantageous and what is the reverse, and it therefore serves to declare what is just and unjust. And it is the peculiarity of man, in comparison with the rest of the animal world, that he alone possesses a perception of good and evil, of the just and the unjust, and other similar qualities; and

> the association in these things which makes a family and a polis." [48]

In other words, politics arise from the human need to co-operate in order to live better than one ever could—if one could—as an individual. Language makes it possible.

But never mind Aristotle: politics was the boundary line *I* drew between intelligent humans and all the rest.

A *New Wilderness* researcher handed me an article by Frans B.M. de Waal on coalitions among the chimps at Arnhem. At that point, de Waal, a young Dutch ethologist, had only published a few papers. While he had excellent credentials, he had nothing in the way of a big reputation, certainly nothing to match that of Goodall, Fossey, or even Galdikas. Yet two years before he had published a book in the US about the relationships among these chimps in Arnhem called *Chimpanzee Politics: Power and Sex Among Apes.* We got a copy.

On page four, de Waal declared: "Entire passages of Machiavelli seem to be directly applicable to chimpanzee behavior," especially in their drive to get and hold power, a startling claim.

Niccolo Machiavelli, a very sophisticated Florentine, was an out-of-work former civil servant when he wrote about political strategies and tactics in *The Prince.* He aimed the book at the Medici. They had just returned to power in Florence after a Republican interregnum. They'd arrested Machiavelli, and tortured him, but let him go. Machiavelli had then retired to his estate where he toiled by day on the farm and communed with the greats by night, writing all sorts of works that were more popular in his day than *The Prince*. But *The Prince* had staying power. People have been reading it for about five hundred years, and likely will be reading it five hundred years from now because it is so insightful about the ways and means and uses of political power.

At first, Machiavellian seemed to me to be a ridiculous adjective to apply to chimp behavior. But given de Waal's credentials, his insistence that chimps display intelligent strategies and tactics had to be taken seriously. Written for the public, but with a wary eye on his colleagues, de Waal did much more than just describe what he saw chimps do at the Burgers Zoo. His book was also an argument about the evolutionary, or biological, roots of complex human political behavior. He showed that chimps do what we do, and sometimes do it better. He said that chimps constrain their violence, and reconcile in order to keep the peace and that we could learn from that.

Reading facts and arguments is one thing: seeing the chimps in action is quite another. I sat in a dark editing room watching raw footage shot at the zoo by cinematographers Peter Fera, Bert Haanstra, Chris Ploeger, and Didier van Koekenhert, listening to the chimps' screams and hoots and pants, watching them rough each other up, and hold each other tight, the big males kissing each other (and the babies) their arms flailing out in gestures of supplication, their teeth bared in anxiety, their fur raised in the fury and the glory of power.

We made our first chimp show from this footage. It explored the relationships among the colony's members. They had dramas, reconciliations, heartbreaks, and perpetrated great deceptions. The alpha male was expected to protect the weak from the stronger: in return he got his just rewards (sex). He had to terrify his competitors while acquiring his status, but he had to make alliances with the losers to retain it. He had to bully his way up the greasy pole, then curry favor by sharing the perks to stay on his perch. That seemed very familiar.

Each chimp, like each of us, had his/her own totally unique face.[49] The more I watched that footage, and matched faces to names, and names to the events detailed in de Waal's book, the more I began to see them as characters, as people. They lived in difficult, tight quarters, dominated by a few big males, and three older females who could be cranky or sweet, depending on circumstances. The juveniles were just as obnoxious as your average human teenager, always bounding in where they weren't wanted, aggravating their elders, terrorizing the young. The adult males seemed to spend their time quelling the juveniles and subverting the alpha male in order to try to have sex with any ready females. They hid from the alpha and from each other, bowed down to him, and joined up against him. This bowing down was almost identical to the way humans once abased themselves before feudal lords. The females did the same, sneaking away to have sex, squabbling among themselves, patching things up, protecting their children from each other, from the juveniles and from the big males, taking sides depending on their own interests and on who was up, who was down.

Shakespeare would have loved this material. These chimps were crafty, calculating, sneaky, ugly, petty, kind, consoling, lustful, ruthless, innocent, not so innocent, sweet, funny, sad.

The oldest male was named Yeroen. There was a younger male named Luit, an even younger and bigger newcomer named Nikkie, "the moving force behind all the developments" according to de Waal, plus young Dandy. Mama, the oldest female, had been the first alpha of the colony in its early days when when there were no big males, but when Yeroen joined the group he eventually displaced her, until Nikkie and Luit came along. Yeroen was dethroned by Nikkie, and Luit, who became the next alpha. But Yeroen did not take this lying down: he would not bow to Luit for a long time and soon Nikkie began to challenge Luit, too. The demoted Yeroen soon unfurled a clever strategy. He made common cause with Nikkie, or Luit, playing one off against the other. Luit was shaky on his throne and so he made friends with the females, especially Mama, who began to support him in his squabbles. Then, when Nikkie failed to support Yeroen as he tried to keep Luit away from a female he wanted, their coalition fell apart. Temper tantrums, fights, bites ensued. For a while, no one was in charge. Eventually, after a long period of instability, there was another big fight and Nikkie emerged the clear winner, the new alpha.

Luit actually shrank as he mourned the loss of power. Nikkie puffed up his fur and became glorious: but he was also arrogant. When Yeroen was alpha, he got as much sex as he wanted. When Luit was in power, Yeroen still got a lot, but not enough to satisfy him. Sex was Yeroen's deal point. When Nikkie took power, he seemed to forget that if he didn't watch it, he could be toppled too. Nikkie behaved as if power was the same as justice which is the interest of the stronger. Yet Yeroen could make sure Luit was elevated again, or, if that didn't work, there was another lad waiting in the wings for Yeroen to turn to—Dandy. This was a triangle that could turn into a square: it involved wordless yet subtle renegotiations over the terms of power.

The footage was ambiguous about what happened next. Luit seemed to drop out of the picture. Why hadn't Luit made a new coalition with Yeroen to regain his rights to the females? (It turned out he had, but it didn't last.) Come to think of it, where was Luit? First, he was there in the footage, and in the book, then he just....wasn't. The reporter in me thought: if this was *Macbeth,* or *Hamlet,* Luit would be dead. You've got to *kill* the king, never just wound him, as one corporate raider I knew used to say. That phrase kept circling in my head.

I asked a researcher to call the keeper at the Burgers Zoo to ask after Luit.

Oh, said the keeper, Luit died.

What? Why? Asked the researcher, with me standing to one side yelling: When? *How???*

It turned out Luit had been murdered in the sleeping cages one night late in the summer of 1980. The next morning, the keeper found the place awash in blood. Luit was a mass of gashes and wounds, with several toes missing. More important, Luit's scrotum had been bitten right through by Nikkie and/or Yeroen. His testicles (and missing toes) were found on the floor. They may have attacked him while he was asleep. Dandy must have been worried that something bad was going to happen because he had made it clear for some time that he would only sleep in a cage by himself. Though Luit had bled out badly by the time the keeper found him, he still wanted to keep his frenemies close and stay with Nikkie and Yeroen. He died though the vets struggled to save him. Shortly after that, Nikkie was attacked by Puist, one of the leading females who had been particularly close to Luit,[50] when they were all let out into the big yard. Soon a new triangle emerged: Nikkie, Yeroen, Dandy.

To my surprise, there was not a word about this regicide and its aftermath in the 1982 edition of de Waal's book, which touted the peace-making ability of these chimpanzees—so very intelligent, so very deceptive—who, according to de Waal, could plot and scheme, deceive and attack, yet also calculate just how far they could push each other and just how much could be forgiven. At first I thought that when the murder occurred the manuscript must have been too far along for de Waal to make changes. But then I realized that this murder had happened two years before publication. In journalism there is a rule: you never bury important facts, and you certainly don't leave them out altogether.

We're gonna tell it the way it happened, Stephen said when I conveyed this news to him, and we're gonna shoot Part II. So I wrote a shooting script, and we hired a cameraman to go over to Arnhem. His orders were to be there as the chimps emerged from their quarters and to watch for the dynamics among Nikkie, Dandy, Yeroen.

Meanwhile, I wrote the script for the first show. I included a description of the murder and the nascent coalition between Yeroen and Dandy that might lead to Nikkie's downfall. But when the researcher checked these facts with the keeper, we heard that Nikkie's reign was already over. Dandy was up and Nikkie was down. In fact, Nikkie was dead. One day not long before, Nikkie had run outside with the other chimps screaming behind him, and charged straight for the moat, where he drowned. He'd behaved as if he knew what his fellow male chimps would do if they caught up with him. Nikkie had run for that moat before, but had survived because it had been covered the first time by a thin skim of ice. This time, no ice. Was this a suicide or a mistake? We thought it might be a suicide and others did too.[51]

So that also went into the script, along with the upshot: Dandy had become the new alpha. Our second show would be about whether Yeroen would use yet another young male—the growing Wouter—to keep Dandy in line.

The more I thought about it, the more it seemed to me that de Waal must have left the murder out of his book because it was contrary to his thesis. On the other hand, I thought the work he did publish was commendable. Nobody before him had been so insightful about how chimps hold their societies together. No one else had made it so clear that you can't draw a line between their kind of social intelligence and ours. But de Waal was also a scientific consultant to our show. I was worried that when we called to check our facts with him there would be a massive fight. I wanted the responsible producers to hear what was said.

We all went to Chuck's office and got de Waal on the speakerphone. Stephen told de Waal if he didn't agree with anything in the script, he could take his name off the show, but Lorne Greene would say what we knew to be true no matter what. The truth is we were covering the politics of the chimps at the Burgers Zoo the way we would have covered a political assassination and its aftermath. We were telling a very 'human' story about politics, sex, jealousy, murder, and despair. I had completely forgotten by then that I once thought of politics as the boundary dividing humans from all the other animals.

While Chuck took notes, I read de Waal the words Lorne would say, sentence by sentence.

He was unhappy, he wanted to know where we got what I had written, but he confirmed the facts. He didn't take his name off the show. We hung up.

Slam the cursor forward thirty years, to the obverse of the Reagan Eighties. Obama, the rhetorical president, was being subverted at every turn by a House full of libertarian Tea Partiers. Bankers had become villains, and the gated communities were plastered with for sale signs. Edward Snowden was about to show us that the freedom-touting New Media moguls in Silicon Valley had shared all our secrets with the US intelligence community with barely a whisper of resistance. Snowden would become the new Gecko. And I had an appointment with Frans B.M. de Waal, in Atlanta.

De Waal's work had shown me that Aristotle was wrong. Human political behavior is neither unique to our species nor dependent on spoken language. It is rooted in (but not defined by) biology. This idea thrilled the right wing, just as an older generation's arguments about aggression had done. Chimpanzee politics had even begun to influence the human kind. On the flap copy of the latest edition of *Chimpanzee Politics,* published in 2007, *Business Week* was quoted to the effect that Newt Gingrich, the former Republican Speaker of the House, had been a fan for years. Inside, de Waal mentioned that the book became required reading for freshmen congressional representatives.

More to the point, his early work had changed the way I thought about whether humans are uniquely smart. In the Arnhem chimps I saw myself and my friends, and many politicians in their huffed-up, puffed-up glory. De Waal had even changed the way I behaved toward my children. Thanks to him, I understood that no matter what their parents do, boys and girls will develop with certain tendencies and at different paces. He showed me that as a feminist, my task was not to re-litigate biology-as-destiny, but to ensure that the start point is not the end point.

In the years since *New Wilderness,* de Waal had also advanced the field, created new techniques, brought ethology into the lab, and the lab to ethology. He had moved on to the comparative study of the intelligence of bonobos, chimps, and macaques. He was even testing elephants. His results were so interesting that they were changing the questions about what intelligence is, what culture is, and the nature of the interaction between them.

As Russon had made clear, among psychologists the word culture has acquired a very broad meaning. It encompasses much more than art making, storytelling, singing, and dancing. Culture can mean opening a nut in a certain way, with a certain kind of stick. It is both a byproduct and handmaid of sociability, like an invisible, external, communally-shared brain that augments individual memory and learning for many different species, from primates to cetaceans. In other words, culture is group intelligence.

De Waal's recent experiments confirmed the pattern Russon had found about how new ideas spread among primates. He showed that the use of a new tool will be adopted much more quickly in a captive chimp community if chimps learn it from another chimp rather than from a human.[52] He had shown that chimps will synchronize their actions to work together to make sure everybody gets a treat. Another paper described chimp and bonobo gestures and facial expressions as the being the roots of human language, as a kind of language of their own. The facial expressions are the same between the species, but the gestures are different. They are more nuanced and flexible among bonobos than among chimps, but for both,[53] they convey meaning.

This work paralleled a track opened many years earlier by University of Pennsylvania psychologists and partners Robert Seyfarth and Dorothy Cheney. Instead of trying to teach animals a human language, they had looked at how animals communicate naturally. They had taped the specific calls certain monkeys make when faced with specific dangers.[54] When they replayed these calls to a group, the correct escape behaviors ensued. In 2011, another group took Seyfarth and Cheney's thinking a step farther. Working with bonobos, they tested whether single calls or *call sequences* convey meaning. They found the call sequences were what mattered.[55]

But I was not looking forward to this meeting. It still disturbed me that de Waal may have held back vital information at the beginning of his career. Not wanting to do him an injustice, I had gone over his publication record to be sure about when he first published on Luit's murder. In April of 1984, he submitted an article about the Arnhem colony to the US journal *Ethology and Sociobiology*. It described coalitions and reconciliation among male chimps as being different in kind from those displayed by females. The article made clear it was based on data gathered up to the "first half of the summer of 1980," in other words, just before Luit was murdered. It did *not* mention Luit's death.[56] In September of that year, he submitted to the same journal an article describing Luit's murder. However, he didn't revise this article until 1986 which is when it was finally published. In other words, though our show was shown to millions and his name was attached to it, he kept Luit's story from the world of science for six years. Even in the 1986 article, he made no mention of Nikkie's 1984 death in the moat.

So I had questions.

I had emailed de Waal, by then the C. H. Candler chair of Psychology at Emory University as well as Director of the Living Links Center at the Yerkes National Primate Research Center, and asked to see him. On the phone I explained I was working on a project about intelligence. I was truthful, but I didn't tell him the *whole* truth. I didn't mention *New Wilderness* or that I had worked on the program.

I could sense wariness in him. Perhaps he was sensitive about the 2012 Ig Nobel Prize in Anatomy he'd just been awarded (along with Jennifer Pokorny) for demonstrating that chimps recognize pictures of their friends' bums.

Why not meet in Minneapolis, he suggested, he was going there on the publicity tour for his new book, *The Bonobo and The Atheist: In Search of Humanism Among the Primates.* In this 2013 book, he located the origin of morality deep down in the primate order.

I would also like to see the chimps at the Yerkes Field Station, I told him.

His schedule made that impossible.

Well how about Yerkes at Emory, then?

Oh no, way too much red tape and bureaucracy for that. There had to be supervision, the PR department had to be involved. No, it would be better to talk in his office at Emory.

So I found myself walking across the Emory campus on a clear, cold Monday morning. I found the Psychology building just down the road from a row of sorority and fraternity houses. De Waal was waiting in his office, dressed in shirt sleeves and dark sweater, dark pants, sensible leather shoes. His hair, once a big puff of dark curls, had gone silvery and thinned out. On the wall between the windows, where he could see it, he had hung a drawing of the chimp enclosure at Arnhem. The wall beside his chair displayed headshots of his subjects.

I sat down at the end of his u-shaped desk and began to ask him questions about his personal history: I wanted to know what shaped him, but I was also stalling. You never know when the time has come to ask the tough question. Too soon and you get less than you came for because you are shown the door. Too late and you're out the door before it's out of your mouth.

He said he was born in 1948, in the Netherlands, in the town of Den Bosch, in the southern, Catholic part of the country. His parents had six children, all boys. His father was the director of a small bank. His mother was a housewife.

Rich, then? I asked.

Not at all, he said. They lived in the temporary housing thrown up at war's end. It was shelter, but not much more. When he was four, they moved to a real house.

He was always interested in animals. In the backyard, he created a little aquatic zoo full of eels and fish. High school in the Netherlands was more like an undergraduate degree with early specialization. He studied biology, but due to a boring teacher he developed an aversion to it. He thought about going into math or physics but his mother insisted he'd like biology better since he'd always kept animals. In 1966 he enrolled in biology at Katholieke Universiteit van Nijmegen. Four years later he had his degree but still had doubts.

Biology had begun to go molecular, zeroing in on viral and bacterial genetics, on cracking the DNA code. Soon, some biologists would triumphantly transform the characteristics of one kind of bacteria by moving in pieces of DNA from another. But that wasn't what he liked to do: what he'd liked most as an undergraduate was a job he took on the side to make money. He'd worked in a psychology lab doing cognitive tests on two young male chimpanzees. But to study primates, he couldn't stay where he was (though in Europe people don't generally move from one university to another as they move up the degree ladder)."My university specialized in ecology, physiology, organic chemistry."

Rijksuniversiteit Groningen, on the other hand, taught animal behavior. So he went there. At Groningen, the focus was on birds and rats, ethology in the Tinbergen tradition. However, he soon heard that at Utrecht, Jan van Hooff was working on the facial expressions of primates. So off he went to Universiteit Utrecht to do his doctorate on aggressive behavior among macaques. And that's how he was introduced to the chimps at Arnhem. Van Hooff's brother directed the Burgers Zoo which was owned by the van Hooff family.

"It had the first large colony of chimps in the world," de Waal said. While most zoos kept chimps, they were thought to be too aggressive to be kept in large groups. But Arnhem had space for them—two acres of space with trees to climb and grass to laze about in, all bounded by a moat. Most of the adults and adolescents had come from other zoos, four had been bought from animal dealers, several had been born at Arnhem.[57] He was twenty-five when he began his postdoc at Arnhem under van Hooff. He stayed for six years. There were twenty-five chimpanzees in residence when he published his 1982 book, *Chimpanzee Politics.*

In those days "everyone did aggressive behavior," he said. "Lorenz claimed it was instinctual and people would always have wars. He had served in the German Army. So many Jewish psychologists here in the US and in Canada thought it was a whitewash of war. So Stanley Milgram did that experiment to show aggression could be manipulated by authority..."

When he started his doctoral thesis on macaques, he was still "under the influence of this thesis." But. "I soon saw macaques were occasionally aggressive, but mostly playing and grooming." At Arnhem, he saw it was the same way with the chimps.

"I discovered they reconciled after fights," he said.

He thinks he noticed reconciliation because his childhood had prepared him to think differently about aggression. He grew up in a household with six boys and "never saw aggression as problematic. At the end of the day, we were still playing together."

"I noticed in chimps the fights brought parties together and we could measure that. They were more often together afterwards than before the fights."

He also observed that the fights themselves were mainly much ado about nothing. "It is bad when it injures them, and that happens on occasion. But mostly it is screaming and yelling and that's all it is."

Though his professor knew these chimps very well, he was concerned about de Waal's views on their strategic behavior. As he wrote *Chimpanzee Politics*, his professor "got cold feet. I wrote of planning and strategies, he wanted me to go slowly on this."

His professor thought he had come to these conclusions because he was a free thinker and predisposed to such views. Yet in the end, he helped de Waal publish the book anyway. De Waal thought he would get hammered in Europe for "being so unorthodox." He wasn't concerned at all about the reactions of the behaviorists in the US. "The US barely existed for me..." But he thought he would be accused of extreme anthropomorphism by his European colleagues who would take umbrage at the notion of Machiavellian political patterns in chimp behavior. Some people advised him to use scientific terms instead of plain language.

"A scientist said I should call reconciliation post-conflict contact. I said if it acts the same as a human in the same circumstance, give it the same label."

He also argued the same behavior came from the same origin. He believed that humans and apes are too evolutionarily close to argue otherwise. "In fish you could argue that the behavior got there in a different way."

He wrote the book in Dutch, but it was translated into English by his British publisher, Jonathan Cape. I had to admit I had always been curious about how such a young researcher with only a few published articles had landed a book contract with such a prominent publisher. (Harper & Row published it in the US.)

"It was Desmond Morris's publisher [in the UK]," he explained. "Jan van Hooff was a friend of Morris's. Morris promoted it to the publisher, who took it."

To his surprise, when the book was published there was very little criticism.

"There were not many complaints from primatologists in general—they said they were thinking along these lines. In general, it was well received. There was a book by Don Griffin (*The Question of Animal Awareness*, published in 1976) that argued against the behaviorist crap and said look at the animals as thinking animals. In the US, people were ready. Many had read that book. It fell on fertile ground."

Others had also been looking at social behavior and developing a concept called social intelligence, or the social brain. Many were interested in how sociability might make animals smarter. A Japanese group had been making strikingly similar observations about triangular power relationships among male chimps in the field. But de Waal hadn't followed others' work closely while working at Arnhem from 1975 to 1981. He was marking his own trail.

I asked how he became interested in Machiavelli and his famous treatise, *The Prince.* He hadn't studied politics, just biology, right?

"I'm not sure why I read *The Prince,*" he said. It had something to do with the way the chimps used their intelligence to dominate, and his dissatisfaction with the aggression hypothesis of Lorenz and Ardrey— that most of our intelligence "is occupied with dominating others." He thought that was too narrow a view of intelligence and didn't reflect what he saw. "A female who mediates a fight between two juveniles also needs intelligence to do it the right way. It narrowed down all the other intelligence we have." Yet when he read *The Prince,* he recognized the same patterns of behavior he had seen in the male chimps in their dominance struggles, followed by reconciliations, and among the females who defused arguments between juveniles.

Because *The Prince* deals at length with the uses of deception, it seemed the right moment to ask the hard question. Why had he not written about Luit's murder in *Chimpanzee Politics*? I asked.

He drew back in his chair. He looked away. His hands folded and unfolded. His body said he didn't like this question, though he gave me the same well polished answer I had already read in the epilogue to the 2007 edition of *Chimpanzee Politics.*

"I didn't feel *Chimpanzee Politics* should end on this note," he said. "A bad ending. It would make a bad ending....I would have had to re-think the whole course of the book," he added. "It was submitted for translation around the time he died."

That answer seemed way too pat. A different possibility began to form in my mind. His professor was the brother of the man who ran the zoo. The zoo had been opened with fanfare by the famous Desmond Morris, a friend of his professor, the same famous man who had promoted de Waal's book to his own British publisher. Maybe de Waal didn't want to embarrass the professor and Desmond Morris with a changed set of facts. Or maybe his professor told him not to publish on Luit's death because it would harm the zoo.

Did anyone ask you not to print it? I asked, thinking that's it, he was helping the zoo. Hah!

There were no constraints on him, he said, although that *might* have happened and it did happen to others. "There are many places that would not permit you to publish if something bad happens," he said.[58]

But at Arnhem? I asked. Did you sign a confidentiality agreement?

"I didn't sign a contract," he said, and spread his hands. And shrugged. "...It happened. But I thought conflict resolution was the most important thing that happened there." Besides, he added, when he had got around to publishing on it, "instead of making it an additional topic, I made it a central topic."

How about Nikkie? I asked.

Nikkie died after he left Arnhem, he said.

I didn't know what to make of that answer anymore than the first, so I just left it there.

Back to *your* story, I said.

He had arrived in the US in 1981, the year before his book came out. At a meeting he had met a man named Bob Goy, a psychologist interested in how sex hormones "organize" a baby's developing nervous system. (Goy showed how the introduction of sex hormones during development can change sexual characteristics regardless of what an animal's chromosomes say about its gender.) Goy was Director of the Wisconsin Regional Primate Center. "He said you have to come to Wisconsin..."

This was fortuitous because de Waal needed a job. Good academic jobs in Holland were not easy to find at that time. He went to Madison for a year to see what it was like. He found he had access to a lot of primates, there was money for travel, and that Madison wasn't even that different culturally from Holland. He was in a relationship, but not married and had no children. It was an ideal time to try something new, so he got married and he and his wife came to the US.

In Madison, he worked with Rhesus macaques and capuchins. He was also able to work with bonobos at the San Diego Zoo as well as the chimps at Yerkes. In other words, he had access to four species. For the first time, he was able to compare and contrast how each species makes peace, shares food, and resolves conflicts. Yet he wasn't entirely happy. He wasn't a professor at University of Wisconsin. He had assistants but no doctoral students. He felt he had a specific approach and "you want to carry that on" and for that he needed doctoral students. He was offered a position on the Madison campus teaching biology, but then he got an offer from Emory to join the psychology department and the staff at Yerkes.

Until that point, he had never done cognitive experiments. He had always worked as an ethologist watching the [almost] unimpeded behavior of captives in zoos. He had counted this kind of display versus that, subjected these data to statistical analyses, and made inferences. But he wanted to get a better handle on cognition. In other words, he wanted to be a cognitive ethologist.

"All my work is about social cognition, but to test it you need to do controlled experiments," he said, echoing Suzanne MacDonald. He felt he could not advance his arguments about ape social/political intelligence until he did experiments that others could replicate.

So he came to Emory in 1992, after ten years in Wisconsin. At that time, there were 200 captive chimps at the Yerkes Primate Center, some of which were held at its Field Station just outside of Atlanta. There were two large compounds within the station, each big enough to hold twenty chimps. "I also brought my own capuchin colony with me," he said. He was pleased to say the San Diego Zoo had recently "adopted" all these capuchins so they were no longer 'with' him.

"They were so easy to work with. They are small, you can go in among the group, put them in the test chamber, they're friendly, cooperative. I could do six or seven tests a day. With the chimps it is very slow. Depending on their intelligence, to get them to come in, you open the door, call, hope not too many come in....They need time to get used to indoors. They are very volatile....But when they do something, they are very intelligent."

His ideas about the biological origin of morality came out of his interest in empathy. He became interested in empathy when he went to a conference where someone described how they tested for it in human children. He and his students soon produced a series of papers on empathy, reciprocity, cooperation, and imitation among chimps, some of which were published in *Nature* and *Science,* the most prominent general science journals. They were well received. "People believed it."

Most of his experiments were done chimp to chimp, or chimp teaching chimp. Unlike Russon's work observing imitation by rehabs, his could be replicated by others—or not.

"Michael Tomasello (co-director of the Max Planck Institute for Evolutionary Anthropology, a developmental and comparative psychologist) and Daniel Povinelli (who teaches biology at University of Louisiana, Lafayette), got negative findings all the time, one long story of negative findings and they concluded imitation didn't exist in the primates, not even the apes. We said if an ape doesn't imitate you, there are at least five reasons for it, they're not motivated, distracted... you can't conclude the capacity is absent. But they said we humans have a theory of mind that apes don't have...Joan Silk (a professor in Arizona State University's School of Human Evolutionary and Social Change)...writes that something special happened in the human lineage. I am not sure what that miracle might have been, but we argued mainly on the ground of different testing. We test human children while they sit in their mother's lap, they are being talked to, and they're dealing with their own species. They say they test chimps exactly the same way. But the chimp is not on the lap of its mom, is not talked to, and is dealing with a different species, us."

There had been a study of eye tracking, for example, which showed that chimps rarely follow a human's gaze. However, chimps do follow another chimp's gaze. So, the species of the experimenter matters greatly.

"For a long time they claimed ape imitation is not *true* imitation because the experimenters' goals and methods were not understood," he said.

So he and his postdoc worked out a new comparative test. They gave chimps a chance to learn how to get something good to eat out of a puzzle box after watching a human get the food out first. "Not good at that," he said. Then they made a chimp the teacher of other chimps.

"Vicky Horner, my postdoc, she did things chimp-to-chimp and we got excellent imitation. The whole imitation debate is now dead," he said. "After we did that, we don't hear it anymore."

But the debate had lasted for ten years. And when it died, other boundaries between the intelligence of humans and the rest of the primates were proposed to take its place. "As soon as you demonstrate it, something that is supposed to be unique to humans but isn't," he said, "...people keep making these claims. A lot of claims I ignore, for example, theory of mind. It says nothing. Because it is a set of capacities, not a single one. I believe more in empathy, you feel as one with someone else, you make their feelings your own—rather than theory of mind."

He beckoned me over to his computer screen so he could show me videos of his experiments. I stood behind him as he found the first one. It had been done at Yerkes in the cognition room, behind glass. There is a box with a transparent lid. The mother chimp is shown how to lift this lid and slide in a stick and pull out a piece of melon while her daughter watches. "Naïve chimps don't even get to the point of opening the lid," he said. The daughter chimp watches as the mother gets the melon ball out and eats it. She smells her mother's mouth. "It's very different from me demonstrating to the chimp who is behind bars. Here with the model, the same box, she is smelling what her mom eats. Then her mom is sent out. She is already waiting impatiently for the food in the box."

And that was true. The young chimp picked up the lid, banged at the box, impatient for the food to appear, and got that melon out and into her mouth as soon as it was put into the box from the other side by the tester.

He showed another video demonstrating how expert chimps taught inexpert chimps two different methods of getting M&M's. He described the passing on of these techniques as *cultures.* "We are creating different cultural traditions," he said, "lifters versus pokers." That paper was published in *Nature* in 2005.[59]

I sat down again. So what does this all mean, I asked. Is there no difference at all between chimp and human intelligence?

He is certain that there are differences. "Chimps cannot fly," he pointed out, meaning they don't build airplanes. The differences in the acquisition of language will also hold up, he said. "But it is always hard to conclude they lack a capacity."

So how far back in evolutionary time do our capacities go? I asked. I was thinking of that list of unique human attributes taught to me in school, especially the one concerning our understanding of death and our need to make a ritual around it. I remembered his description of the chimps locked up in their cages at Arnhem after Luit's murder. They were totally silent until Luit's body was removed from the building.

He sat back in his chair. Tool use, so far as he can tell, goes way, way back. "I think it is way older than we think. Even the octopus has tool use...absolutely everything you look at is older....Empathy. My most controversial statement in the mid-1990s was that it exists in mammals. *Psychologists* decided it was cognitive. As if you *decide* to become empathic," he said with sarcasm. "In the human literature, we know there's a lot of automation in empathy. In the 1990s tests were done in which subjects were shown facial expressions of humans for 100 milliseconds—so test subjects are not consciously aware of it. You think you are looking at a landscape. But there are images of faces within it. If you see frowns, you will frown more. So that is a mood transmission going on without your knowing. Jeff Mogil [the E.P. Taylor Professor of Pain Studies and Canada Research Chair in studies of pain at McGill University] does studies on empathy in mice—showing emotional contagion....People were upset when I said it's in *chimps.* Later studies showed it in dogs, crows, mice, and elephants. I don't think insects have empathy but they do respond to pheromones of distress and they sometimes do save each other...."

So, do you mean brains may not be necessary for intelligent behavior? I asked him. I was thinking about the tiny number of neurons in *C.elegans,* and the utter lack of same in a slime mold. And what *is* intelligence, anyway? I asked.

"I'm not sure cognitive behavior only applies to brains," he said, slowly. "Somehow an animal has to select options to make the best of a situation...*that* is intelligence, the capacity to select the best course of action. You would assume experience needs to play a role."

Well, I said, thinking, great, finally, a useful definition. So tools go way back, fine, I said. But how far down the evolutionary bush do we need to go to find the origin of numeracy, for example? What about the capacity of *C.elegans* to calculate competitors against food supply and decide to *dauer* (hibernate) or not?

At first he thought that this had nothing to do with numeracy, but might just relate to a sense of fullness.

I explained that there are two variables the larvae assess, the number of competitors versus the number of bacteria around. I also brought up the slime mold that anticipates shocks and always solves the maze by growing itself along the shortest route to the food supply, no brain to be seen.

"If experience and learning comes in," he said, "it's intelligence. If they do it from day one, I'm not sure it's intelligence."

I wondered what he knew about Lynn Margulis's theory of evolution by merger and acquisition going all the way back to the earliest cells. She made cells sound intelligent.

He had met her and her daughter, he said, but he didn't know much about her work.

Well, can a cell be intelligent? I asked.

"To ascribe intelligence to a cell," he ruminated. "...the human brain *is* just a bunch of cells...."

He shrugged.

There was a pause. A lull. I don't know why, but I decided I had to ask him once more about his failure to publish quickly about Luit.

Luit's death, he said, just shifted something in his brain. It was at that point that he *really* saw conflict resolution as absolutely vital because "without it, someone could die." And since he named this chimp behavior as reconciliation, and published that book, there had been about 200 other studies in the area, into something no one had paid attention to before.

A justification, I thought, but not an explanation.

He wanted to explain how he has moved on with his study cooperation. A student of his would be defending a thesis on it the very next day. Instead of doing the normal sort of study of two animals cooperating, or failing to, the student had set up a cooperation experiment with the entire group of chimps at Yerkes. That meant the chimps had to deal with "freeloaders," those who don't participate but get treats anyway and with competitors struggling for dominance who don't want to help or share.

He showed a video of three chimps sitting together in front of a fence. They could only get a treat if they pulled on a string together. "They need to watch each other, and they do," de Waal said. They got the treat.

"See the alpha male and the dominant female try to steal it?" he said. I did. As soon as they lost interest and walked away, the three chimps started to pull on the string again.

Then he ran a video of elephant experiments he had done in Thailand with Joshua Plotnik (director and president of the NGO called Think Elephants, devoted to helping conserve them by figuring out how they think.) In this experiment, two elephants had to cooperate with each other in order to haul over a platform with buckets of peanuts on top. There was one rope with two ends that wound around the wheels of the platform. If only one of them pulled the rope, the rope would come free and the platform would not move. "They learn that the partner needs to be there doing something. And they do it quickly."

To make it more difficult, they brought one elephant up to the place where the rope lay waiting, but kept the partner back. "To see if it waits for its partner," he said. "And they wait. First time. They are thinking about cooperation. Recently in Thailand we did the mirror study."

The mirror study is one that allegedly determines whether an animal has a sense of self. Can it recognize itself when it sees its image reflected? Elephants recognize themselves in the mirror (as do dolphins.)

"The people who work on corvids (crows, jays, etc.) can't stand it," he said.

Why?

Because ravens and crows and parrots don't recognize themselves in mirrors, he explained, and yet they are very smart in other ways. They make tools. They recognize the faces of humans who bother them and remember them for years, so why don't they know themselves? Conversely, why can magpies recognize themselves?

"Self recognition correlates with a bunch of other capacities," he said. "They can take the perspective of someone else. Dogs can't do it. "

But my dogs have known when I'm upset, I said, and tried to comfort me. They know what time it is, which is surely more abstract than recognizing one's own face in the mirror. Bella, my daughter's dog, always nudged me at precisely 12:30 every day to remind me it was time for lunch. And what about the very clever Honky, our first dog, who escaped daily from our old walk-up apartment to go to the subway at rush hour? He was so good at pretending he was starving, though his belly swished on the ground, that people crossed the street to the butcher shop and bought him sausages. And really, does self-recognition in a mirror define a meaningful intelligence boundary? My granddaughter had to be taught to know herself in the mirror, but she answered to her name and made her needs known long before that. I wanted to know how far down the evolutionary bush this ability to know oneself actually went. Two brainless slime molds in their plasmodium phase will somehow decide to merge in order to get to food faster and then decide to disassociate. How do they figure out who is who?

Lots of animals need to know what time it is, he answered. Chimps and bonobos build nests every night. Give any chimp nest materials, and they will make one. Someone in England had discovered there is an advantage to sleeping in a nest. One felt safer, and one slept better. Also, one got fewer ticks and parasites.

So is a nest an invention, a tool shaped by culture, or some sort of innate capacity? I asked.

He thought there must be an innate component, since all chimps do it, but it requires some experience. At which point he said he had to go to the bathroom and off he went.

I sat there waiting, thinking. It seemed to me that his work showed that the border between what is cultural and what is not, between what is learned and what is just there at the ready waiting to be shaped by experience, must be very porous. Do wasps learn to make their complicated nests, or do they do it perfectly right off, because a program runs them? And does it matter if there is a program? Isn't that what culture gives us? Lots and lots of programs, lots of ways to do things that we don't have to think about because somebody else thought about it and convinced others to try it long before we came along. And finally, is being born ready to react to circumstance something other than intelligence? If so, what is it?

When he returned, he showed me the video of the second part of the elephant test. Sure enough, the elephant waited for its partner so each could grab one end of the rope and pull on it together. They hauled in the peanuts.

Time's up, the dog in me barked. It was noon. Stephen would be waiting with the car loaded. We had almost a thousand kilometers to drive to get to Florida for a conference on comparative cognition.

Are you going? I asked de Waal.

He gave me a quizzical look, as in, why on earth would I?

It's mainly the bird people, he said.

I packed away my notebooks. I had asked my questions about withholding information, but I still didn't know what to make of de Waal's answers, and for that matter, what to make of the ethics of this whole field. He and his colleagues wrote about the evolution of morality deep in the primate order, about how empathy is alive and well among the apes, and had very effectively shown that humans and bonobos and chimps and orangutans are almost the same in capacities to think and feel. Yet they gleaned their insights mainly by studying apes imprisoned for life in places no one wanted to show me.

I left without telling him that we'd met over the phone during the making of *New Wilderness.* While waiting to tell him here, in these pages, is morally ambiguous, I was sure he'd understand it.

6

On the Beach

It's always worthwhile going to a conference, I said to myself, jaws clenched, as we sat—trapped—on a freeway in the Middle of Nowhere, Florida. All your burning questions could be answered at the Conference on Comparative Cognition. The bird people had been making headlines in the last few years, showing that some species of birds make tools, know, and remember mean people, and are not just stereotypic in their behavior as per Lorenz. Hitchkock's *The Birds* was beginning to seem like a documentary, as were the First Nations' stories of the clever trickster Raven. Yet I had serious doubts about what they could tell me about *how* they select the best options in the circumstance—de Waal's definition of intelligence. The ape people were still fighting old battles, lab versus field, imitation and learning versus programs. Would the bird people—as de Waal had called them—be more enlightening? Never mind, I said to myself. You have a fall back mission—to find out more about the Great Ape Trust.

It was Heidi Lyn (*not* a bird person,) who lured me to this meeting. I had found her name on papers co-authored with Sue Savage-Rumbaugh and listed on the Great Ape Trust/ Iowa Primate Learning/Bonobo Hope Sanctuary website.

She had even run the place for a brief time. So I had looked her up. I found she was teaching psychology at The University of Southern Mississippi and, to my surprise, also working on the side as a consultant to the Institute for Marine Mammal Studies in nearby Gulfport. She specializes in language, communication, and intelligence in both great apes and marine mammals, especially dolphins. This was almost too good to be true (even *beshert*—Yiddish for meant to be—as my mother would say).

I'd been interested in cetacean smarts ever since *New Wilderness's* first show on killer whales. I'd recently re-read some early papers by John Lilly, the first to systematically study dolphin intelligence and communication. Lilly, a graduate of Caltech and an MD, had done early research on human brain physiology in a fruitless hunt for the location of the conscious self, then moved on to human perception (which he studied by floating in the dark in a tank of warm water to deprive himself of stimuli). By the 1950s, Lilly had become fascinated by big-brained dolphins—*Tursiops truncatus*—which are similar to us in their brain-to-body ratio. He found a way to record their calls and clicks under water and to analyze the physics of these sounds. They make use of frequencies way higher than the human ear can apprehend. He calculated that their capacity to receive and process information is roughly twice that of humans.

Unlike most MDs then or now, Lilly was comfortable with the new and rapidly evolving computer science and information theory of his day. He wrote about intelligence as the manipulation of information: he spoke of bit rates, processing capacity, random access memory. His dolphins were taught to make the same sounds as their human trainers, to mimic spoken

English, and he demonstrated their numeracy. But then he went too far: he asserted that he had taught his dolphins to speak English and that he had come to the conclusion they are smarter than we are and also nicer.[60]He reported using LSD on them and LSD and ketamine on himself, and he investigated dolphin sex in a very unusual way.[61] For these and other transgressions (like his communications with aliens), Lilly had been abruptly drop-kicked out of mainstream science. But not before he let his dolphins go, arguing it was ethically improper to keep his intelligent "friends" imprisoned in what amounted to a concentration camp.

A generation later, Louis Herman of University of Hawaii became the leading figure in the study of humpback whale behavior, dolphin intelligence, and communication. Like Sue Savage-Rumbaugh teaching bonobos to communicate with lexigrams, Herman taught dolphins to respond to artificial, computer-generated signs and signals. Early on in his work, two of Herman's dolphins were stolen out of his research station by former lab workers and released into the Pacific. They believed, like Lilly, that dolphins are persons. Herman thundered that they had probably been killed most horribly because they weren't equipped to handle the dangers of the Pacific, and he soon acquired other dolphin subjects. However, those dolphins had long since aged and died and Herman had retired. So I had been at sea about who to go see about cetacean intelligence. Heidi Lyn was the first person I'd found who had not only studied dolphin and ape intelligence but was trying to compare them.

On the phone, she'd told me she was trying to make a new version of the Primate Cognitive Test Battery so she could administer it to dolphins.

The *what?* I asked.

She referred me to an article published in *Science* in 2007[62] by Esther Herrmann and Michael Tomasello of the Max Planck Institute for Evolutionary Anthropology in Leipzig. This was the same Tomasello who'd been unable to replicate de Waal's experiments on imitation. Herrmann, Tomasello, and three colleagues had created tests for physical and social intelligence that could be given to pre-literate children, chimps, and orangutans permitting direct comparisons. Their tests allegedly measure understanding of space, quantity, causality, social learning, communication, and "theory of mind."[63] These cognitive skills, according to the authors, are required by social animals that forage for food.

The orangutans and chimps they tested ranged in age from young to old: the average age of the chimps was ten, six for the orangutans. The human children were about two-and-a-half, verbal, but not yet able to read. They found that the human children did as well as the chimps on the tests of physical intelligence but children were better than both kinds of apes at social tasks. Chimps were better than orangutans and children at causality involving tool use, and making a judgement about quantitites. But neither chimps nor orangutans were as good as the children at the theory of mind test which involved following the gaze of the tester. "We found that the children and chimpanzees had similar cognitive skills for dealing with the physical world, but that the children had more sophisticated cognitive skills than either of the ape species for dealing with the social world," they wrote.

They argued that humans, after splitting from the common ancestor shared with apes, had developed specialized and unique social/thinking skills that enable us to live in groups and facilitate learning, making us what we are today. In other words, they'd created yet another bar dividing humans from all the rest. This is what de Waal had complained of.

Lyn had been part of a group lead by William B. Hopkins at Yerkes who thought these results were misleading. They believed, like de Waal, that the children in the Herrmann/ Tomasello study were better at the social tests *because they had been raised and tested by humans in a human context,* "whereas the apes were raised in standard sanctuary settings." They set out to use essentially the same tests on human-raised apes, compare them to zoo and sanctuary raised apes, and to children. "Enculturated apes," as the authors called the human-raised, had been brought up in "socio-communicatively rich" environments.[64]

The study group included the bonobos Kanzi and Panbanisha living at the Great Ape Trust in Des Moines, and four chimps at the Language Research Center at Georgia State University. All had been exposed to the human cultural immersion practices of Sue Savage-Rumbaugh and her husband/colleague Duane Rumbaugh. Some but not all of these apes were "language competent," meaning they had not all learned to communicate using lexigrams, but all were familiar with being tested. The other seventy-nine apes included seven bonobos living at the Jacksonville Zoo and Gardens, and chimps at the Yerkes National Primate Research Center, either living in its labs or at the Field Station. The average age of the apes was about twenty, making direct comparisons with children iffy.

Surprise!

The enculturated apes did better than zoo or lab apes on both the physical and social cognition tests. They did as well as Herrmann's human children on six out of the eight tasks, and did better than children on relative numbers and attention, though in this study, children were better than the chimps at causality. Lyn's group argued that the enculturated apes' intelligence had been "modulated" during development by their human environments. "Our data indicate that yes, immersion in a highly social environment rich with opportunities for social learning *does indeed increase intelligence.*" [italics mine.]

It seemed to me that this primate cognitive test battery had flaws similar to human IQ tests. What the battery measures is competence at doing tests for capacities human beings prize. Human-raised apes had been taught how to do the right things. But did that make them smarter than those raised in different ways?

> "...our collective results suggest that the differences seen between cognitive skills in humans and other apes on the tasks employed in this study are less of a reflection of true biological differences between species but rather are largely a result of the different socio-communicative cultural environments in which each species is reared. Clearly there are species differences in the cognitive abilities of apes and humans. However, one must use caution when drawing conclusions regarding species differences...."[65]

Lyn hoped to create a new version of this test battery to compare humans, apes, and dolphins. "Like an IQ test," she said.

Well, I said, I really want to see how *that* works. Can I come down there? Can we also talk about the Great Ape Trust?

There was a long sigh on the end of the phone.

"It's a nightmare," she said.

But she had agreed to meet me at the Conference on Comparative Cognition in Melbourne, Florida, by which time she hoped to have her test working for the dolphins living at the Institute for Marine Mammal Studies in Gulfport, Mississippi, where I could also visit. The conference organizer, Debbie Kelly, a professor at the University of Manitoba, said of course I could come to the conference for a few days. So I'd signed up.

Then I got an email from Ed Wasserman, Stuit Professor of Experimental Psychology at University of Iowa. He said he'd heard I was coming to the conference and would be happy to talk to me.

I'm used to having to ask nicely to get senior scientists to share their valuable time with me, not vice versa. And there was something about his name that jangled the memory cells. Hah! A man named Wasserman had served on the board of the Great Ape Trust but quit after Savage-Rumbaugh was reinstated in 2012. I asked by return email if he was that man. And did he have any papers from his time there?

Wasserman said he was that Wasserman and he had been a collaborating scientist at the Trust, as well as a board member. He sent me the one paper he'd worked on there because that is all there was.[66] We arranged to meet at the conference. I assumed we'd talk about what *I* wanted to talk about: the Trust. It was really beginning to bother me that no one there responded to my emails or calls. Who was taking care of the apes? One lexigram-literate ape had died. The apes still alive would soon have to pay for their food and involuntary lodging by entertaining the public. That reminded me of the so-called "mentally defective" children who had been locked away in Ontario's 'training schools,' imprisoned, yet made to work for the privilege.

The conference hotel sprawled beside the Atlantic above a glorious beach east of Melbourne, Florida. The day's lectures had already started when I found my way to the main meeting room. A group of older men sat in the first rows. A lot of young women sat further back. A few middle aged women were scattered throughout. My notebook records that the seven-minute talks did not fly by. It records my increasingly impatient commentary as, one after another, the researchers presenting their work failed to indicate what questions they were asking that no one had asked already, or what prior claims they were disproving, or how their work advanced understanding of how thinking works.

The methods seemed old hat. Measurements, statistics, cheap digital video. A lot of studies on dogs: one asked if dogs care about relative sizes of bones. Another asked if dogs explore their surroundings as energetically and fearlessly as wolves.

I listened carefully when it was Heidi Lyn's turn to give a talk. She wore smart rectangular glasses, a stylish outfit, with her hair bound behind her head as befits a tenure track researcher. She was interested in how Kemp's ridley sea turtles find food: their methods of navigation in the open ocean are not known. So she'd built a big, six-armed maze and baited it with food and then watched as the turtles tried to navigate through it. But she was still unable to say that they employed any strategy in particular.

I was having trouble staying alert when, hallelujah, along came an older Japanese scholar from the Lab for Adaptive Intelligence at the Brain Science Institute of RIKEN, Japan. Like de Waal working with chimps, he had found that monkeys synchronize movements with a partner monkey. He had set two of them face to face in front of an apparatus that permitted them to push buttons. They had synchronized their button-pushing almost immediately. When one of them doubled or halved its speed, the other followed. He had decided this "motor synchronization" was the same phenomenon we see when people automatically match their movements in crowds (he showed funny videos illustrating this with examples). He decided it was unintentional in monkeys too and therefore an old phenomenon, so this group level behavior had evolutionary meaning. Why do they—we—do this, he asked.

No answer.

I was fighting sleep again when a woman of late middle years spoke about ants rescuing each other from danger. De Waal had mentioned this, too. It has been known for 140 years that some Mediterranean ants try to save each other (if they are related) when they fall into the pit of a predatory insect called an antlion. It's a matter of pheromones—chemical distress signals—that alert a rescuer to a nest-mate's peril. Karen Hollis, Professor of Psychology and Education at Mount Holyoke College, and a colleague in Paris, had reported on this behavior[67] in 2009 and attracted a lot of media interest. In their opinion, this smart ant rescue behavior had evolved as the result of an arms race between ants and antlions. Hollis told us she had decided to see if these rescue behaviors also occur in ants in the US. She had tested some she found right outside her office window in South Hadley, Massachusetts. Again, she found that family members do rescue each other. They do one kind of movement when a sister is stuck under sand, another when she has fallen into an antlion pit trap, and they even attack the antlion directly with jaws and stingers. I thought I heard her say empathy is as possible in ants as it is in rats. Did she mean it? Or did she mean enemies force the emergence of smart behavior? And if she meant the latter, why hadn't the antlions changed their behavior to counter what the ants do?

At the next break, I found her sitting at a round table, talking to a colleague. Pardon me, I said, but what did you mean when you compared the empathy of ants to rats? Did you mean to say both are altruistic, or neither?

She said she had meant to say that she didn't think a certain rat study claiming to show altruism in rats was well done. She did not think ants are capable of altruistic feelings.

At first I thought she must then consider ants to be organic robots, their behavior triggered by the incoming chemical signals. But I was wrong. She thinks ants learn. But she doesn't think they *consciously* decide to sacrifice themselves the way we do when we endanger or even kill ourselves on behalf of another. *That* is altruism and it is a cognitive function, she said.

De Waal had said the exact opposite.

If a frowning face displayed for milliseconds in a pastoral video can trigger a frown, as de Waal had pointed out, human altruism might not be 'cognitive' either, I thought. Also, I'd been reading about brainless robots that invent altruism to solve problems. I asked her if she knew that work.

No, she didn't. She reached over and grabbed my name tag to see where I came from. With a sniff, she dropped it again.

Ed Wasserman walked into the hotel lobby at the appointed time and plopped himself down in a big club chair on my right. He has a *face.* It's as long and narrow as a greyhound's. His body was long and narrow too, swimming in a white cotton shirt and green cargo shorts.

I asked if we could please talk about his time on the board of the Great Ape Trust. He brushed that aside.

"I have an agenda," he announced.

What would that be? I asked.

It turned out that he wanted me to trumpet the virtues of comparative cognition science to the world. This was the twentieth meeting on this subject to be held in Melbourne. He and several of his colleagues had founded the Comparative Cognition Society in 1999 with only thirty-one members and it had grown quickly. The 250 people at this meeting amounted to a goodly number, but what kind of future awaited the young graduate students who had devoted themselves to this science? Dark. Bleak. He and many of the rest of the Society's American founders were no longer being funded by the American science granting agencies.

Why is that? I asked.

There had always been money issues, he said, but not like this. Former US Senator William Proxmire, a Democrat from Wisconsin, had made a career out of his monthly press releases bestowing his infamous Golden Fleece award on some poor *shlemiel* of a civil servant, or an agency, whose work, in Proxmire's estimation, represented a senseless drain of tax dollars. Scientists were often on the list. But that was just embarrassing. Several years ago, Wasserman and his colleagues had been cut off by the National Science Foundation (NSF), the go-to organization for science grants in the US (which had itself won a Golden Fleece award after Proxmire retired and Taxpayers for Common Sense took up his baton). Then he and his colleagues were also cut off by the National Institute for Mental Health (NIMH), another formerly reliable funder of their work. Now, unless their projects could show direct application to humans and/or human mental health, they might as well go whistle, because they wouldn't get a dime.[68]

And yet, the public is fascinated by this work, Wasserman said, so why should he and his fellow scientists be cut off? Their work had shown that there is much more variety to intelligence than that which humans display, that there are many different species who think and decide in their own fashion. They had held a Comparative Cognition Society executive meeting to discuss their plight. He had proposed that they invite and even pay for some media people to come to this conference so they could get their story out, so the public would see what they're doing, and press for money to flow their way. Coincidentally, it was at that moment that I had contacted Debbie Kelly to ask if I could come. So a decision had been made: he would buttonhole me.

I advised him to get in touch with some newspapers. He growled that most media just don't bother with follow-ups, not even the illustrious *New York Times*, he could tell me stories...

I said that the Canadians who'd given papers so far had all acknowledged Canadian federal government research support. So what is the problem on the American side?

"I don't want to talk about anything negative," he said firmly.

I threw my hands up.

Because, he explained, he had other ideas about how this interview should work. He wanted to tell the good news story—that his science is ushering in "the Second Copernican Revolution in which man is finally removed from the center of life on earth."

Was this a revolution? It didn't feel like one. Proofs and anecdotes about animal intelligence had been proposed and discounted by scientists since Darwin's day. More to the point, most of the comparative cognition researchers still seemed to compare animal behavior to a human standard, as if humans continue to occupy the smarts pinnacle. On the other hand, revolutions do roil with factions, and there was plenty of that going on, too.

The funding agencies just didn't get it, he continued. "I'm appalled to know that even the NSF insists on societal (meaning human) relevance....This is to accommodate Congress."

Congress was in gridlock and government spending had just gotten a haircut via something called the Sequester. But lack of money wasn't the real issue either, he said, it was that the NSF's funding panel that used to give grants for animal cognition and learning had been disappeared. Its files had been moved to a panel that had to do with evolution "and there is no science money coming through that panel that is mainstream animal cognition," he declared. "We aren't whining because we were never successful," he said. "We had long running laboratories, and then we hear 'we aren't funding your science anymore'...."

Problems at the NIMH had begun in 2001 when E.Fuller Torrey, an MD, wrote an article in the *Washington Monthly* asking why the NIMH funds studies into pigeon perception, yet spends so little on its true mandate, human mental health, specifically drug studies to aid humans with disorders like schizophrenia or bipolar syndrome. Wasserman studies perception and learning in pigeons and in other animals. This issue caught the attention of a "redneck" member of the

House of Representatives, Randy Neugebauer, a Texan billed by one journal as "the most conservative member of the House." First elected in 2003, he joined the Tea Party caucus in 2010. "My grant was selected for de-funding," Wasserman told me.

There was a fight. It was "painful and disrupting." Nevertheless, he kept his funding.

"But… the Animal Liberation Front trashed my lab," he said.

He should change his name to Job, I thought. Do you mean PETA? I asked.

No, he didn't. He explained that the Animal Liberation Front, founded in Britain in 1974, is far more extreme in its opposition to zoos and animal experiments. They do awful things to stop what they disapprove of. They had even "disinterred the remains of a woman who supplied products to medical labs," he said. [Actually, her farm supplied animals to labs: her family was told to get out of this business if they wanted to get her body back.] "They do break-ins, start fires….They attacked my lab in 2004. They actually destroyed it, there's a video online of them hammering away at my computers."

Later, I found that video on the Animal Liberation Front's site: their spokeswoman, her face available for scrutiny by any policeman who cared to look, claimed credit for the whole thing. The video opened with two people wearing black bandanas across their faces, black hoodies over their heads and carrying black backpacks. They swiped their way into the locked psychology building at the University of Iowa. The camera followed as they pushed around stacks of plastic bins on trolleys. They tore through the labs like Sherman through Georgia. [The ALF website says that they tried to destroy the entire psychology department at University of Iowa, but missed the labs on one floor.] They picked locks and opened doors and dumped rats from various cages into plastic take-away bins. They held wriggling rats with electronic ports sutured into their heads right in front of the camera, as if to say, see? Mean buggers work here. Then they went to work on a row of computers with their sledgehammers. Those were Wasserman's computers.

So you'd better explain why your science is so important, I said, thinking if I just hear him out about this, maybe he'll tell me about the Trust.

"We ask the grandest of all questions," he said, sitting back in his club chair, waving a noble hand. "What is it to be human? How do we fit? Are we exalted? Unique?" Comparative cognitive science has two important aspects, he continued, one romantic and the other practical. The romantic side has to do with the big questions: are other beings intelligent? And what *is* intelligence anyway? The practical side has to do with knowing how vision or learning works so we can aid the impaired, the autistic, or the ADHD child.

But then he arrived at the moral problem that sticks to this work like a limpet mine—is it right to imprison and do tests on intelligent living beings? Can their wrecked lives be justified by the science produced? (Or by the entertainment they provide?[69])

"The more we learn of animals, the more we appreciate they are intelligent, adaptable, like us," said Wasserman. "Then, if that's true, some people believe that there is no reason, no moral excuse for studying them. The very fact we contribute to understanding our proximity to animals, so the pushback comes, which is, we should not study them."

"It's not a comfortable place to be," he said.

So why wasn't more science produced at the Trust? I asked. [Sue Savage-Rumbaugh later sent me a list of the papers and book chapters produced by scientists who used the resources of the Trust between 2004 and 2013—I counted only 31 papers.]

"It sure didn't run like *my* lab," was all he'd offer, though according to him, $25 million not just $15 million had been poured into the place. Talk to Heidi on that, he said. He must have said talk to Heidi fifteen times.

Later, several talks were given in celebration of Alan C. Kamil of the University of Nebraska—Lincoln, the conference's Master Lecturer. One in particular caught my attention though it wasn't about birds. In fact, it wasn't even about vertebrates. Jennifer Basil of City University of New York spoke about a mollusk called the chambered nautilus and how she tested its intelligence in a watery maze. She described how quickly they learned to find the exit, how long they remembered the route, and how they used the geometry of the room around the maze tank to orient themselves. They are not social at all, she said. Nevertheless, they are so very, very smart that the species has survived three Great Extinctions.

But how does it learn, I wondered? And does that mean culture and intelligence have nothing to do with each other?

Finally, Alan Kamil rose to give his address. He is a big, burly bald man with a gray beard, not the sort you would think had spent much of his career watching birds, exploring how hummingbirds, jays and nutcrackers learn, remember, and solve life problems. It was Kamil who surprised everybody by discovering the smarts of certain jays (corvids), finding them to be almost as clever and social as *Rhesus* macaques. This had led him to ask questions about what—if anything—is similar about their environments. Or is this a case of convergent evolution?

In a book called *Behavior Mechanisms in Evolutionary Biology*[70] he told a story from the early days of his career when he watched the very territorial behavior of male Anna's hummingbirds in the field, to the surprise of his psychology colleagues. He'd gone to Hawaii, where these birds live, and set up a net trap. He'd watched as a male hummingbird successfully spotted the net, avoided it, and then used that net *as a tool* to force a competitor off his turf by driving *him* into that net. Tool use by a hummingbird! Yet he hadn't written it up. He called it an *anecdote*, comparing it to Goodall's observations of a young female chimp who used her relationship with a big old male as a "social tool" to keep an older female in line. Apparently, Kamil still believed that observations of natural behavior count for less than stripped down lab tests of the same species. He called bringing the field to the lab *validation*, a radical idea in the 1970s, when there was a great divide between psychology and biology. "Psychologists said it's not psychology if it's outside the lab. Biologists said learning is not important," he explained.

So he had studied Clark's nutcrackers in both lab and field. They stash their hoards here, there, and everywhere and somehow can find them again even when they're all covered by a blanket of snow. Nutcrackers, he observed, are social, but also selfish. They hide what they're doing from their fellows, and move things around if they think they've been seen while caching. Which was all interesting, fascinating, even funny, but I wanted to hear what he had to say about *how* they remember. He referred to some papers he'd done on the geometry of their hunt-by-pecking technique. Then he got to his main point, which had nothing to do with intelligence per se. He wanted to talk about how comparative cognition folk—psychologists by and large—might benefit from the tools of another discipline.

He had had the happy fortune, he said, to head a biology department for years, though he had only taken one undergraduate biology course himself. So he had informed himself about his colleagues' work by going to their lectures, by listening to them talk over coffee, and by reading their papers. That is how he had discovered the value of what biologists do, specifically molecular studies, genome sequencing comparisons, and most important, epigenetics, the study of how gene expression is regulated and modulated by areas of the genome—control regions— once dismissed as junk. These control regions are influenced by the environment, and then act upon, regulate, rewrite, and rearrange genes which are passed on generation to generation.

As an example of epigenetic interaction, he cited fish called cichlids. Cichlids' reproductive behavior is socially regulated, meaning the dominant male in a territory changes color and puts on a display that pulls in the females. Dominance can be determined by which male gets to a territory first.[71]

"You can take an animal and make it dominant and you see an increase in the production of a gene called EGR1" he said, which triggers a cascade of other interactions which change the physiology of the male fish, producing the dominance display. This interplay between the environment, gene expression, and behavior means "no more talk of nature versus nurture as separate things," he said.

So are we humans like cichlids? I wondered. Is rational, intelligent decision making just a story we tell ourselves, even as our genes pummel our behavior into line according to the circumstances we find ourselves in? Was this another way of saying the behaviorists got it right?

Kamil urged the students in the room to inform themselves about biology. The biologists have methods we could use to answer our questions, he seemed to say. In effect, he was admitting that the methods used in comparative cognition studies needed serious refreshing. There might well be a Second Copernican Revolution going on, I said to myself, but Kamil thinks it's happening elsewhere.

Heidi Lyn had promised to meet me in the hotel bar while the rest of the conference celebrated Alan Kamil at dinner.

When she appeared, I almost didn't recognize her. This Heidi Lyn wore jeans and a hoodie, her long brown hair hanging free, her freckles scrubbed of makeup.

She climbed on a rickety stool, where she twisted and stretched, trying to ease a muscle cramp in her back. Her son, who is autistic, had been leaning against her for hours, she said. She pushed her hair behind her elfish ears and sighed. There'd been a problem getting her dolphin cognitive test battery going, she said, she was sorry to say it wasn't going to be ready to show me.

Never mind, I said, I'll come to Mississippi anyway. Why don't you tell me about the Trust instead?

Lyn's eyes rolled at my question. She said Sue Savage-Rumbaugh was her mentor, and had made a great contribution. That's why she hadn't gone public before. But she was torn between wanting to be kind and loyal, and finding a way to protect those apes.

And that was why I couldn't let the story go either. In 1985, a federal law was passed in the US requiring any study of vertebrate animals supported by federal funds, or carried out in an institution receiving federal funds, to have each study plan cleared by an ethics committee before it began. Every university in Canada and the US has such committees (though a 2012 comparative study suggested these committees are commonly overloaded with the scientists doing the review work, not independent people). I found one online report to the USDA filed in 2010 by the Trust, a basic form about the kinds of experiments being done there. It reported that two animals were being bred, and thirteen animals were being studied, but not by means that could cause pain or do harm. And that's all it said. But harm is a relative concept: did the USDA ask whether Kanzi's life as a study subject caused him harm or pain? What about baby Teco?

She offered bits and pieces of a strange tale. As she talked, her head fell forward, again and again, as if she could make her concern and frustration disappear by banging it on the drinks table.

"My husband says I should write a book," she sighed.

I got many disconnected fragments of Heidi Lyn's story in that bar. But I had to wait until we got to Mississippi to gather them into something like a narrative.

7

Dr. Doolittle and the Dolphins

On our way to Gulfport, Stephen and I hashed these questions out in the car. Is intelligence shaped by externalities, or determined by the brains we're born with? Is intelligence utterly plastic, or is it culture-specific? Is it a property of groups who live together? Or is it driven by the forces of evolution acting on individuals? We were still arguing when we reached Longboat Key where we were staying for the weekend with friends.

At twilight on our first day there, we went to watch the birds gather and settle for the night among the mangroves on Sarasota Bay. On the Gulf side of the Key, a pink wash slid across the water as the sun plopped down in the west. The bay side was already indigo. Hundreds of birds cried and cawed and fluttered and honked, hovered and circled and came in to land feet first. The largest, the pelicans, took over the top story of one big mangrove, their weight dragging its boughs down, and tucked their heads under their wings. Then came smaller grey gulls who settled on branches in a layer below. Then came white egrets with hot pink feathers tracing the edges of their wings. At the bottom, and to the left, in a mangrove less favored than the big one, the smaller species and the lesser pelicans took their places, one layer above the next, each giving the other just enough space or a warning cry to shove over.

Their cooperation in the mangroves seemed to have nothing to do with their daily competition. These birds all fished the same waters, yet here they were, cooperating, tolerating, sleeping together, defending each other.

It was a repeat of what I'd seen on the beach earlier at Melbourne. I'd walked there on my first morning in the hotel the way only a winter-worn, sunshine-deprived Canadian ever walks on a cold, blustery beach—without a warm jacket. The sun was on its way up. The biting wind made me want to cuddle down in the sand like those birds up ahead which hadn't bothered to wake up and go fishing. They had arrayed themselves in a loose square, right there in the center of the beach, neither at the water's edge, nor at the beach's highest point where people walk, but smack in the middle. They lay only a few yards below a scraggy patch of sea grass and salt-sprayed shrubs proclaimed by a wooden sign to be an off-leash dog area.

There were at least three different species snuggled together. There were big seagulls, their heads tucked under their wings. They lay in the center of the square. Smaller birds, much smaller, were arrayed on its outer perimeter. Why would birds go to sleep on the sand right near where people walk and dogs run free? And wasn't that a cat under that bush, scrunched up and ready to spring? At first I didn't get it, then— bam!—the pattern made sense. The quick lift-off birds were on the outside, the bigger ones that need a stiff breeze or a running start to get airborne were on the inside.

Oh My God, I said out loud, it's a seabird version of a Roman military square. They can all sleep yet will be warned in time to react to danger. It was so *smart.*

A collective kind of smart.

It was as if Nature was yelling in my face: intelligence is dynamic: it doesn't come from *either* competition *or* predation *or* being social *or* solitary, it arises from all these circumstances. Recognizing intelligence in other animals only when they behave like us, or trying to remake their intelligence into something more like ours, won't help us figure it out. As Suzanne MacDonald had pointed out, we need to understand how others get along in life. And all the boundaries need a major rethink.

The next morning, a voice on our radio, tuned to NPR, caught my ear. I turned the sound up. It was Sue Savage-Rumbaugh. She was telling an interviewer about how, many years earlier, she had come to understand the importance of culture—and the boundaries thrown up by culture. She had been trying to get Lucy, a human-raised chimp, to breed with a strange male chimp brought in for that task. But Lucy was so human-oriented that she masturbated to magazine pictures of people. Savage-Rumbaugh could see that Lucy was terrified of this male chimp, especially when he held out his hand to her, a come-on gesture. Lucy was caught between two worlds, Savage-Rumbaugh said, belonging in neither, like a mythological creature.

Savage-Rumbaugh sounded sane, empathetic, wise. But what empathetic person would introduce one intelligent ape—who she knew thought of herself as human—to another ape, who didn't, to get more apes to study?

Stephen and I found our way to the Institute for Marine Mammal Studies in an industrial/residential suburb of Gulfport. The Institute is right beside an industrial canal. Its waters were dark with silt and who knows what else. On the opposite shore, a tractor stood guard over mounds of broken concrete and piles of rolled steel tubes. The Institute's front lawn featured three cracked wooden statues of dolphins, tails down, rostrums up, like land-locked mermaids caught in a patch of roses. There was a large, shed-like building with offices and a much bigger display pavilion. A visiting class of schoolchildren was touring inside, dunking their hands in a shallow cement trough to touch the nurse sharks swimming there, and oohing and ahhing over everything else. In the covered walkway between the two buildings, caged parrots nattered at each other. Beyond, there was a grassy space around the dolphin pools and low bleachers set out for people to watch them

Heidi Lyn walked in wearing a puffy down vest, jeans, clogs, hair down. She gave us the penny tour.

What was I expecting? Not this. The name suggests a private think tank. While research is done at the Institute, it looks more like a mini-version of Marineland, though it is not nearly as grand as the giant marine entertainment dome that once stood near the beach downtown—until it was destroyed by Hurricane Katrina. One of that dome's former directors, Dr. Moby Solangi, had founded this place in 1984. Later, he had been involved in the rescue of dolphins washed out into the Gulf by the hurricane. But for thirty years before that, he had captured wild dolphins and sold them to entertainment parks. He would tell me later, when he popped into Lyn's office to check me out, that he had sold animals to Canada's Wonderland and that Marineland is one of the best operations anywhere, never mind what the *Toronto Star* had to say.

Four dolphins were living in the pool complex. I had never been so close to dolphins before. They were beautiful: so incredibly fast, so smooth, so agile, so powerful, so curious about us. I was thrilled to watch them fly through the water. But that water! It was the bright aquamarine of human swimming pools, not the life-dense brine of the sea, or the silty water of river mouths, where wild dolphins make their livings. They'd tried to use real sea water at a dolphinarium in Harderwijk, the Netherlands, Lyn said, when I asked, but got complaints because people think it's dirty. Walk boards separated and connected the big main rectangular pool and two smaller ones. Another pool lay hidden behind a twenty-foot-high blue tarp. A dolphin named Apollo swam behind it: he had arrived at the Institute with an infection and until he was healthy he had to be kept isolated from the others.

Two adult dolphins chased each other in the main pool, racing under the surface at dizzying speed until we got close enough: then they slapped the water hard to spray us. There was nothing playful about it: they were pissed off. They want their buckets of fish, Heidi explained. In the smallest pool one lone dolphin bobbed up, and then thrashed up higher so he could see us better. He fell back in with a sploosh. Then bounced up again. And again. An array of toys (a foam letter H, a soccer ball, rings, a football) floated on the surface, but apparently looking at us was more fun than playing with them.

Stephen couldn't bring himself to leave until he'd asked Heidi a question that had bothered him since *New Wilderness* days. He said: why didn't anyone follow up on John Lilly's early dolphin communication work? He meant: why hadn't any modern researchers tried to decode what dolphins say to each other instead of trying to teach dolphins to talk like us?

Heidi looked at Stephen as if he was nuts: first, she said, many, many researchers have studied dolphin communication, starting in the 1940s, and that work had never stopped. A man named Vincent Janik had been doing a lot of work on dolphin signature whistles. He'd shown that they amount to a unique personal label. Studies done in Sarasota had shown there is more communication among kin than with non-kin. And Peter Tyack had been doing work on communication since the 1970s, too.

But Lilly was after more than signature whistles, Stephen pressed. Lilly said dolphins exchange dense amounts of information. He wondered about whether these sounds constitute a language, and about deciphering it.

While she conceded Lilly's early work had been groundbreaking, Lyn still said he deserved no credence. "He gave his dolphins *LSD*," she said, forcefully.[72] He had this *woman* who stayed in his lab who had *sex* with them, she added.

By this point we were standing by the main pool. But the small dolphin in the small pool kept leaping up, higher and higher. He reminded me of our last dog, a Bichon Frise called Loki, who used to jump five feet straight up in the air so he could see, and be seen, by whoever was knocking on our front door. This dolphin desperately wanted our attention.

That's Chance, she said. Chance had stranded on the beach in Alabama in 2012. Katrina had killed many marine mammals, but BP's terrible Deepwater Horizon oil spill in the Gulf of Mexico in 2010 had killed many more by far and was still killing them. In the pavilion, a board listed the numbers of the mainly newborn dolphins found dead annually. Year after year, the numbers of the dead kept climbing. Chance, who was about four, had stranded for reasons unknown (though it might have been due to the BP spill). When he was found, he had big open sores on his body, a growth on a spinal disk that prevented him from moving properly, and a damaged pectoral. Though he whistles under water constantly, he doesn't seem to echolocate at all. Apollo, who is three, had probably got separated from his pod and then got lost. He too had stranded.

The Institute's trainers were trying to teach Chance and Apollo how to swim through the narrow underwater gates connecting one pool to another. It's a hard job. Dolphins don't like to swim through narrow spaces: they have to learn to trust the human trainers who show them how. They nevertheless found Chance in the big pool from time to time: he probably jumped out and skidded on his belly across the walkway.

Bo and Buster, the adults in the main pool, were around thirty-three: they had been caught in the wild when very young and so had been in human hands almost all their lives.

These were some of the dolphins she intended to test with her modified version of the primate cognitive test battery. But even before she figured out what to test, she had to figure out how. Should she present them with a floating platform and offer them a device they could push with their rostrums to answer questions? Whatever equipment she made, the dolphins would have to be trained to use it. It would take time to turn these individuals into a group, and time to make sure the adults tolerated the young newcomers. And even before that, they had to learn to come when they were called.

I wrote all this down in my notebook but I was thinking instead about the tiny living space available here for these huge, smart animals, the boring nature of these clean and featureless pools compared to the complexities of the Gulf. I was thinking of them breaking out and skidding on their bellies across the lawn and down into the canal, whistling free at last as they went. Thirty years in what amounted to a bathtub for God's sake.

As if she read me, she began to explain that the older ones would never survive in the wild if released. The two younger ones had been at death's door when they were found. While dolphins will carry sick pod-mates for a long time, Chance wasn't just sick, he had a spinal malformation. He was named Chance because when he was found, he had precisely none.

Stephen left and we went to her office which was small, spare, next to nothing in it except for a desk and a couple of chairs.

She sat in one of them and shoved her hair behind her ears. That's when I realized she looked like my friend Marion from the gifted class. It wasn't just that spray of freckles across Lyn's nose or the way her ears bowed under the weight of her hair, it was her mix of vulnerability and toughness, her willingness to strike out for new shores.

She was born in Dallas, Texas, in 1970 to parents who were both musicians and teachers at Southern Methodist University. But when she was two, and her big sister was three, her mother, who had apparently felt overwhelmed by the marriage, escaped it. She put Heidi and her sister in the car, and they camped for nine months as they made their way across the country to San Francisco. There her mother met a man from the Philippines. And off they all went to live in the Philippines—until that relationship fell apart. When they returned to the US, they went to live with her mother's parents in Green Bay, Wisconsin. But then they moved again to Saugatuck, Michigan, an artist colony/vacation town for people from Chicago where Lyn's mother opened a restaurant on the main floor of a house. They lived upstairs. Heidi was ten, her sister was eleven by then: they soon became very good students. Her sister went to University of Michigan to study economics, and she went to University of Pennsylvania to study computer science and linguistics.

When Lyn was in her junior year, her mother called to say she was sending her sister to Hawaii for a two week Earth Watch tour as a graduation present, did Heidi want to go for the second week?

"It changed both of our lives," Lyn said. After that trip, her sister decided she wanted to work with animals, not with money. "My sister grew less happy with her job (in a big insurance company that paid her very well) and applied for one at Chicago's Shedd Aquarium and another at the Brookfield Zoo that paid something like six dollars an hour. The zoo hired her and she stayed for ten years. Now, she's a dog trainer as are many of my friends."

When Lyn got to Hawaii for her whale watch experience, one of Louis Herman's postdocs, Adam Frankel, was managing the tour. He asked her what she was studying. When she said linguistics and computer science he said she had to visit Herman's dolphin lab, the Kewalo Basin Marine Mammal Laboratory that he founded in 1970 in Honolulu. It turned out they needed each other. She had been looking for a project that would be the basis of her senior thesis. They were happy to put her to work.

Did they want your linguistics? I asked, thinking surely that must have seemed attractive to Herman. If you want to decode a dolphin's language, a linguist could be helpful. But Herman was teaching them an artificial language, Lyn explained.[73]

Herman, like Savage-Rumbaugh, had immersed his dolphins in the human world to teach them human means of communication. He described their training process as akin to the fifteen years of education in a human culture that human children get.[74] In the 1980s he had done studies of the capacities of two dolphins to follow instructions given in two invented languages that allowed for the creation of novel sentences. One dolphin learned to link computer generated sounds to specific objects and actions, the other learned a language of visual signals. They were able to perform as requested by their trainers using these made-up languages with almost no errors.[75] But this work had been attacked by a colleague named Ron Schusterman. Herman had responded, and Schusterman had responded to Herman, producing a series of papers and counter papers that Lyn called the flame wars.

"Ron claimed Louis' stuff (the strings of symbols) were not sentences, but sequencing," said Lyn.

By the time she arrived in Honolulu, Herman had moved away from the language studies to a broader exploration of intelligence. He came to define intelligence as "flexible behavior."[76]

> "By flexibility I mean the ability to organize and carry out behaviors that are appropriate to new situations, new contexts, or new events, and that are not necessarily genetically constrained or part of the species' naturally occurring repertoire of behaviors. Flexibility is demonstrated then by the animal's ability to go beyond the boundaries of its naturally occurring behaviors or the context of its natural world."

An interesting definition, I thought. It did not exclude slime molds that solve lab mazes and anticipate being shocked, though *C. elegans*, and a lot of humans, might not make the cut.

Herman moved on to study dolphins' physical memories: could dolphins invent and execute a number of actions and then repeat them? They could. One of them invented and performed a series of thirty-six different actions, then repeated them perfectly.[77] To some degree, this had to do with practical considerations, according to Lyn. There was no grant money being awarded for animal language studies by then and language papers had a rough time getting published anywhere. "It was much easier to step away from language," she said. "You become a more successful scientist when you do."

And that, according to Lyn, was mainly due to Herb Terrace who publicly announced his failure to teach Nim Chimpsky any syntax.[78]

Lyn reported for work in Herman's dolphin lab in 1995. Unlike most of the others there, she was not a volunteer. She was not required to pay Earthwatch, which in turn made grants to Herman's lab so he would permit Earthwatch volunteers to help out. Herman paid her out of his budget. But it was a difficult environment. "There were no super long term workers there."

Lyn fell in love with Herman's dolphins, and stayed on as a volunteer to avoid falling out with him. He asked her to return as a graduate student. But when she saw a video of Sue Savage-Rumbaugh working with her language-immersed bonobos, Kanzi and Panbanisha, "talking English and Kanzi responding in lexigrams, my brain just exploded. The video showed the tests, and them walking in the woods..." Kanzi would request to go to a particular location, and the experimenter would ask him to take the group there, and he would. "All these things were clearly referential," she said.

"I had worked in a day care center and watched the development of language," she continued. "It is odd. It isn't learned like anything else—it's from nothing to full blown explosion in eighteen months. Just amazing. I wanted to be Dr. Doolittle and she was doing it. So I wanted to figure out what she was doing and bring it back to dolphins."

She applied to do her graduate work with Sue Savage-Rumbaugh. She had never met her, but she had heard plenty.

She was accepted as a graduate student. Sue Savage-Rumbaugh and her then husband Duane Rumbaugh still had a lot of research money. "They had a huge National Institutes of Health project for the Language Research Center. They said they had enough put aside to pay a stipend...It was $600 a month." But who could get by on that? "They said just come on down and we'll figure something out." In return for the stipend, she worked part time in their lab and helped take care of the apes. They put her up in a trailer on the campus along with their computer programmer, a man named Ryan. She found herself sharing a double wide with Ryan and his dog, her office on one side, Ryan's on the other.

"Kanzi and Panbanisha would walk by and say hi. The bonobos were housed in a new building right next to the trailer."

And so how did you find Sue Savage-Rumbaugh back then? I asked.

She sat back to consider that.

"When I first met her...she was a difficult task master, like Lou. Same attitude....Things that were apparent—I could never know what reaction I'd get from her. It got worse, or I got more sensitized to it as a pattern. I was there four years, four-and-a-half. I left there before I finished my PhD."

Why?

"It got really upsetting, there were lots of things I was not comfortable with."

Like?

"We had a grant from the NIH [National Institutes of Health] with Bill Fields as the PI. [The PI is the principal investigator, the person responsible for overseeing the experiment, characterizing the results, and managing the money.] He had no bachelors degree." [He got it in 1999.]

Well, why was that wrong?

"My name was removed because I did not have a PhD. I was 'abd,' all but for dissertation..."

She and Sue Savage-Rumbaugh had had a big fight. Why? It centered on Lyn's concern that Savage-Rumbaugh was going in the wrong direction and this would endanger her funding. Lyn heard rumors that people at the NIH wanted Savage-Rumbaugh to change her methods. But for her next set of experiments, Savage-Rumbaugh wanted to explore the effect of love on apes. Her idea was that she and Bill Fields would act as parents for Panbanisha's son. But Lyn didn't think they had worked out anything like a proper scientific study method. It was the same with Savage-Rumbaugh's idea to study art and music with apes.

"No plan. Or it changed. I sing. My family was musical. Panbanisha would sing along with "Cheese Burger in Paradise" [Jimmy Buffett's song] and it was awesome." Panbanisha didn't actually sing: she pointed to the right symbols on the lexigram that indicated the words of the lyrics as the music played. There had been no music studies done with apes at that point, and it would have made sense to analyze what Panbanisha was doing. "But Sue banned popular music. Only indigenous allowed. We said why not kids' music? She taught them to blow into recorders. Apes have little control over their breath. To get them to blow out is difficult, but Kanzi and Panbanisha were blowing into recorders. Peter Gabriel came and played with them there. But the analysis was not finished...there has been no publication based on this material since."

"So I decamped to take care of my grandmother in Florida." [Her grandmother passed away five months later.]

Duane Rumbaugh told her he was worried that she wouldn't finish her thesis if she left. "I said I promise you I will, but if I stay here, something will happen and it will be bad."

Lyn left Georgia with one dissertation plan, but once she arrived in Florida, she changed it. She had been talking with another graduate student about the kinds of vocabulary mistakes the apes made when they were asked to point to specific words on their lexigrams. She thought their mistakes were not random. She'd ask for lemon and they'd point to lemonade– wrong, yet related. The Language Research Center at Georgia State had fifteen to twenty years worth of other researchers' similar test data. She went through it, transcribed it, and analyzed the results looking for patterns.

She found that when they made mistakes it was because they picked words associated with the word in question, or words that sounded like that word.

"In humans, these types of mistakes have been studied and are known ..... As far as I could tell, it was the same kind of errors as with human speech."

Her thesis was eventually published in the journal *Animal Cognition* but not before it was rejected by five other journals first. "I sent it to *Science"* she said, "but they wouldn't review it. They said it wasn't *important* enough."

In the meantime, Savage-Rumbaugh had acquired "suitors" for her kind of research.

One was Ted Townsend. He had seen Duane Rumbaugh on television news and found out he was from Iowa. Townsend had said if the Georgia State apes came to Iowa, he'd support them in a center for life. "But he didn't put it in writing," Lyn said.

In 2001, Heidi Lyn was still living in Florida working on her dissertation when she learned that Sue Savage-Rumbaugh had been in negotiations to work with Penny Patterson (keeper of Koko the gorilla) and Lyn Miles (who worked for years with a male orangutan named Chantek who used American Sign Language to communicate). But then Ted Townsend came along with a counter offer.

"There was a meeting I was part of with Paul McCartney," Lyn said. She covered her face with her hands and murmured again about writing that book.

McCartney had given a concert in Atlanta. He had let Sue Savage-Rumbaugh know that he wanted to come and meet the lexigram-speaking, recorder-playing Kanzi, the Bonobo Genius.

"But Penny and Lyn were here. And Ted Townsend was coming. There was a meeting at Sue's house while McCartney was visiting with Kanzi."

So you met McCartney?

"I had left. There was too much drama," she said. "But the other guys did. I think the plan was to get everybody together. But I think Penny and Lyn were blindsided by Ted."

Apparently, they hadn't realized there might be a competitor for Sue Savage-Rumbaugh and her apes.

"He presented his thing....It's a soap opera," she said, her head hanging low. "He (Ted) wanted Sue's animals. He wanted all four ape species."

How is it scientists get to own these animals? Or does somebody else? I asked.

"Nobody knows," she moaned. "One, Matata, is owned by the Congo but was loaned to Emory and then to the Trust.... Sue has tried several times to claim ownership, most recently against the Trust, the day before I took over as director in 2011."

Townsend's plan to set up the Trust was announced in 2002. It took two years to finish the specially designed buildings on land donated by the City of Des Moines and an energy company. The first apes arrived in 2004. By 2005, Lyn, and others were being flown to the Trust for grant meetings. The scientists working there were being paid very well, "it was all roses, chocolate, and champagne," she said. Yet once the apes moved to the Trust, it was hard for outsiders to get permission to come in and gather their data. There were always scheduling problems. And then the grants never came.

Why?

Lyn believed it was because the NIH had begun to move away from funding Sue's kind of research. But there was also the matter of the Trust's Senior Research Scientist, William Fields. "He was a security guard at Georgia State," she said. "Six months later he was Principal Investigator on one of her grants. He's got a BA. He finished it before Iowa. He is on a bunch of her papers."

Patricia Greenfield of UCLA, and a linguist, Tom Givon, had won grants to do analyses of videos made of the apes' communicative gestures before the apes left Atlanta. Lyn worked on these projects with them, though mainly from home in Florida where she was living with her boyfriend of six years. But she never forgot about applying what she'd learned about ape communications to dolphins. When she went to a marine mammal conference in Vancouver she met a researcher named Diana Reiss, then at the New York Aquarium of the Wildlife Conservation Society on Coney Island. Reiss mentioned that she was working on a keyboard system in New York that allowed language studies with dolphins.[79] Would Lyn like to help? She grabbed at the chance.

She went back to Florida. Not too long after, she split up with her boyfriend. "I had the house sold in twenty-four hours and was on the way to New York City."

I must have looked shocked. She explained that she and her boyfriend had had a conversation about children that "was the deal breaker." That very evening, Reiss told Lyn she should come down to New York because something could be worked out. "I met my husband there."

The Aquarium hired Lyn, and she also got an adjunct teaching job at Hunter College where Reiss had an appointment. Lyn commuted from her cousins' house in New Jersey to Coney Island every day. Everything was going well until there was turnover in the upper ranks of the Wildlife Conservation Society. Then a dolphin died. Within two months, the Society had sent the other dolphins elsewhere which wreaked havoc with Reiss's dolphin study. Reiss and Lyn switched their focus to another of Reiss's studies, comparing belugas to walrus, hoping the dolphins would be returned. It became clear they would not be when the administration sent the belugas away too. Lyn left at the end of four years.

"We had a whole plan for a keyboard, we had the bids." But without dolphins, why build it? "Diana and I agreed that if a keyboard ever happens with either one, we're both part of it. She is hoping to start one at the National Aquarium in Baltimore. I hope to do it here."

She saw a post doctoral fellowship position advertised on the marine mammal message board by Vincent Janik, professor of biology at The Centre for Social Learning and Cognitive Evolution, and the Sea Mammal Research Unit in the School of Biology at St. Andrews University, in the UK. Janik had already shown that some of the whistles made by dolphins are actually identifiers, like names. He had also found that dolphins learn and imitate each other's signature whistles.

"He wanted a postdoc with experience in training dolphins one-on-one," she said. However, the dolphins he studied were not at St. Andrews. The UK effectively bars the captivity of dolphins through very strict regulations "...a huge set of requirements for them and none could meet it, and it was very controversial."

Janik had studied wild dolphins with the Sarasota Dolphin project in Florida, and captive dolphins in Germany. Then he put together a plan to work with a very large colony of captive dolphins held in a very large dolphinarium at Harderwijk in The Netherlands. She worked for him there from 2006 to 2008.

"My job was to train them in match-to-sample task and to help with the husbandry." In the middle, she got pregnant, which had been her plan. But just after she gave birth to her son, in 2007, she got a panicky call in the Netherlands from Sue Savage-Rumbaugh who had just been kicked out of the lab for the first time.

Savage-Rumbaugh wanted to know if Lyn would consider being the director of the place if she could get herself taken back by the Trust.

Lyn later heard versions of what had happened from the staff, and another from Savage-Rumbaugh. Apparently, Savage-Rumbuagh had wanted to go to the Democratic Republic of the Congo and the Trust had offered to send her. But things went wrong.

Savage-Rumbaugh did not return to the lab until 2008. When she was allowed to come back, it was as a scientist, not as the research director. Bill Fields, her former protégé, was firmly in charge. Lyn said if I wanted to know more about that period, I could try others who were there at the time, including her friend Rob Shumaker.

I told her I had already contacted Shumaker. He'd agreed that I could come to the Indianapolis Zoo, where he now works, to see the orangutans he'd brought first to the Trust and then taken to Indianopolis—if it was warm enough for them to be outside. I wasn't allowed to see their winter quarters. He'd already refused to say a word to me about what happened at the Trust.

"Nobody talks..." Lyn said.

In the meantime, Lyn's postdoc had ended in the Netherlands. She went home to Michigan with her husband and young baby and spent the next six months looking for work. She went to an evolution of language conference in Barcelona. There she ran into William Hopkins of the Psychobiology Division of the Yerkes National Primate Research Center at Emory. She'd known him for years. He said he had an unfilled postdoc position. She applied, and after an interview, she got it. When she arrived at Hopkins' lab at Yerkes, Duane Rumbaugh told her that his former wife, Sue, needed to sell her beautiful house and fifty acres quickly. He would arrange financing for Lyn if she would buy it. She and her husband and son lived in that house for two years while she worked out of Hopkins' lab.

"His lab," she said, "is the reason I'm doing the cognitive test battery." For their first study in response to Herrmann and Tomasello, she had arranged to try the test battery on the bonobos at the Great Ape Trust. She went there three or four times over the next two years, rarely seeing either Savage-Rumbaugh or Bill Fields. The apes were basically fine, although, as usual, Kanzi was hugely overweight and Panbanisha was too. Working with Hopkins changed the way she does science: it was less about inspiration, more about efficiency. Hopkins had accumulated the MRI scans of ninety chimps and could churn a new paper out every four months.

"He had a trailer at the Yerkes property. He had a white-board with lists of publications in review, or awaiting publication, there were thirty on the wall at any given time. At other places, it would be one or two, no organization, no plan. Now I have several studies in data collection stage, several in analysis, others in writing, others in review, others in press. If any slots on the board are empty, I get nervous."

As her postdoc with Hopkins was drawing to a close, a friend she'd met at a Comparative Cognition Conference sent an email about an assistant professor position opening at Southern Mississippi. She applied. She went from the medical appointment, where she got her son's autism diagnosis, to the plane, to the job interview. She was hired in August, 2010. "The job was meant to be."

She and her husband found a house online. They negotiated a lower price. By chance, their house turned out to be in the best school district for children with autism.

"And finally, after my fourth postdoc, a tenure track," she said.

By my tabulation, Lyn had spent almost seven years doing postdocs. That is a long time to be a free-lance scientist.

So how did you end up with your appointment at this Institute? I asked.

"My mother needed acupuncture for her dogs," she laughed. Her mother's vet "was also Moby`s vet." Lyn had never met him but she had heard about the Institute. "I was hired on as a consultant."

Moby is also "an interesting person," she said.

She'd worked with a lot of interesting people.

When Ted Townsend announced that he was no longer going to support the Trust, things there got very interesting indeed.

"Bill Fields told me I couldn't gather data at the Trust unless I brought money....I said I couldn't get money without data.... Then I got a call from a person who heard that December 31, 2011 was the date of no more money left, and Kanzi and Panbanisha would be homeless."

This must have been a terrifying prospect. It wouldn't just be terrible for Kanzi and Panbanisha. It would backwash on anyone who'd ever had anything to do with the apes at the Trust. Many people doing primate studies already knew what had befallen Nim Chimpsky after Herb Terrace was finished with him.

Nim's story had a dreadful arc. He was born at the University of Oklahoma's 'chimp farm,' and was torn from his birth mother's arms at age two weeks. (She had already had six other babies taken from her and there would be five more taken from her after Nim.) He was sold to Herb Terrace of Columbia University. Terrace gave Nim to a human family with six kids so Nim could be immersed in human language culture. The mother of this human family had once been Terrace's graduate student and lover. Terrace asked her to be Nim's mother and to teach him American Sign Language as part of his study of language acquisition by chimps. When she proved incapable of treating Nim as a science project (she breast fed him!), Terrace took Nim back and moved him to the empty mansion formerly used by the presidents of Columbia University. Nim lived there under the care of a female graduate student/pseudo-mother, in a bucolic setting with all the comforts. This student also had a brief affair with Terrace: after it was over, she was shown the door. Nim was then trained to communicate using ASL by a succession of students who lived with him. But after years of serving science, Nim began to display normal, aggressive male chimp behavior.

Most adult male chimps are much stronger than a human male. By the time he was five, Nim had bitten all of his teachers, severely in some cases. Then he bit one of his female teachers on the face so hard that his teeth almost tore right through her cheek. It took the woman three months to recover. Herb Terrace became frightened that he would be sued. He called the Oklahoma chimp farm from which he had purchased Nim and they agreed to take Nim back. Nim was bundled onto a private plane, unconscious.

When he awoke, his life as a human was over. He endured the abrupt loss of everyone who had ever loved and cared for him. Instead of his own pet cat to cuddle and hump in a mansion on an estate, he had to live in a cage with chimp strangers who terrified him. If he didn't behave, he was zapped with a cattle prod. Still, he made friends with some of the students who took him out of the cage from time to time and into the fields for picnics. But when the Oklahoma facility got into financial trouble, Nim was sold to a biomedical research facility in upstate New York. There he was kept in a much smaller cage and used as a model in medical experiments. After his student friends in Oklahoma made a big fuss about that, Nim was sent to a sanctuary in Texas. But there were no other chimps there. He lived in a cage, mostly alone, for the next ten years. Finally, the Texas sanctuary changed ownership and a few more chimps came to join him at last. He died of a heart attack at age twenty-six.

This awful story had been told in a powerful documentary, *Project Nim,* directed by James Marsh and based on a book by Elizabeth Hess. It was shown at the Sundance Film Festival in 2011 and then appeared on television. It is a scalding indictment of scientific incompetence, heartlessness, arrogance, ruthlessness, and unethical behavior. Terrace acted as if being a senior scientist gave him carte blanche to mess with the life of an animal he thought smart enough to learn a human language, as well as with the lives of the students he supervised. Terrace is recorded on camera saying that his affairs with his grad students *of course* did not interfere with his scientific judgments. But Terrace's former graduate students had different ideas about that. They flatly disagreed with Terrace about whether the language learning experiment was a success or a failure. Terrace had claimed in *Science* that he shut the project down after realizing that Nim had only mastered about 125 words and acquired no syntax. The students said that Nim had learned thousands of words, not just a few. And most of his keepers cried on camera about what had been done to Nim, displaying something like slaver's remorse.

Any researcher who played a role in a similar fate for Kanzi, the Bonobo Genius and the Trust's other bonobos would earn a terrible reputation.

When Lyn heard there was no more money at the Trust, she called Savage-Rumbaugh.

"Sue said Ted was pulling out, but there was money for six months."

Lyn talked to the then president of the Trust's board to be sure that was the case. He said Sue was correct.

But then the staff protested Sue's behavior and demanded her ouster. The board promised to investigate and asked Sue to step aside while they did so. The board president asked Lyn to come in as temporary director.

"The day I was to fly in to Des Moines, he said 'look at the front page of the *Register* and see if you still want to come.'"

What she saw was a story detailing accusations made by Sue Savage-Rumbaugh about the staff. Savage-Rumbaugh alleged that they had slashed baby bonobo Teco's feet. "She said they had held him down to cut his feet...because, she said, he was walking bipedally and they did not want that scientifically published. He is two-and-a-half. She has him in total immersion in human culture....When I was on my way there, Sue was convinced that Bill Fields was stealing the bonobos. She had her sister drive a truck back and forth in front to make sure no one was stealing the apes."

So were people trying to steal the apes?

No, said Lyn.

But Savage-Rumbaugh thought otherwise. On December 6, 2011, Savage-Rumbaugh's lawyer, William C. Zifchak, of Kaye Scholer LLP in New York, sent a letter to the federal court judge supervising a case that had been brought to the court by the Trust in February, 2010. He asked for an emergency status conference. The purpose of the case before the court was to resolve conflicting claims about the ownership of some of the bonobos. In 2009, the Democratic Republic of Congo had asked for one female—Matata—to be returned. Sue Savage-Rumbaugh then claimed direct ownership of at least two of the bonobos, both given to her by the Japanese Monkey Center that had at first lent them to her when she was at Georgia State. As well, Townsend's withdrawal of funding, according to her lawyer's letter to the judge, nullified the agreement by which Savage-Rumbaugh had donated them to the Trust in the first place. If the Trust, also known as the Iowa Primate Learning Sanctuary (IPLS), was to become defunct, Savage-Rumbaugh wanted to be the receiver of these two, as well as the other bonobos, through her new charity, the Bonobo Hope Initiative, Inc. Through her lawyer, she accused unnamed persons within the Trust of undermining her fundraising for Bonobo Hope, and claimed that certain unidentified members of the staff had slashed baby Teco's feet. Her lawyer also mentioned that she had seen ape packing crates on the grounds, suggesting that someone planned to remove the apes and shut the place down.

Lyn spent only a week and a half in Des Moines, though she was listed as Director for a month and a half. During that time, the staff came to her, one by one, with their concerns and stories about Savage-Rumbaugh. As Lyn could not give them assurances that their concerns would be dealt with, one by one they quit. Soon there was only one keeper left. The board of the Trust found the staffers' complaints to be groundless. Savage-Rumbaugh was reinstated in her role as scientist.

There were seven animals still in the care of the Trust at that point, and the question was, could one lonely staffer manage them all? The answer was yes, because Savage-Rumbaugh and her sister were also acting as caretakers. Still, this would only work if no one got sick or took a vacation and no science got done and there were no public tours. Lyn couldn't stay to help because she had a job in Mississippi and the Trust had no money to pay her. "I said I will resign as Director. And I was out by January 31, 2012. I went back to Mississippi to teach..."

Time went by. Then Lyn got a phone call that Savage-Rumbaugh was ill and couldn't take care of the apes. It was April, 2012. The USDA was examining the place to decide whether or not to grant a license to allow the public in to view the apes for a fee, a way to raise money for their upkeep.

How would that work, I asked.

It is pathetic, Lyn said. The rules in the US are such that it is considered okay to keep a male ape in an area only five-feet by five-feet square.

Lyn thought something had to change. But no one had anything set up.

"I called in May....They said Sue got treatment, she's back in the lab, full of energy."

And just like that, the crisis seemed to be over, at least so far as Lyn was concerned.

But in August, 2012, Lyn got another call out of the blue from the president of the Trust to say they were out of money, they couldn't house the apes. "Be prepared," he said.

She had already talked to Moby Solangi and others about bringing the apes—and Sue—to Gulfport if worst came to worst. And so she waited. But again, nothing happened.

"I call him again. He says Sue went all in," she said. Savage-Rumbaugh had come up with $200,000 and it was in the Trust's account.

Then, Susannah, the one caretaker who'd stayed, posted to her Facebook page that her time taking care of the apes at the Trust was over. She had been told that Panbanisha had made 'accusations' about her...She had been told she was going to be fired. Soon Lyn "got an email that some former caretakers were going public...Then they did go public, and [the blog] *Chimp Trainer's Daughter* got the emails. That was the third time Sue was kicked out of the lab."

Sue Savage-Rumbaugh was asked to step aside by the board of the IPLS so it could investigate the former staffers' allegations.

One day late in the fall of 2012, while in a meeting, Lyn realized she had missed several calls and text messages about something going on at the IPLS.

"I pulled up Facebook to see that Panbanisha had died of a respiratory illness....Respiratory illness is the number one killer of apes....All the apes were ill at that point, several seriously."

By then, Savage-Rumbaugh had been reinstated by the IPLS board. The board had decided that these newly published complaints were actually the same as those made by the staff previously, and there was nothing to them, then, or later. Lyn wasn't sure what kind of investigation had been done. Some board members resigned. Sue and her sister went on the IPLS board.

The Trust did get its animal display license from the USDA. Lyn thinks that by then she had already noticed a picture on the website of Savage-Rumbaugh holding baby Teco amid a group of people. This is against all regulations for managing apes because the risk of infection if they are exposed to crowds of people is very great.

Lyn was in despair about all of it, and just hoped for basic miracles. "That they will be able to feed the apes and not have the electricity cut off."

As to her own research, she was determined to build a keyboard for use by the dolphins. But language studies with animals were "the pool of death" for grant applications. That's why some people she knew were doing crowd-funding to raise money for their experiments. She thought another way to go would be to set up a rescue center. Or even run a dog care facility.

Why?

You can get money for that and there would be students, she said. "I am saying you could make enough to self-fund," she said.

But who would publish something you self-fund? Who would think it was reliable? I asked. And what about oversight committees, I thought, what about ethical review of the study protocols?

As long as you're affiliated with a university, and do good research and are supervised by an ethics committee, she said, that would be enough.

If the need to crowd-fund is a measure of scientific credibility, the Second Copernican Revolution is on the rocks. But come to think of it, that's what happened to the First Copernican

Revolution, too. Few took seriously Copernicus' findings that the planets move in elliptical orbits around the sun until Galileo took a run at the same idea about 100 years later. And Galileo had to recant. The psychologists and ethologists had certainly made progress since *New Wilderness* days—by observing and demonstrating that many species do smart things, by getting out of the lab and into the field, and vice versa, and by combining the tools of their disciplines in new ways. But they aimed much higher: they wanted to unravel the mystery of the nature of intelligence, not just to expand the list of who displays it. And in spite of Wasserman's claims, humans still remained at the dead center of their cognitive universe. Fifty years after Goodall's first reports on tool use by chimps, they could no more agree on what intelligence is than Kanzi could blow a Benny Goodman riff on a clarinette.

Part II
Evolution Is A Universal Turing Machine

The Smarts Mind Map

Scientists often say that they stand on the shoulders of giants. What follows on the next two pages is a map setting out names and the relationships among those who have contributed to the study of intelligence (natural and otherwise). The earliest thinkers are on the bottom of both pages and the most current contributors are near the top. Some names are larger than others, suggesting relative contributions. But that's just how things look from the vantage point of now: thirty years ago, or 30 years on, it would have been/will be different. The biologists are mainly on the left and the philosophers, mathematicians, computer scientists and hard scientists are on the right. It starts with Aristotle, who wrote on everything from biology to politics to theatre and had the biggest set of shoulders in science and philosophy for a thousand years. Darwin and Turing get equal billing because Darwin's theory changed the study of life forever, and Turing, who invented digital universal computers, invoked Darwin's evolutionary process as a metaphor for how to teach a machine to think. The work of both, taken together, makes it possible to evolve intelligent robots. James Lovelock, a chemist, and Lynn Margulis, a microbiologist who brought merger and cooperation into our understanding of the evolutionary process, showed us that all living things are part of larger systems which form the self-regulating, dynamic circulatory system of this planet. Those named near the top of both pages have used these insights to reframe our world as smart. In the last thirty years, these microbiologists, zoologists, primatologists, computer scientists, ethologists, plant scientists, philosophers of mind, engineers, neuroscientists and psychologists have redefined the nature of intelligence, who and what displays it, and have figured out how to manufacture it, busting the boundaries between all the disciplines.

Smarts Mind Map

Jennifer Mather
Suzanne E. MacDonald James Shapiro Susan Murch
Heidi Lyn Monica Gagliano Anne Russon
Marthe Kiley-Worthington
Frans B.M. de Waal Anthony Trewavas
Alan Kamil Richard Dawkins Kanzi
Herbert Terrace Louis Herman Sue Savage-Rumbaugh
Dian Fossey Jane Goodall Birute Galdikas
Sydney Brenner Duane R. Rumbaugh
Louis Leakey
Barbara McClintock Jacques Monod Francois Jacob
Lynn Margulis
Donald O. Hebb John Lilly
Konrad Lorenz Nikolaas Tinbergen Karl von Frisch
Wolfgang Kohler Robert Yerkes Charles Spearman Lewis Terman
Ivan Pavlov Jean Piaget Jagadish Chandra Bose
Jacques Loeb Herbert Spencer Jennings
Hans Driesch Wilhelm Roux
George Romanes Francis Darwin Francis Galton
Charles Darwin
Sir John Scott Burdon
Jean-Baptiste Lamarck

Leonardo da Vinci

Aristotle

Chris Eliasmith
Edward Snowden
Malcolm McIver
Stevan Harnad Ray Kurzweil Larry Page Sergey Brin
Stefano Mancuso Tomonori Kawano
Frantisek Baluska
Toshiyuki Nakagaki
Andrew Adamatzky
Julian Jaynes
Dario Floreano
Terry Sejnowski
Stefano Nolfi
Inman Harvey
Geoffrey Hinton
Daniel Hillis Valentino Braitenberg
James Lovelock Carver Mead

Norbert Weiner William Grey Walter
John von Neumann
Max Newman Claude Shannon
Alonzo Church
Kim Philby
Alan Turing
Ludwig Wittgenstein
George Boole Martin Heidegger
Georg Wilhelm Friedrich Hegel David Hilbert
Charles Babbage
David Hume
Rene Descartes Gottfied Leibniz
Thomas Hobbes
Francis Bacon
Niccolo Machiavelli

8

Original Sins

Slide the time cursor back to the late spring of 2012. The zeitgeist was in full rebellion mode. In Quebec it revealed itself as the Maple Spring. Students ruled the streets in Montreal night after night, thousands upon thousands of them banging drums, taunting the cops with come-and-get-me red squares fluttering on their t-shirts as they protested proposed hikes in tuition fees they said they shouldn't have to pay at all. It felt like 1968 in France and in the US all over again what with millions on the march as well in North Africa, the Arabian Peninsula, and Egypt. The formerly powerless young were making coalitions to square off against dictators, those puffed-up human versions of Nikkie the Chimp, trying to hang on to power at all costs. Several would be pulled down by groups of democrats armed with iPhones, videos, Twitter, and Facebook (in Libya, there was a little help from NATO). And the government of Quebec would be replaced too.

Wherever it flows, the zeitgeist often twists old ideas into new configurations shared everywhere and all at once. Bad ideas, no matter how erroneous and harmful, hang around, seemingly forever, waiting for a chance to roar once more. But good ideas live on too. They may be forgotten for a time, but they will be rediscovered (as you will see) because they reveal something true.

I googled Professor Stefano Mancuso who had given a TED talk about plant intelligence. I thought his work, suggesting that plants display it, was really interesting. I was looking for his email address because I wanted to arrange an interview with him at his lab in Florence, Italy. But up came an item that said Mancuso was going to lecture at The Turing Memorial Summer School of Cognitive Sciences to be held at the Université du Québèc a Montreal that July.

The title for these sessions was "Evolution and the Function of Consciousness." I looked over the abstracts. Other speakers were going to talk about the evolutionary meaning of consciousness (is it necessary, or is it an accident?); invertebrate cognition and behavior; artificial intelligence; and robots.

Well, well, I thought, look at that: all sorts of people I should listen to. Maybe the lectures on artificial intelligence will shed light on the natural kind.

I contacted the organizer of the Institute, Stevan Harnad, a professor of psychology and Canada Research Chair in Cognitive Sciences at Université du Québèc à Montreal. (He also has an appointment at the University of Southampton.) Harnad said I could drop in if I wanted to. He is interested in what he calls "the hard problem" (the how and why of consciousness), and was going to talk about something called the Turing Test.

I am ashamed to say that was the first time I'd ever paid attention to the name Alan Turing, though he is the Charles

Darwin of computer science, the point of origin for computers and ideas fundamental to artificial intelligence. He is therefore the original sinner behind our governments' burgeoning capacities to stick their noses into everything we do via computers, iPhones, cell phones, and Wi-Fi. As soon as I paid attention, wham! It was like sticking a straw in the wind and sucking down a hurricane. It was the centenary of Turing's birth (June 23, 1912). TV documentaries and dramadocs popped up on various networks, including the UK's Channel 4. "The Imitation Game," a movie still in development, had already started its promo. (Starring Benedict Cumberbatch and Keira Knightley, it won top prize at the Toronto International Film Festival in 2014, and a nice group of Oscar nominations.) The *New York Times* carried a review of a nonfiction book by George Dyson called *Turing's Cathedral.*[80] It wasn't really about Turing's life: Turing's definitive biography, *Alan Turing: The Enigma,* was written by Andrew Hodges, a mathematician at University of Oxford who also manages a website devoted to Turing's life and ideas.[81] Dyson's book is a history of the first generation of digital computers from which all our smart machines, search engines, and networks descend.

As Dyson explains, our 'Smart' era began in 1936 when Alan M. Turing arrived at Princeton University to do a graduate degree in mathematical logic under Alonzo Church. There, for the second time, he met John von Neumann, who had settled into Princeton's Institute of Advanced Studies along with Einstein and other major thinkers on the run from Hitler.

Von Neumann was from Budapest, the son of an investment banker. His formerly Jewish family had a purchased title and had lived in style: he was different from Turing in almost every way. Von Neumann had published his first mathematics paper at age eighteen[82] and by 1936 was internationally renowned. He had an eidetic memory that allowed him to recount word for word anything he'd read, and he read widely. In 1930, he had been invited to teach part-time at Princeton (at a huge salary increase from what he then earned at the University of Hamburg). Then, in 1933, he took a permanent position at the Institute of Advanced Studies, before the Nazis began to clear all Jews out of teaching positions in German universities. "He felt at home in America from the first day," Dyson quotes the physicist, Eugene Wigner, who was invited to Princeton at the same time.[83] In Princeton, von Neumann and his wife had a big house with servants, and threw grand parties. They lit the place up.

Von Neumann later presided over the development and use of America's first electronic computer with stored memory, dubbed Electronic Numerical Integrator And Computer—but better known as ENIAC—for some of the calculations necessary to build the atomic bomb. Von Neumann's report on how the next computer system should be designed—the Electronic Discrete Variable Automatic Computer, aka EDVAC—became the basic engineering schema for computers for the next forty years. But as von Neumann admitted himself, he wouldn't have got there without Turing. Though von Neumann is the practical hero of Dyson's book and Turing gets much less space, it is clear Turing is the *sine qua non.*

According to Hodges' fascinating book,[84] (which I rely on here), Turing disdained America and Americans and was not given to throwing parties. When Turing's mentor at

Cambridge, Max Newman, asked Alonzo Church to take him on as a student, he explained that Turing had done groundbreaking work alone and without supervision. Hodges describes Turing as having had a high whinnying laugh, and more than his share of eccentric behaviors (including: chaining his tea mug to a radiator to keep his colleagues from using it; riding around on his bike in a gas mask to ward off pollens in hay fever season.) In his later years, he became a competitive long distance runner. He was also gay in a time when male gay sex was a crime.

Turing's story seemed to exemplify Francis Galton's theories about the British families that harbor genius. Alan's father Julius, who became a high ranking official in the Indian civil service, came from a Scots merchant family which sported a baronet but had fallen on hard times. (Turing's nephew eventually inherited the title). Turing's mother came from an important Protestant family in Ireland. Her father became the chief engineer of the Madras and Southern Mahratta Railway in India. While they were based in India, Turing's parents made sure that he and his brother were raised in Britain. When their parents had to be in India, the two boys remained in England, cared for at first in Hastings by an Army couple who had children of their own.

Alan was eventually sent to an independent public (meaning private) boys school called Sherborne, in Dorset. According to Hodges, he rode his bike sixty miles to get there for his first term's opening day because most major rail traffic was stopped due to a general strike. He was brilliant at mathematics and science but did not shine in classical subjects and was described as dirty and disorganized in his personal habits. At one point, it was suggested he had to aim at being "educated" (as opposed to being a science specialist) if he wished to stay in a public school. In the sixth form (the last before graduation) he became friends with Christopher Morcom with whom he shared many ideas about mathematics and science. He deeply loved Morcom, who soon won a coveted mathematics scholarship to Trinity College, Cambridge. But then, in 1930, Morcom died of tuberculosis caught from tainted milk.

After Morcom's death, Turing, in his grief, was full of questions about the nature of spirit, its whereabouts, whether it leaves the body at death, or even during sleep, and most of all how it informs matter. He told Morcom's mother that her dead son Christopher was still around in some fashion because he felt that Christopher helped him. He had by then become very interested in relativity, quantum mechanics, and the principle of indeterminacy, noting that a fully determined universe (Newton's version) must conflict with free will, but that an indeterminate universe might explain it. He was interested in how mind is produced by the brain, how mind directs matter, how matter is predictably shaped during development. He had no patience for any orthodoxy, even that which assumed that the human mind/brain is of a different order from all other animals. These ideas would shape his work for the rest of his life.

Turing entered King's College, Cambridge in 1931 with a mathematics scholarship of eighty pounds sterling a year, after having failed twice to win one to Trinity. King's was a small but rich college, not quite as prestigious as Trinity for mathematics. There were only eighty-five entrants in Turing's first year. Cambridge had long been, and would remain, ground zero for mathematical and biological genius. In a place where interesting minds were ordinary, Turing was marked as interesting.[85]

In the 1930s, Cambridge (especially Trinity and King's Colleges) was also ground zero for British social revolutionaries–from pacifists to Communists. Few were willing to lay down their lives for King and country, and all the other old verities were up for grabs too. A number of Cambridge students were recruited to the Communist Party and some joined the underground Comintern. Turing was interested enough to join an organization he described as "politically rather communist," and to become a reader of the left-wing *New Statesman*. He also considered taking a vacation to visit Russia to see things for himself in 1933. Hodges nevertheless insists that Turing was not politically committed.

A small subset of Turing's contemporaries at Cambridge later became spies for the Soviets and betrayed their country. That list includes: the infamous Kim Philby who went on to head the Soviet counterintelligence section of Britain's Security Intelligence Service, later known as MI-6; Guy Burgess; Donald Maclean; Anthony Blunt; and John Cairncross. Run by officers of what became the KGB, they were recruited into a network of brilliant, mostly upper class and sometimes gay students who would likely be privy to secrets concerning science, technology, and government policy when they took their places in the civil service and academe.[86]

Philby recruited himself by asking a particular don well known for his Marxist opinions how he could help the cause. He was introduced to a man in Paris who introduced him to the Comintern underground in Vienna and later ran him as a spy. Philby also gave his handler the name of seven others who might like to serve. Guy Burgess, another member of the Cambridge spy circle, or the Cambridge Five as it was later known, was a brilliant gay man who was also a notorious drunk. He later held raucous parties in his flat in London's Bentinck Street, (in a house leased from Victor Rothschild where other members of the Circle either stayed or visited) featuring powerful men in compromising positions with working class boys.

The members of the Circle first met each other in arts clubs and debating societies such as the Ten Club, the Moral Science Club (where Turing gave a paper), and in a not-so-secret secret society founded in the early nineteenth century known as the Apostles. Turing was not on its list though two of his closest Cambridge friends were—David Champernowne and Robin Gandy. Another friend of Turing's, the mathematician/physicist Alistair Watson, was the Apostles' secretary. The Apostles had included luminaries such as John Maynard Keynes. The society's alumnae, called Angels, wielded enormous political and social influence in Britain and its empire. David Champernowne served in Churchill's office during the first years of World War II providing statistical analyses of policy proposals. At meetings of the Moral Science Club and the Apostles everything was open to inquiry.

According to Hodges, the pure mathematics Turing studied at Cambridge had by then become increasingly concerned with mathematical principles rather than with numbers, with the *logic* of the relationships beneath the surface of physical phenomena. This shift in the nature of mathematics paralleled and proved useful to the work of physicists who were reconsidering everything about the foundations of physical reality: relativity, indeterminacy, and quantum mechanical theories required new mathematics to express new ideas. Turing immersed himself in these problems as an undergraduate, reading Bertrand Russell and von Neumann, who had struggled with these questions before him.

After graduating with distinction in 1934, Turing continued to study mathematical logic at Cambridge under the eye of Max Newman[87] who introduced him to three central problems that had been set a few years earlier by a leading German formalist mathematician named David Hilbert. These problems concerned mathematics as a consistent system, for example whether or not a finite mathematical system can predictably deal with an infinite number of problems.

In particular, Turing worked on Hilbert's Entscheidungsproblem —the decision problem. As Hodges describes it, Hilbert asked: is there a definite method which in principle can be applied to any assertion and be guaranteed to produce a correct decision as to whether the assertion is true? Hilbert was sure there was such a system but had not proved it. Turing showed there is no such system. Alonzo Church at Princeton, who had been working on the problem too, solved it with his own new form of calculus and published his findings just as Turing was completing his own proof. But Turing's method was completely different from Church's. Turing's paper so impressed Max Newman that he asked Church to take Turing on as a student and helped get it published. Turing's paper on what he called computable numbers and the *Entscheidungsproblem* was published[88] in 1937. By then, because of another dissertation he'd written the year before, Turing had already been made a Cambridge Fellow (an appointment of several years duration which included rooms to live in and financial support).

Turing had approached the decision problem from a fresh angle.[89]Max Newman had set Turing to thinking about whether there could be a mechanical process to answer Hilbert's decision question. So Turing imagined a computing machine (before Turing, all computers were people, after Turing, all computers were machines) with an unlimited supply of paper or a tape of unlimited length divided into separate squares which operated under simple rules. Turing asked: is there a class of problems such a mechanized computer would be unable to solve? So far as Turing could see, such a computer—a universal machine– could be used to solve any problem so long as it could be expressed in algorithms.

Algorithms are simple rules of operation such as: slide the tape one space to the left if the sum of A and B equals C...

Turing's imagined computing machine would use the simplest information system: it would have a code built of 1s and 0s and blank spaces (generally referred to as digital). He proposed that the tape of infinite length divided into squares bearing these data could be scanned by the machine. The machine would also have the capacity to: move its scanner one position to the right or left; change its state of "mind" by injecting a new program or set of rules. It would have a memory. It would be able to erase its memory. It would have the means to print out results.

A Turing machine set up in this fashion would be able to do more than calculate numbers, it would be able to process any sort of *information* so long as that information and process could be reduced to a simple numerical code and an order of operations.

Turing's paper was a theory and a guide for how to build a real, universal, digital computer, the kind of machine we use all the time now but which most of us simply do not understand. It was also an elegant solution to Hilbert's Entscheidungsproblem: he demonstrated logically that there must be a large class of problems that cannot be reduced to a rule—an algorithm—in advance of trying the problem, and decided (solved). Some problems would have to be attacked by trial and error. This conclusion led him to see that there must be sharp limits to mathematics, but also that there could be myriad uses for his imaginary computing machines. They could bootstrap themselves through increasingly difficult problems *because they have memories.* In effect, a Turing machine would be able to learn.

Turing was not the first to imagine a binary system (made of 1s and 0s) that could be both a means of calculation and also a code—that idea had been considered first by Francis Bacon, then by Thomas Hobbes in 1623, and by Gottfried Leibniz, according to Dyson. In 1714, Leibniz published an extraordinary thesis. He said that there are multiple possible universes, one of which is ours. Ours is filled with myriad mental particles—monads, or little minds— which are the "local actualization of a universal mind." According to Leibniz, our universe is optimized so that a minimum of laws leads to a maximum diversity of results.[90] Substance, he argued, arises from relations: he foresaw "a universal symbolistic in which all truths of reason would be reduced to a kind of calculus." These ideas are not far removed from how some modern biologists and physicists see phenomena now: scientists at CERN hunting for the Higg's boson imagine it as the minimum particle underlying all of the universe's unimaginably diverse phenomena. Sometimes called the God particle, this boson, this quantum of nothing/something, is believed to insert mass or substance into this universe.

The idea of a calculating machine wasn't new either: there were lots of electro-mechanical machines available in Turing's day that could perform specific functions like adding and subtracting, etc. The ancient Greeks made analogue computers. The Antikythera, made around 100 BC, was found among the artifacts of a shipwreck. It was recently rebuilt. It consists of clockwork parts whose gears turn wheels that relate the positions of planets and stars to various other phenomena.[91] The Jacquard loom is a kind of programmable computer. So is the player piano. But these early computers could not be used to solve every kind of problem, only the specific ones they were created for. Charles Babbage designed the first digital computing machine to perform sequences of arithmetic operations under punch card control in the early nineteenth century, but it was never built.

According to Dyson, Turing's 1937 paper was read and recommended by von Neumann to just about anyone working on the first American computer. He told them that Turing's work was fundamental.[92]

Turing stayed at Princeton long enough to get his PhD. While he was there, as the coming war with Germany seemed inevitable, and then imminent, he became very interested in ciphers and de-coding, and in building enciphering machines. This work extended the ideas of his doctoral thesis which dealt with a series of 'machines' able to bootstrap through problems and even consult an outside "oracle" (a human) from time to time. He hand-built an electric multiplier to generate codes in Princeton's physical laboratory's machine shop. When he returned to Cambridge, he offered himself to the Government Code and Cypher School (now called GCHQ), the British government's hush-hush cryptographic/cryptanalysis headquarters.

The School was under the control of SIS—now MI-6—which was in turn under the control of the Foreign Office. By February 1938, Turing was taking courses there and taking the German military's mechanical coding device problem—called Enigma—back to Cambridge to work on. In August, 1939, the School moved about fifty miles northwest of London to a mansion called Bletchley Park. Turing worked there part-time until September 1939 when war was declared. After that he was there full-time until 1944.

The secret work done at Bletchley Park literally saved Britain from losing the war. Max Newman, Turing's Cambridge mentor, and formerly a pacificist, became a leading figure there too in 1942, helping to invent the machines that mechanized the decoding of Germany's mechanically enciphered high command communications. The Cambridge Circle spies were aware of Bletchley's importance. Kim Philby tried to get himself a position at Bletchley Park, but failed. John Cairncross arrived there as a German language specialist in 1942.

Turing was put in charge of hut eight (the word "hut" is no exaggeration as the images on the Turing.org.uk. website make clear). His self-appointed task from the start was to find a way to reliably break the German Navy's U-boat communication code. The Navy's version of the Enigma machine was much more secure than that used by the German air force or the army. Turing is said to have taken on this task because everyone considered it unsolvable and he could have it all to "myself."[93] And it had to be solved. By 1940, German U-boats were hunting down and destroying the merchant marine ships hauling food and supplies from North America to Britain.[94] The British navy had to be able to find and sink those U-boats.

Enigma, an electro-mechanical coding machine, had been invented at the end of World War I, used commercially, and then been adopted by all the German military services. Messages were typed on a keyboard connected to an array of three rotors each with twenty-six electrical contacts to represent each of the twenty-six letters of the German alphabet. Reading Enigma-coded material required an Engima machine working in reverse plus knowledge of the start points for the rotors that coded the message. The start points changed every day, sometimes three times a day. For top secret communications, especially naval U-boat messages, a fourth rotor position was added to the machine along with a plugboard to further scramble ten letters. To break an ordinary Engima message by brute force trial and error (meaning trying every possibility until one that works is found), at the rate of one thousand tries per second, would have taken over three billion years. To break the four rotor Enigma would have taken 200,000 tries per second over a 15 billion year period. Even if one got incredibly lucky once, the start points kept changing. So the Germans were sure these codes could not be broken.

The Government Code and Cypher School had bought itself an Enigma device in the 1920s which gave Turing and the people in hut eight an understanding of the problem. Then, early in 1940, three Polish mathematicians who had escaped to France gave them a second boost. The Poles had broken some German messages using an Engima machine that Polish customs had confiscated in 1928. They had first reduced the complexity to a series of subsets of possiblities and built what they called a bomba, a machine that mechanized the search through these subsets by electromechanical trial and error until an arrangement of letters similar to coded text caused the machine to stop. Solutions could then be found using very clever statistics and comparisons. Turing was sent to France to interview them. British versions of the Polish bomba, called Bombes, were soon built and run as Engima emulators. Eventually thirty-six Bombes ran at the same time, ticking through possible configurations.

After other small breaks (an Enigma machine was captured from a sinking U-boat and a German weather information coder sent the same message in both the Enigma and the known meterological cipher), Turing's group got quite good at breaking Enigma messages quickly. For a time in early 1942, Turing's group was able to reliably break each day's U-boat messages and have the decrypts sent to be translated from the German into English by lunch. Because of that, the Allied merchant ships were able once more to reliably bring food and supplies to Britain. The decrypts sent around to military units for planning actions were disguised as being from a source called Ultra. The truth about Ultra became the Allies' most closely guarded secret of the war.

Then the Germans changed their system. For a time, Bletchley was in the dark, until a German operator made another mistake. But the problem turned out to be twenty-six times harder than before. They needed to speed up their decryption machines.

In addition, in late 1941, Hitler and the German high command started to use new digital coding devices for their teleprinter communications which employed five-holed tape in a method of automatic encipherment. The devices were called the *Geheimschreiber* or the *Schluselzusatz*, one made by Seimens, the other by Lorenz. The British called them Fish, their product Tunny. They reduced letters to 1s and 0s with a digital key so they could be scrambled much more thoroughly and quickly than Enigma messages. With a key applied to the enciphered text, the original would be revealed. Speed in breaking these communications was vital so that the information encoded was still useful. There was no way an electro-mechanical machine like a Bombe would suffice.

Turing and his colleagues got a toehold on Fish when another mistake was made by a lazy German coder who repeated himself with a key only advanced by one character. This repeat was caught by a quick-witted chemist/mathematician, William Tutte, who worked in Turing's hut (and spent his post war career in Canada at University of Toronto and then University of Waterloo). Initially, portions of these scrambled messages were compared to each other by hand, then on machines called Heath Robinsons developed under the guidance of Max Newman. But the Robinsons were electro-mechanical and not fast enough to deal with the volume of signals. Soon, the first electronic version of a Turing machine was built to speed up comparisons.

Electronics were brand new. They were understood only by a few engineers, some of whom worked at the British Post Office. The first Bletchley Park electronic machine, with a memory made of vacuum tubes, was called Colossus: it was built by a Post Office engineering crew led by a man named Tommy Flowers. Eventually, the tapes to be read by the Colossus machine grew to be 200 feet long, scanned at a rate of 5000 characters per second. And soon there were ten Colossus machines working at once. Each machine was programmed with a plugboard; their memories consisted of 2400 large vacuum tubes and their states could be changed by toggle switches. They could distinguish encrypted German from random alphabetic noise, making them "the distant progenitor of the search engine," as Dyson puts it. They made it possible to decode the full range of German communications.

If the Germans had known that the British were able to decrypt Enigma and Tunny messages in close to real time, they would have changed their codes much more frequently. But they believed in the power of their machines. The Soviets weren't supposed to know how the British got their spectacularly good information about German intentions because their communications were quite leaky: the British had found that out when decrypting German signals.[95] But John Cairncross exposed some of the secrets of Engima to the Soviets in 1942 when he began smuggling out actual decrypts from Bletchley Park to give to them. The most important decrypts he gave them concerned the positions of the Luftwaffe before the battle of Kursk, part of the last German offensive on the Eastern Front, and a turning point in the war. The Soviets had ignored most of the previous warnings the British gave them because they hadn't known the source: with the actual decrypts along with unit identifications in hand, thanks to Cairncross, the Soviets moved on the German air force and caught many of the planes still on the ground. It is hard to imagine that Cairncross didn't tell his handler about Turing and Newman and the machines at Bletchley Park. And if he didn't, Philby must have.[96], [97]

The Americans definitely knew about Turing. In December 1942, Turing served as a liaison among the British code-breaking establishment, its American counterpart, and American industry. According to Hodges, the British were way ahead of the Americans and everybody else when it came to decryption but they needed American help to build more Bombe machines to keep up with the sprawling requirements to decode ever more information. To get that help, they had to share the secrets of Enigma decryption. They gave that job to Turing. While he was in the US on that assignment, he started work as well on a way to encipher speech while visiting at Bell Labs.

This wartime sharing of methods and means laid the ground rules for the Cold War and post-Cold War sharing of signals intelligence (sigint). The first agreement was written between the US and Canada in 1943. Another was written between the UK and US (called UKUSA) in 1946. Ultimately these agreements morphed into the current five-nation sigint consortium involving Britain, the US, Canada, Australia, and New Zealand known as Five Eyes. (A secondary set of agreements include a larger number of governments that share signals information in a system called Echelon.) The Five Eyes governments' secret signals intelligence agencies (NSA in the US, GCHQ in Britain, the Communications Security Establishment, or CSE in Canada, the Australian Signals Directorate and the Government Communications Security Bureau in New Zealand) mine worldwide global Internet traffic, telephone, and cell-phone communications, satellite communications, etc. There is a Five Eyes shared signals communication system called Stoneghost.

None of these agreements was revealed in public until decades after they were signed. The names of the organizations involved in the work, their budgets, their installations, who they listened to, were kept secret from the public paying the tab and apparently even from one Prime Minister in the "dominions," as Britain's former colonies were referred to in the early documents. These early agreements stipulated from the first the requirement for absolute secrecy about the organizations doing this work: anyone having knowledge of this information-gathering system was forbidden to engage in any kind of dangerous activity that might lead to their capture.

The burgeoning Five Eyes intercept system was briefly exposed in 1975 in the investigations that followed Watergate. Up until then, the NSA was not mentioned in public. The Senate's Church Committee mainly focused on wrongdoing by the CIA and swept the NSA's intercept system under its rug, but the Pike Committee's inquires about the NSA led to secret recommendations about accountability that were leaked to the *Village Voice.* Then Bella Abzug, chair of the US House subcommittee on Government Information and Individual Rights, a subcommittee of the House Committee on Government Operations, held hearings into the NSA. NSA officials refused to answer important questions. But when the subcommittee tried to issue contempt citations, its overlord committee refused to authorize them and that was the end of that. Another piece came into public view in the 1990s when a former National Security Agency employee named William Hamilton raised Cain about software he'd created for the US Department of Justice to track cases. He alleged a new version had been turned to another purpose. It had been sold/fobbed off on allies (Israel, Canada,

Singapore, Jordan) so they could be spied on as they used it.[98] Reporters digging into this story (like me) found it to be like a ball of wax that stuck to everything else, but explained nothing.

But in the meantime, out of the public eye, the Five Eyes consortium moved far beyond snatching and decoding the military signals of other countries, and into the interception of their own citizens' communications as the Internet expanded and cellphones became ubiquitous. Growth became exponential after 9/11/2001.

Then, in late 2012, Five Eyes swam into public view when a low level Canadian military man, Jeffery Deslisle, pleaded guilty to handing over untold volumes of Five Eyes' material to the Russians for money. He'd been doing it for years because he was unhappy in his marriage. He was finally arrested after being paid in cash in Rio de Janeiro by his handlers. Exactly what he gave the Russians was smothered when he admitted his guilt, but the court was assured it was extremely damaging to the US.

This system would emerge again in the middle of 2013 when Edward Snowden blew the whistle in *The Guardian* and the *Washington Post.* It would quickly become clear that nothing escapes Five Eyes—NSA alone tracks the locations of five billion cellphones *a day.* [99]

But in September, 1944, Turing's sigint progeny were still only twinkling monads in the sky when he began to work at Hanslope Park, a facility also belonging to the Security Intelligence Service (later called MI-6). It fell under the direction of Section ix (Soviet Counterintelligence). Philby had headed that Section since the previous year. By then, too, John Cairncross had also moved to SIS.

At Hanslope Park, Turing worked on a new project of his own devising. Over the next two years, he and an engineer named Don Bayley worked out and built a secure voice scrambler for use in telephone communications, code-named Delilah.[100]

Meanwhile, back in the US, a group of mathematicians and engineers raced to finish the ENIAC. Started by researchers at the University of Pennsylavinia, ENIAC was the first American universal digital electronic computer and it went into action in the middle of 1945.[101] It had been conceived as a machine to quickly calculate ballistic tables, a task overwhelming the human computers trying to keep up with the demands of war. It was turned to the calculations necessary to build the atomic bomb, under the direction of von Neumann. At the end of the war, the Americans tried to keep the bomb technology secret (they failed due to Alan Nunn May and Klaus Fuchs who handed detailed plans to their Soviet masters) but did release this computer technology to the public.

The British, by contrast, covered everything done at Bletchley Park with the blanket of the Official Secrets Act. The Bletchley Park team members, over 10,000 workers, most of whom had no idea about what they were really working on, were told that they could not discuss what they'd done there, ever. Though Turing was awarded an OBE in 1946, very few knew why. After he finished Delilah, which, according to Hodges, failed to interest anyone in the British government though it was years ahead of existing technology, Turing was asked to work at the National Physical Laboratory. There he designed a much faster version of a universal electronic computer called the ACE, for Automatic Computing Engine.

He wanted to use the ACE to explore models of the action of the human brain. But the money, the equipment, and the cooperation necessary to build it were slow in coming. Turing's frustration went off the scale when an engineer at the Laboratory proposed a scaled down version. Turing left on an unexpected 'sabbatical' to Cambridge, where he resumed his Fellowship, after the director of the Laboratory (Charles Galton Darwin, a grandson of the original) agreed that Turing *should* take himself off. The ACE was completed by 1950—but not by Turing.

Instead of returning to National Physical Laboratory, in 1948 Turing joined the faculty at the University of Manchester where Max Newman had set up a Royal Society-funded laboratory to build new electronic computers. This is when Turing either mind-melded with Leibniz's Universal Mind, or tapped into a zeitgeist from the future, pick your favorite Germanic metaphor. He began to think hard about how to make a machine version of human intelligence.

Newman also wanted to build what he called a brain.

In 1949, the same year McGill psychologist Don Hebb published his seminal *Organization of Behavior*, Turing attended a seminar organized by chemist/philosopher Michael Polanyi (father of Nobel Prize winner John Polanyi) and several other leading thinkers, including Max Newman.[102] They met to consider "the Mind and the Computing Machine." Like the psychologists who measured other animals' cognitive powers against a human gold standard, this group too seemed to think they could and should model an intelligent machine on the human mind.

They took note of the fact that intelligence persists even after big hunks of cortex are cut away and that consciousness cannot be located precisely in one set of neurons or another. They asked: how could consciouness/intelligence arise in a machine unless specified in some way? They spoke of the theory of neural networks, and about how the modeling of neurons would need to relate to reality. Turing offered the idea that random operations can be made to become regular after a certain prevailing tendency has shown itself, an idea taken up in robotics years later. His suggestion of a machine modeled on neurons led to a discussion of model cells, which lead to a diagram of star-shaped arrangements, another set of ideas taken up many years later. The observation that new and unspecified properties might emerge from coupled relationships, such as a neural network, was extremely insightful. It would not come into its own in computing for another forty years.

In a paper called "Computing Machinery and Intelligence," published in the journal *Mind* in 1950, Turing took these ideas and flipped them upside down. In this article, Turing turned the can-we-model-the-human-brain-to-make-a-thinking machine question into a very different one: can a machine think?

Turing appeared to recognize that most specialists in psychology then (and now) evade the question of the nature of intelligence by focusing on intelligent behavior alone and in extreme cases by assuming that all behavior is just a response to a stimulus. Turing adapted their method of ignoring what can't be understood in favor of what can be. He suggested that if a machine could fool a person into thinking it is a human, *it did not matter whether it thinks as we do, or not.* If it walks like a duck and quacks like one....

Enter the Turing Test.

His article proposed an imitation game or test in which an interrogator is trying to figure out which of two hidden humans is male and which is female. The rules of his game permitted the interrogator to ask them questions and for one of them to lie, but the interrogator could not hear, see, or touch them. Their interactions would consist of typed questions and answers. So, Turing asked: could a machine fool the interrogator as well as a human could? This was both a test of logical principles, and also something much more intriguing. He devised it to draw "a fairly sharp line between the physical and the intellectual capacities of a man." While at some point one might be able to make artificial skin that is indistinguishable from a human's, Turing argued, "we should feel there was little point in trying to make a 'thinking machine' more human by dressing it up in such artificial flesh."

> "...May not machines carry out something which ought to be described as thinking but which is very different from what a man does?"

While he was willing to admit that one day someone might be able to make an organic thinking machine from the cloning of a single human cell, he limited his test to what everybody was then interested in, "a digital computer." His proposed digital computer was comprised of a store, an executive unit, and a control. The executive unit carries out the operations. The store —like the memory of a human computer—contains the information, including the table of instructions (the program). The information in the store is broken up into packets of small size. Numbers are assigned to the place in the store where these packets are held, etc. The control makes sure the instructions are carried out and in the right order.

He asked: if this kind of machine can mimic any other kind of similar discrete machine, can it also fool a human into thinking it is a human? He wrote:

> "I believe that in about fifty years' time, it will be possible to programme computers, with a storage capacity...to make them play the imitation game so well that an average interrogator will not have more than 70 per cent chance of making the right identification after five minutes of questioning....I believe that at the end of the century the use of words and general educated opinion will have altered so much that one will be able to speak of machines thinking without expecting to be contradicted."

He made short work of possible religious objections such as that thinking is a "function of man's immortal soul." Too restrictive of the Almighty, he said. God could put a soul wherever it liked, including in a machine. He didn't bother to refute the idea that man is the best thinker of all, because, he said, it was not worthy of the effort. He insisted on keeping it simple: are there any problems a discrete machine could not answer whereas humans have no limits? He had already demonstrated that such a machine has limits, and, he asserted, it had never been demonstrated that humans have none. In fact, he said, humans have a limited number of brain cells most of which would not be involved in higher level thinking. Then he spat out the kernel of what later came to be called the Singularity—that point when all the world's thinking machines together will have greater capacity than all the human minds piled together. Turing said: no human would ever be a match for a group of learning machines: "...there would be no question of triumphing simultaneously over all machines. In short, there might be men cleverer than any given machine, but then again there might be other machines cleverer again, and so on."

He carefully sidestepped the pit trap of whether machines could have consciousness.[103] Sixty-five years later, people are still trying to define conciousness. Yet he obviously believed a machine could have it. Again, he used the psychologists' evasion. Don't waste time with that which cannot be shown. Get on with what you can do.

> "According to the most extreme form of this view the only way by which one could be sure that a machine thinks is to be the machine and to feel oneself thinking. One could then describe these feelings to the world, but of course no one would be justified in taking any notice."
>
> ...I do not wish to give the impression that I think there is no mystery about consciousness. There is, for instance, something of a paradox connected with any attempt to localize it. But I do not think these mysteries necessarily need to be solved before we can answer the question with which we are concerned in this paper."

And on he went demolishing possible arguments, even including one regarding extra-sensory perception, until, finally, he came to his startling, out-of-the-box main point. Though a machine is different from a nervous system, *a machine could be taught to teach itself, improve itself, and develop—like a child—through experience.*

> "We may hope that machines will eventually compete with men in all purely intellectual fields. But which are the best ones to start with? Even this is a difficult decision. Many people think that a very abstract activity, like the playing of chess, would be best. It can also be maintained that it is best to provide the machine with the best sense organs that money can buy, and then teach it to understand and speak English. This process could follow the normal teaching of a child. Things would be pointed out and named, etc. Again, I do not know what the right answer is, but I think both approaches should be tried. We can only see a short distance ahead, but we can see plenty there that needs to be done."

The point was to build an intelligence from the bottom up.

The first path Turing laid out was eventually followed by IBM which built a chess-playing computer called Big Blue. The idea of teaching a machine about the world through a machine version of sense organs— like Herb Terrace and Sue Savage-Rumbaugh's immersion recipe for making chimps language adept—reached back to Scottish Enlightenment philosopher David Hume. He had insisted that human thought/reason emerges entirely from feelings or sensations. But it also flashed forward to Google and its driverless car and to the iPhone's helper Siri which both learn as they interact, and to military contractors building autonomous robots. Turing was pointing to the idea that to behave intelligently, a machine would have to interact with the world.

Only a year later, William Grey Walter built robot turtles on three wheels with "sense organs" (light detecting sensors) mounted on their heads. These organs allowed them to follow a moving light. He exhibited the turtles at the Festival of Britain in 1951. Turing went to see the show.

For his remaining years, Turing kept on building intellectual bridges between living things and machines, between mathematical logic and the patterns of bodies to get at the how and why of organic systems' mathematical relationships. He was fascinated by the Fibonacci numbers displayed in the anatomy of so many organisms, from animals' digits to the branching veins in leaves. Where were these patterns specified? Why do they repeat? How? He began to spend more and more time in a lab he set up in a room in his house, examining the behavior of chemical reactions which seemed to be sufficient to explain basic cellular movements. He studied chemical compounds that lock into self-organized and repeating reaction/diffusion interactions (now called the Belousov-Zhabotinsky reaction, it explains the "streaming motion" of a slime mold in its plasmodium phase). He argued in an article published in 1952 that such reactions were sufficient to explain how an embryo unfolds into a functioning body.[104] In short, he was trying to grapple with the problem of design when there is no Designer. Where is the pattern specified?

Turing had offered his hand in marriage to Joan Clarke, a mathematician in his hut, but had second thoughts. He told her he leaned towards being interested in men. Though she was fine with that, and would have married him anyway, eventually he withdrew the offer. He was wholly a gay man. For years he had been very careful about who he told, and who he didn't tell. Gay sex was illegal and gay men were prey to blackmailers and worse. Also, he liked *young* men. He even made a pass at a young boy refugee for whom he had taken responsibility. He seemed proud of being gay, at least according to Don Bayley, his colleague at the Security Intelligence Service's Hanslope Park.[105] Being gay was tolerated at King's, and it had been okay in the security services too. But then the Cold War took hold.

In 1950, the Americans became obsessed with 'perversion' as a weakness that could lead to subversion. Gay men and women were marked as dangers, just like those who had ever shown interest in Communist groups or Marxist ideas. Those of either persuasion who served in the military or in the diplomatic or intelligence services were hounded by public hearings and by intelligence agencies, sometimes to their deaths.[106] Nevertheless, Turing retained his security clearance, consulted with GCHQ, and continued to have affairs with men.

In 1951, Guy Burgess who worked out of the British Embassy in Washington, was warned by Kim Philby, by then MI-6's liaison in Washington with the CIA, that their fellow spy and

Cambridge friend Donald Maclean was about to be arrested in London as a mole for Moscow. Burgess went to London to warn Maclean and they both fled to Moscow, to Philby's horror. When MI-5 searched Burgess's flat, they found documents that could only have been given to Burgess by Cairncross, who by then was working on defense matters at the UK's Treasury.[107] When he was confronted, Cairncross admitted that he once gave the Soviets confidential papers, but not to being a spy, and he was not charged. Cairncross soon left Treasury and went off to work for international agencies: he didn't make a full confession until 1964. Philby was declared unacceptable in his liaison role in the US by the FBI due to his close relationship to Burgess and Maclean. But he was defended by his friend James Jesus Angleton, head of counterintelligence at the CIA who had had many drunken lunches with Philby where he talked on and on about what the CIA was up to. Philby was recalled, investigated and interrogated by MI-5 but didn't break. He was retired with a golden handshake of 4000 pounds sterling but was closely watched. Eventually, his MI-6 friends would take him back into the fold and send him to the Middle East where he worked as a journalist, though he still reported to MI-6, until he was exposed in 1964 and defected to Moscow.

And so, in 1951, it became extremely dangerous to be a gay Cambridge Fellow who had moved in Communist circles during the Thirties, had sexual interests similar to Guy Burgess's, had worked indirectly under Philby, and from whose mind the burgeoning machine sigint codebreaking system had been born. Especially if one still picked up very young working class men on the street and took them home.

In December, 1951, Turing's house was broken into and certain things were stolen. He went to the police. There were fingerprints found that matched those of a person already in custody, and soon the story came out about how and why that young man had burgled Turing's house. He had been told about Turing by another young man, Arnold Murray, who Turing had picked up and had an affair with. The police interviewed Turing, who told the truth, right down to who did what and to whom. He gave a five-page written statement specifying details. Turing was charged with gross indecency under the criminal law. His brother, a lawyer, told him to plead guilty since he'd already confessed.

In 1952, positive vetting of security personnel replaced the previous regime of once-over-lightly-because-we-knew-your-Dad-and-you-gave-up-those-bad-ideas-you-had-as-a-lad-right?

After charges were laid, according to Hodges, Turing told Don Bayley he could no longer consult with GCHQ, though one of his GCHQ colleagues appeared on Turing's behalf at his trial, held in March 1952, testifying that he was a national asset. The trial was covered only by a local paper and a regional edition of *News of the World.* Turing was given the option of jail or probation with chemical castration achieved by taking large doses of a synthetic female hormone. He took the hormone for a year as required. He grew breasts, and, according to the Channel 4 docudrama *Codebreaker* (premiered in Britain in November 2011, starring Ed Stoppard as Turing) it is likely that his testicles shrank, that he became impotent for a time, that his thinking grew fuzzy, that his attention began to wander. Yet according to Hodges, in the summer of 1952 in the middle of this course of 'treatment' he took a vacation in Norway (which borders on Russia) because he'd heard at Cambridge that there were all male dance clubs there.

A young man he met in Norway tried to come to the UK to see him the following March, 1953, but there was a problem and somehow the police drove the young man away before he got to Turing's house. In other words, Turing was being closely watched. Later on that summer, he went on holiday to a Club Med at Corfu, Greece, right across a small strait from Albania. There he got numbers and addresses of several more young men.

All these behaviors violated the terms of the UK/USA agreement. So did MI-5 wonder if Turing was a spy? His name does not come up in former MI-5 executive Peter Wright's book, *Spycatcher*, which lists Britons suspected of being spies in the early 1950s. By 1951, MI-5 was certain that Philby was a member of the Circle, though it could not prove it. For years the questions would circulate: how big was the Cambridge Circle? Was it just three? Five? Eight? Was there more than one spy circle? Was there another at Oxford? The known spies did not work at the leading edge of science, though the Soviets were obviously interested in nuclear research and had only built their first computer in 1950. Peter Wright accused Alistair Watson, a friend of Turing's since his Cambridge days, of having been a Circle recruiter and then a spy himself. Wright alleged that Watson became a very high level spy when he went to work at the Admiralty's research station in 1953.

In the same book, Wright described how, as a senior member of MI-5, he had kept a poison antidote kit in his safe in case defectors under MI-5's protection were poisoned by the

Soviets who had a history of assassinating spies who became dangerous or expendable.[108]

In early June, 1954, well after Turing finished his course of 'treatment,' he was found dead in his bed with a half-eaten apple beside him. Potassium cyanide was found in his home lab and in his stomach. Without testing the apple, the coroner ruled Turing's death a suicide.

But was it? Turing had not appeared to friends to be despondent and had appointments in his calendar for the following week. His mother was convinced he had somehow either inhaled or ingested the cyanide by accident. On the other hand, friends later remarked that Turing was known to eat an apple in bed every night, and to be obsessed with Disney's Snow White and the Seven Dwarves. He and his friend David Champernowne had gone to see the Disney film together when it came out in 1938 and he often quoted bits of its dialogue. So was there a conspiracy to kill him and make it look like suicide? Who would want him dead? The Americans? The British? The Soviets?

Hodges suggests that Turing was told something terrible when he visited a fortune teller at Blackpool at the end of May, 1954, and the Channel 4 docudrama suggests that he hinted at his decision to kill himself to his Jungian analyst. But there is another piece of information to consider. By the spring of 1954, Kim Philby, who was still under careful watch by MI-5, was rapidly deteriorating under the strain. He had no work, was drinking heavily, had a crazy wife, and a large family to support. The KGB worried that he would break and spill the beans. One of the Cambridge Circle's former spymasters, Yuri Modin, a youngish, fair-haired man, was ordered to go back to Britain and get money to Philby to buck him up. Modin was told not to directly contact Philby but to find someone else to hand him 5000 pounds in cash. Modin presented himself as Mr. Greenglass, a Norwegian, at a public lecture given by Anthony Blunt. Blunt had been a member of the Circle but had not been active for many years (and would not be outed for many more). Sidling up to Blunt, Modin signaled that he needed a meeting. At that meeting, he asked Blunt to deliver the Soviet cash to Philby urgently.[109]

That meeting happened a week after Turing's death.

Is it possible that Mr. Greenglass was the same young 'Norwegian' who had tried to contact Turing the year before, but been chased away by the police? Had Turing been approached again at Blackpool and asked to provide services? But said no?

Turing died two weeks before his forty-second birthday. There were members of the press present at the inquest in the front row. Nothing about Turing's homosexuality or his previous criminal charge was mentioned in the papers. His suicide was put down to the sort of inexplicable thing that very brilliant people sometimes do. According to Turing's niece, Turing's brother told his children that if anyone asked if they were related to Alan Turing, they were to say that they did not know him.[110]

And why have I told you Alan Turing's ambiguous story? So you understand that the earliest ideas about how to create human-like intelligence in machines, and how to use machines to decipher intelligence, were born in the same mind. Because these phenomena were linked at birth, scientists in what were once distant disciplines like psychology, mathematics, microbiology, plant science, robotics and computer science, are engaged in a joint enterprise. By examining how intelligence works in living things, they are finding ways in which it can be generated in machines. And vice versa. This will soon extend the list of who and what displays intelligence ad infinitum.

9

Philosophy of Mind Using a Screwdriver

I found the right lecture hall at Université du Québèc à Montréal by following a trail of the Summer School's posters. They featured the bust of a man with a computer motherboarded skull.

The lecture hall's desks filled quickly with about 100 students and speakers. Stevan Harnad stood down in front—a tall, stooped man with grey stubble for hair, a black t-shirt floating free of his spare frame, his cheekbones sharp enough to cut paper. He's a vegan—he won't eat flesh because he thinks that all animals are situated in this world via feelings (which non-organic machines like computers can never have).

I looked around in the classroom for students wearing the red square, the emblem of revolt. There was only one. That's when I noticed there was also only one person preparing to take notes with a pen on paper—me. Everyone else had smart phones, Notebooks, iPads, laptops, some had several different kinds of smartness. It's not that I don't have smart machines: I do. It's that I don't trust them. When I write something with my hand on paper I remember it: when I type it, either it doesn't impress itself wherever it is that my brain stores new things, or it doesn't hang around for long. At Université du Québèc à Montréal I became vividly aware of living in two eras at once: before this Smart one and after.

I had by then acquired a group of phrases used by experts to define what intelligence is. I could reel them off, not that they satisfied me.

- Intelligence is the ability to learn and remember.
- Intelligence is flexible behavior.
- Intelligence is the ability to adapt to change.
- Intelligence is the ability to choose the best course of action.

But I was also sure there is more to it than these definitions implied. In addition, they said nothing about how intelligent behavior is generated by lumps of fatty tissue with electrical currents running through it, as one person so delicately phrased it. I had more questions than I'd started with.

Turing thought machines could acquire intelligence the same way we do, accumulating memories over time, possibly through experience. But isn't intelligent behavior also displayed by groups working together as swarms, for example? Does memory and experience accumulate in a group that comes together and breaks apart? Where? Is cooperative intelligence fundamentally different from whatever it is that goes on in individual minds or brains? Does intelligence ramp up because of the complexities of groups? Or does it get a bigger boost from the desperate needs of lonely individuals?

One thing I did know. My own intelligence has a time limit. I will need help one day to remember myself.

I watch my mother forget to use a knife, a fork, and even a plate at brunch; she has forgotten her brother, her sisters, her parents. Even her husband of sixty-six years has become a stranger who stares down at her from her living room wall. This woman once knew by memory every note of every piece of music I ever played and would shout from upstairs when I used improper fingering. She has a brain still hard at work in her body, she has all her senses, yet her ability to learn new things and recall what she learned throughout her life is fading fast, even as her great granddaughter's intelligence powers up, retaining every new word presented to her, connecting dots so fast that her learning curve, if graphed, would look like a hockey stick.

Maybe that's why I had begun to think of intelligence as a ghostly organism with a natural history, or, if you prefer, a story arc with a beginning, middle, and end.

Since reading the abstract for Harnad's talk I had been trying to pay attention to my own consciousness, to see if I could distinguish my Self from the operations of intelligence. To me, they were separate phenomona. My Self (my conciousness of me?) has been a constant companion since I rolled off a lumpy, flowered couch when I was a baby, everybody screamed, and I discovered that there is a me distinct from all of them. But this Self has changed over time. Once I—my Self—would have felt as if every pair of eyes was upon me as I made my way to a desk in a crowded lecture hall. I would have felt conscious of being clumsy, of being a stranger, of being embarrassed by these things. As a teenager, I would have burst into a full body sweat. It took many more iterations of Self for me to learn that no one cares a fig about whether I am clumsy or not. And even this more sanguine Self has shrunk away (unless a car careens too close to me).

It seems to me that as I am writing this the conscious part of me is like a laidback theater director issuing languid orders while intelligence is a stagehand who does the hard work of moving the sets around. Yet when I first began to write, my conscious Self was always shouting out what to do. Writing was an analytic exercise amenable to outlines. But writing has long since become something else entirely, a matter of letting whatever-it-is-deep-inside say things by using my hands to type words on a screen. Then another gal, a nonunion backstage apprentice, rearranges things for clarity. I am often surprised by what I write. The meaning of a first draft is often reversed by the third. And I can't step outside and observe while any of this is happening. When writing, as with any sport, it is best not to focus on how you do it.

But is that what Harnad meant by consciousness—a self with no location who feels all the slings and arrows and shouts out warnings? Or did he mean the intelligence that hides within and pummels and kneads and rearranges reality?

A man named Inman Harvey, formerly a Senior Lecturer and now a Visiting Senior Research Fellow in Computer Science and Artificial Intelligence at University of Sussex, took the podium after Harnad. He gave a lecture called "No Hard Feelings: Why Would An Evolved Robot Care?" He apologized in advance for dropping anvils on Harnad's argument, since he had come at Harnad's invitation, but then he got right to it.

He first described consciousness as a matter of different levels of attention ranging from hyper-alert to nobody home, depending on the occasion. Understanding that consciousness is highly variable, he said, made one think again about the so-called hard problem of how do we feel and why. When one considered it in a certain way, the hard problem "goes poof."

He described himself as an evolutionary robotocist.

What the hell is that, you ask? I asked too and so had looked up Harvey's [111] website where I found a book chapter called "Robotics: Philosophy of Mind Using a Screwdriver." The way I read it, Harvey wanted to get at the problem of how a human thinks by starting with a much simpler machine, a robot, to see whether any thought at all is necessary to get intelligent behavior. He argued that designing autonomous robots "has an intimate relationship with the study of autonomous animals and humans—robots provide a convenient puppet show for illustrating current myths about cognition."[112]

Cognition, in case you've forgotten, means thinking/deciding and is often used instead of the word intelligence.

Autonomy requires robots to make their way through disorganized environments, he said, just as intelligent animals must do. Building autonomous "situated" robots, Harvey asserted, forces one to choose a particular philosophical stance about the nature of mind and intelligence. Those who build machine intelligence from the top down, which he called Good Old Fashioned Artificial Intelligence—Turing's direct descendents—start with a belief in Descartes' division between body and mind. Good Old-Fashioned AI produces robots controlled, as he put it, by a homunculus, aka a series of programs that amount to a little person within who receives information from the machine's sensors and calculates a series of responses. But this sort of intelligence is not the same as that of living animals because in living things, cognition is not the same as calculation. Thus, Good Old-Fashioned AI had been a disappointment. It is mainly useful for building robots that do the same thing over and over on an assembly line, not for building robots that learn.

He and his evolutionist colleagues had taken a different approach. Their version of philosophy of mind arose in the twentieth century, not the eighteenth, among thinkers such as Martin Heidegger, Maurice Merleau-Ponty (Jean Paul Sartre's colleague), and Ludwig Wittgenstein. For Merleau-Ponty, mind arises from the action of the body in a situation, where we are "organisms and animals first, humans only later." Harvey argued—like the behaviorist psychologists and Turing—that it is adaptive *behavior* that should be our focus. Our cognition is "situated: a robot or human is always in some situation rather than observing from the outside; and embodied: a robot or human is a perceiving body rather than a disembodied intelligence that happens to have sensors." If one starts investigating intelligence from this phenomenological viewpoint, by building a robot for example, it simply doesn't matter whether there is some appropriate representation of reality that we might recognize going on inside the robot as it computes its response to events. It doesn't matter if there are no representations of reality at all. The fact is, his paper argued, intelligent adaptive behavior can and does emerge from brainless robots capable of simple interactions with their environment.

Harvey's paper showed a picture of a simple walking robot. It was nothing but two rudimentary legs with flexing knees and blade runner feet. These legs were attached to each other at what would be a hip and the whole contraption was set on the top of an incline. When built, this robot—sans brain, sans nerves, sans skin, sans muscle—walks, one leg after another, down the ramp. "The designers of course carefully computed the appropriate dimensions. In the natural world, organisms and animals have their bodily dimensions designed through natural selection," he wrote.

Instead of trying to design programs that could run autonomus robots by a series of programs running a series of calculations, Harvey and his colleagues had been inspired by how living animals are perfected, specifically the role of evolution.

The environment throws up challenges to living things which must eat and replicate. They are dynamic systems situated in difficult and particular sets of circumstances making choices in real time. As they go about this business of living, random changes happen within them, usually as their cells reproduce. Something new will result. Sometimes this new characteristic kills them but sometimes it gives them a leg up so they can produce more progeny, each one inheriting the new bit that led to greater success. This part of the evolutionary process is what Darwin called natural selection. Darwin compared natural selection to a careful farmer picking the characteristics he wants as he breeds his livestock or crops, and he insisted it happens slowly. Darwinians see every organism living now as the product of 3.5 billion years of struggle and selection.

So, said Harvey, evolutionary robotocists copy this process. They design robots by *evolving* virtual programs for virtual robots operating in a system in a computer. They aim for a behavior they want this virtual robot to perform (such as flying in a circle, or going quickly to a charging station without hitting a wall in a maze). Virtual robots made of 1s and 0s are programmed with simple abilities chosen at random, encoded as virtual genes, and are set a task. They run these programs over and over as if they are populations, changing an operational rule—a 'gene'—at random, to see what happens. Eventually, simply by chance, a few virtual robots will perform the desired task just a bit better than the others. So their 'genes' are saved and recombined, the rest are dumped, and the new and improved virtual robot populations are run through many more iterations, making more random 'gene' changes to see what happens. Virtual genes when recombined will sometimes produce unlooked for behaviors that perform the task much better than before. The developer may never understand how his virtual robots solved a problem "for the same reasons that it is difficult to design complex dynamical systems, it is also difficult to analyse them," wrote Harvey. Eventually, they end up with a program that gets the desired task done well enough in a virtual setting that it is worth testing in the real world. They build a real robot with that evolved control program and try it out. Computers can run thousands of different iterations of virtual robots through tasks in an hour or two, which is many thousands of times faster than anyone could design, build, and test a series of real robot control systems. As Harvey put it:"...this emphasis on behavior assumes that there are significant internal states, and in my view, is compatible with the attribution of adaptive intelligence."

In other words: Harvey and his colleagues had turned evolution—the theory that says there is no Grand Designer— into a design tool to produce intelligent machines.

Professor Dario Floreano from École Polytechnique Fédèrale de Lausanne in Switzerland took the podium next. He showed images of robots that he and his students had evolved virtually and then built which presented new, unlooked for behaviors. He too spoke of 'genes' encoding simple neural networks. (These are not to be confused with real neurons that go to work in real brains. Neural networks in computers avoid the problem of von Neumann's serial architecture. Neural networks 'learn.') Little more than pucks on wheels, or flying rotors in light cages, these robots solved problems that should have required at least ant-sized brains and societies with cultures, including: how to escape predators; how to navigate; how to co-opt one kind of attribute for another purpose—communication; how to share a task; how to pick a leader.

Floreano showed videos that included two-dimensional virtual robots—bright-colored lines that improved their capacity to move from none, to oscillation, to hunching across the screen like inchworms. Watching these neon-colored lines move, I couldn't help but wonder: how much more could Floreano speed up this kind of virtual evolution if, instead of relying on random rule changes and recombination he made use of *all* of nature's tricks to drive change.

As he spoke, I also thought I heard the zeitgeist howl by, or at least the leading edge of *something* moving fast.

These were dizzying ideas in their own right, but they induced an awful vertigo when considered in conjunction with what another speaker (who builds robots using what Harvey called Good Old-Fashioned AI) had to say about intelligent robots. He concluded his abstract by asserting that robots used to be used to understand the world, and now they were being built "to change it." In that context, as he stood in front of us, he mentioned DARPA (Defense Advanced Research Projects Agency), the advanced research arm of the US Department of Defense. DARPA spends billions (at least $2.8 billion in 2013 according to its unclassified budget: who knows what the secret budget was?) in grants and contracts to get smart people to create what the US military might need in the future. It has held a series of competitions to get industry to build what it wants now: autonomous robots and vehicles.

Autonomous robots are different (but not all that different) from the pilot-less drones that have been bombing suspected Al Qaeda operatives in Yemen and Pakistan since 2001. Drones take off, land, surveil, and drop bombs with no human in the plane, but under the control of distant human pilots in a safe silo somewhere. A human operator in a remote location makes the decisions about where and when to drop a payload while the drone does the deed. Drones monitor the Canada/US border and spook the battlefields of the Middle East. Police departments have ordered them up. In Canada and the US anyone can buy small cheap helicopter type drones with tiny camera attachments in boys-and-their-toys stores and use them to spy on their neighbors.

Autonomy means that the human is taken out of the system and the robot goes about its business on its own. Robots have to be adaptively intelligent to be autonomous, as Harvey pointed out, in order to maneuver on their own in complex settings. Some robots shown on the DARPA site already carry on semi-autonomously. One, nicknamed Big Dog, can carry soldiers' heavy loads while finding its own path across rough terrain, rolling over, righting itself, picking its way until it gets where it's been told to go. Small autonomous robots can roll or crawl or fly into buildings to scout for dangers. Smaller ones can fly in swarm formation.

DARPA, said the speaker, had just put out a call for narrative software.

At first I wasn't certain I'd heard him correctly, but he repeated himself. I checked the DARPA website later: he was correct. At that point, DARPA was soliciting proposals on:

> "(1) quantitative analysis of narratives
> (2) understanding the effects narratives have on human psychology and its affiliated neurobiology and
> (3) modeling, simulating and sensing—especially in stand-off modalities—these narrative influences."

This project would run in three eighteen-month sessions, and the end product was supposed to revolutionize understanding of narratives in human political contexts and lead to new software and devices.[113]

What I read suggested DARPA wants troops to be able to whomp up good narratives by inputting a few facts into their handy tablet software when they are dropped into someone else's territory, stories that might prevent hostile ideas from turning into hostile behavior. DARPA seems to believe a good story can work material changes on brains, behavior, and societies. DARPA's challenge to all comers was to determine the principal components of stories and how they affect human brains. In other words, they want software that will deliver a propagandist's version of what I do—which is to organize facts and spin them into a story that a reader will care about. Journalists like me want to build stories that are as close to truth as it is possible to get. Unlike journalists, propagandists hope to spin stories imbued with Stephen Colbert-style 'truthiness' to advance their side's version of events at the expense of their opponents, so as to shape a political response. DARPA wants to turn propaganda into neuroscience. That they should use public money to do it, and that they believed they could succeed, was terrifying.

I soon learned that one US company, Narrative Science of Chicago, is already selling narrative software based on their patented artificial intelligence platform, which they claim can "move beyond numbers and communicate true insight." *Forbes* is a partner. If you input sports scores their software will write the game results[114] as a story, complete with sports-writing-type verbs. A company spokesman told Steve Levy at *Wired* that we can expect ninety percent of news stories to be written by software in the future.

Smart machines are getting way too close to my particular form of intelligence.

Later, as Stephen drove us back to Toronto, I sat in the passenger seat chattering like a mad thing, blind to the traffic and the countryside. These phrases—dynamic systems, neural networks, evolutionary robotics, *narrative software*—spurted off my tongue. Darwin and Turing had met and merged. Where would this take us? Would future wars be won and lost by autonomous machines? That could bring a welcome end to body bags, right? Or would some of the one percent turn themselves into armies of one augmented by machines and bully their way to new versions of *L'état, c est moi(the State is me)?* Could you have war crimes trials if the crimes were carried out by autonomous robots? Or, if they are intelligent and autonomous would they be considered persons with rights and responsibilities? Would they get the vote? Could they be jailed? Executed? Wait: why couldn't smart autonomous robots decide not to fight our wars but get rid of us instead, as in *Battlestar Galactica*? Would smart software rewrite history?

Christ, what if we're too late to stop this? I shouted at Stephen.

Because, if it is possible to make autonomous, weaponized robots, they will be made, if not by responsible folk (and every nation's military leaders consider themselves to be responsible folk) then by others. And if robots can learn to feel with sensors as my fingertips feel, or see with sensors as my eyes can see, or smell as my nose can smell, or hear as my ears can hear, why couldn't they also devise such good narratives we'd believe whatever they say?

They could render me as obsolete as a secretary.

The next weekend I flew back to Montreal and Harnad's Summer School.

Meant to be, that's what my mother would have said, because if I hadn't gone back, I would have missed Marthe Kiley-Worthington, Malcolm McIver, Jennifer Mather, and their lessons, which are: we are already surrounded by autonomous intelligent organic machines completely unlike us but which make sense of this world in ways we need to study. They allowed me to shove my fear of obsolescence due to narrative software, and the upcoming violent overthrow of everything by smart robots, into the corner for later consideration.

On a bright Saturday afternoon, Dr. Marthe Kiley-Worthington took her turn at the podium in the lecture hall. I'd noticed her the previous weekend and wondered who she was. She didn't fit in among the students or professors any more than I did. She was an older woman with an air of intellectual rigor spiced by a dash of back-to-the-land. She was stocky, wore a bright dashiki and cotton pants, had pulled her hair into a roll at the back of her head. Her skin had that muddy tan people get from toiling for long hours outside. She was educated in the UK but now works as the Director of the Eco-Etho Research & Education Centre in La Drome, France. There she studies the behavior and learning styles of equines and elephants.

Her curriculum vitae set out her fine science training: she had earned a doctorate in ethology comparing wild and domestic animal displays from University of Sussex, an MPhil in Philosophy of Mind from University of Lancaster, done several honorary fellowships including one at Cambridge, and over the years had been a referee for various journals. She had been a visiting professor at University of California, Berkeley, at the invitation of John Searle, a philosopher famously interested in artificial intelligence. So I was very surprised when she said, in an angry voice, that almost nothing of substance had been contributed by science to understanding how other mammals think, whereas 7,000 years of domestication of horses and elephants had taught farmers a great deal. Unfortunately, science had paid this source of knowledge little or no attention.

Scholars, she pointed out, live in isolation from the animals they study so they often miss the obvious. But animal trainers who fail to understand the thoughts and feelings of the large animals they work with very quickly end up injured or dead. Two hundred mahouts (people who train working elephants) have met their maker in Kerala, India, just in the past ten years, she informed us, for want of attention to their elephants' ideas.

She said she wanted to return to a modified anthropomorphism. As she put it, the burden of proof should be on those who say animals don't feel, don't think, don't have a point of view, don't have consciousness. They should have to damn well prove it. She also wanted to bring back into science the power of observation, even the anecdote, to make use of "folk knowledge and belief because it arises from close observation, not from the isolated position most scientists adopt." Bear in mind, she said, "that the simplest explanation is that the animal has the attributes attributed to it.... They accumulate knowledge."

If we were going to get anywhere, she said, we have to assume that animals do perform mental work. While there are similarities between equines and humans (a humerus is a humerus and both humans and equines have them, as she put it) the areas where brains differ are well worth paying attention to. And size has a large impact on world view. "Distance is very different for an equine who can travel 160 kilometers in twenty-four hours. An elephant can only keep up top speed for a few minutes, and how he senses will affect his world view.... An elephant's trunk is used to smell and feel, and they use it for communication. It's like hands that can smell. An equine uses smell for only five percent of its interactions, sight for seventy percent. Smell is almost vestigial for humans."

So, she reasoned, horses were likely to have memories that are visually based, elephants would remember feel and smell. But they are mammals, she pointed out. So they feel emotions as we do, such as fear, panic, hunger. And joy. We can understand them if we choose to, just as they understand us because they have to.

On she went, peeling away the so-called objective standards of the science of comparative cognition like so many old banana skins. Take theory of mind, she said. She pointed out that elephants and equines obviously recognize individuals and their roles in their communities because they have to. Those who break community rules end up as sad isolates because they have difficulty "reading and predicting behavior."

We don't even acknowledge the normal sexual behavior of equines, she said. We've basically made up assertions to support the way we choose to breed them. It is an article of faith that mares will not have sex except when they are in oestrus, never when pregnant, she said. Typically, stallions are kept isolated from the females and the young. Normal breeding methods involve twitching a mare who is in oestrus (wrapping its muzzle to make it stand immobile), so it can "be raped" by a stallion. But in a wild herd, she said, females will have sex with stallions, pregnant or not, oestrus or not, because they are actually wooed. An image flashed on the screen of two horses nuzzling each others' backs.

Intentionality can be seen when lionesses hunt together, she said, each taking a role, each exercising judgment. And intentionality can and is shared with other species. A picture flashed of a group of people working with a group of elephants. *Collective intentionality,* she called it.

And that brought her to imitation: animals learn from each other and from humans "by watching and imitating." In fact, she said, they "perform better if the teacher does it first."

She showed us a video of a woman teaching horses to kick their forelegs in unison like members of a chorus line. After that, up came a video of a baby elephant, a two-year old clambering into the back of a truck. No words were necessary to explain this but she spoke them anyway—the baby wanted to stay close to its human 'mother' who had already climbed into the truck's cab.

In short, she said, they are aware of themselves, their communities, their bodies, and they have intentions, and memories. And even if we're not sure of any of that, it's better to assume they do than that they don't.

Her argument amounted to the simplest scientific rule for dealing with uncertainty and competing explanations—Occam's razor. Generally, an explanation based on the fewest assumptions is the more reliable.

Malcolm MacIver of Northwestern University is an engineer and neuroscientist. He described his work discovering that at least in a species of small electric fish that hunt in the dark waters of the flooded forest floors of the Amazon River system, the mechanics of an animal and its sensory abilities fit together. Fish live in one kind of world, animals on land inhabit another world altogether. The fish he studies can't see. Even if they could, the water they live in is totally black with debris. But they generate a small electrical current, a field all around their bodies. They use it the way a dolphin uses sonar, to feel changes in the electrical patterns given off by things around them as they move, especially the water fleas they eat. They have about 17,000 tiny electrical sensors in their skin that pick up tiny perturbations in voltage, which is enough information for them to flee predators and pick off any prey that gets close enough to grab. McIver has done an art installation by converting the different electrical fields generated by fish to specific sounds, which together sound like a deranged symphony. More to the point, McIver has already moved to commercialize this strange sensory system for application to robots. He has built one "able to navigate through clutter with great agility."[115]

Jennifer Mather also looked a little out of place. Her hair was a white ear-length cap, and she wore a blindingly bright blue shirt that kept catching my eye no matter where she sat in the lecture hall. Like Kiley-Worthington, she too expressed annoyance. She studies the behavior of cephalopod mollusks— octopus, squid, and cuttlefish. She stood at the podium and asked a very important question. How come in ten days of talks there had only been three lectures, including hers, which described what we know about the intelligence of invertebrates? Isn't that weird, she asked, since ninety-eight percent of all animals alive have no backbone?

She didn't say, but didn't have to, that it makes little scientific sense to draw conclusions about the nature of consciousness and intelligence if we draw information from only two percent of living animals. It certainly makes no sense from an evolutionary point of view, since invertebrates have been around a lot longer than mammals. How could we know when and where intelligent behavior originated if we only study it in the most recently evolved forms?

So it was not a question of what we *might* have missed: we'd obviously missed almost everything.

Mather is a professor of psychology at the University of Lethbridge in Alberta which seems an odd place to be if you want to study creatures of the ocean. But she grew up on Vancouver Island playing in tidal pools. She did a post-doc with Martin Wells, (grandson of H.G. Wells), one of the world's great experts on cephalopod behavior. Mather is an advocate for these animals. She was next going to lecture in Paris about these intelligent/sentient animals' right to be free of pain and suffering.[116]

The cephalopods are definitely intelligent and sentient, she said. Though they evolved from shelled mollusks, they abandoned their shells. Without shells, they were able to move with great speed. They developed brains, acute vision, a lensed eye "convergent with the mammalian eye….Sensorily, neurally, physiologically they made big changes," from their shelled ancestors.

"Every species," Mather said, "lives in a slightly different motor and perceptual world…so how are cephalopod worlds constructed? …They are all masters of various means of camouflage."

She showed a picture of an annoyed octopus which seemed to be staring at whoever held the camera. In fact, the camera was not looking at the actual eye but at an eye bar, something the octopus displays when angry, or when it's interrupted.

"Evolution is obviously about getting you to the point where you can survive," she said. "Horseshoe crabs got it right: they've been the same for 315 million years."

The original mollusks once lived in chambered shells, moving from chamber to chamber in a serial fashion. They were efficient, even dominant in the seas. But then the bony fishes came along. For the mollusks it became a matter of adapt to this competition fast, or go extinct. So far as Mather is concerned, cephalopod behavior became intelligent *because* of the rise of the bony fishes that prey on them. In order to evade them while finding their own food, cephalopods had to learn to plan, think, judge risk, explore, remember.

Cephalopod, as she pointed out, literally means 'head feet,' and that is what they are, a head in a sac with a lot of legs, and the heads are smart. Many people who visit zoos have reported that the octopuses seem to know them if they keep coming back. That turns out to be correct: they recognize human faces.[117]

Lab studies, including hers, have shown that they also use tools and select camouflage to confuse specific predators.[118] "Octopuses play, have personalities, and solve problems, capacities which have led to a suggestion that they have a simple form of consciousness."[119]

"They play, but they're not social, yet they have...distinct personal differences," as she put it. They have a centralized brain, but three-fifths of their neurons are in their arms. While they have some central control, each arm also has a great deal of local control of motor outputs. They have no skeleton at all. They use a muscular/hydrostatic system to propel themselves. Their arms can move through infinite degrees of motion. They can pass an item along a single arm from suction cup to suction cup to their mouth. They can use their suckers to place a pincer-like grasp on just about anything.

"My professor discovered that after surgery, when he tied their wounds with surgical silk, they would have untied it the next day."

"And fabulous camouflage," she said, as images came up on the screen behind her. One seemed to be a reef. I could see no animals in it at all, only sand and rock. Then some of it moved. It was a cuttlefish.

"Interestingly," she said, "they are color blind but their camouflage system evolved to fool fish that do have color vision."[120]

Bees, she reminded us, dance to indicate a food location to their hive mates. She thinks cephalopods are equally capable of mental mapping of locations and routes. Her studies have shown that they typically arrive in a new location, make a home for themselves, explore the territory around it over the course of four or five days. They always find home again without retracing their paths. Then they move on to a new location. The species she studied did not hunt in the same place two days in a row. "They are win/switch foragers," she said, "which suggests they have a procedural memory as well as a spatial one."

She's studied octopus behavior in the Caribbean, where they live near the shore. And, she's done controlled studies in the lab at the Seattle Aquarium with her colleague Roland C. Anderson. "They explore, it's the first thing they do in a new environment…a thermometer in a lab tank has a life of about five minutes before they take it apart." She and Anderson did the facial recognition study in his lab. They sent a stranger (a student) into the lab who riled an octopus in its tank by touching it with a bristly stick. Then they sent in another student who fed it. These behaviors were repeated with seven other octopuses, separately, and their marks of distress or interest were recorded. The octopuses quickly figured out who the friend bearing food was and moved toward that friend when she appeared, whereas angry eye bars would appear on their mantles or they would aim their funnels and squirt at the student who had touched them with the bristly brush whenever that student came close.[121]

Octopuses are inventive as well as exploratory. Another study described how an octopus without suitable shelter grabbed a coconut shell, cleared it, pulled itself inside, closed it up, and went to sleep.

When foraging they use their eyes to look for a likely hunting spot and then follow chemical gradients in the water to close in on likely prey. "They'll go over a rock, surround it with the web under their arm, and go under a rock to see what they've caught." The distance to home is a basic indicator of whether they'll take food home or stop and find a place to eat it. They are "specialist generalists" in what they choose to eat and where they'll live. And they use different methods to eat food, learning quickly whether it is more efficient to pull a particular shell apart or drill it open.

She showed an image of a pile of scallop shells outside an octopus's den. Other octopuses of the same species might prefer something else.

"I think it comes from personality. We need to study individuals. Evolution is working at the individual level. The learning, forgetting, more learning cycle is very important."

"In order not to be eaten," she continued, "intelligent animals develop a bunch of strategies." Up came pictures of Caribbean squid. Squid often travel in groups of up to 100 animals, unlike the singular octopuses. "They live in mid-water and are very vulnerable." So squid must do very careful risk evaluations constantly and will leave and return to a hunting location as often as eight times an hour. They also frequently change their appearance, turning their skin from all white to zebra stripes. Mather has spent a lot of time trying to understand their camouflage. Depending on the type of fish that comes by, they use different displays to make them go away. "They have learned the movement patterns of fish fins in the water."

All cephalopods have complex distraction and display systems. Cuttlefish in particular are adept at matching themselves to their backgrounds. It's not exact, but as her images showed, it's close enough to fool almost anybody if they stay still. "Is it conscious of its surroundings in order to make the camouflage?" she asked. "Does it know?" she asked.

Octopuses make what she called dymantic displays, making eyes on themselves where there aren't any, spreading arms to look huge, both of which deter predators with fish perception systems. They also have fine muscular control of the skin and are able to raise small papillae to rumple and dimple it or make it smooth. They manipulate the pigmented chromatophore sacs below the skin surface to expose the sacs's colors (yellow, red or brown/black) and develop more chromatophore sacs as they age. This allows them to create ever more complex mottling effects. In addition to these colors, their skins have light reflectors called iridophores and leucophores. Iridiphores create linearly polarized reflective patterns. Nerves run directly from the eye to the brain and from the brain to the muscles that control chromatophores. "This is a control system that can change in thirty-five milliseconds down to a square millimeter," she said. "They have a complex display."

Perhaps most important, they vary these displays depending on who is approaching, how far away, and how dangerous, and they do it in a randomized way, meaning no predator will ever see the same camouflage pattern again and learn to recognize it.

Squid also do interesting displays both to confuse predators and to send messages to other squids. Males signal to attract females and to send male competitors packing, and females signal availability, specific interest, or disdain. "They can do plaid, or zebra. Zebra is an agonistic display to (ward off) other male squid," said Mather. A female goes through particular color changes when she is receptive, and a male flickers when he is interested and trying to get a female to consort with him. Males compete for female attention, but the female decides which spermatophores she will accept. Like the octopuses, squid have remarkable control of what their skins show. A male squid can be saying 'shove off' to a competitor on one side of its body, even as it produces an attractive display for a female on the other side. "They talk out of both sides of the mouth," she said.

Scholars started asking as far back as 1975 if this color signaling amounts to a means of communication, a skin language. Mather had taken up that question herself. She described these displays as systems that evolved as defenses against predators that were later adapted by squid, who live in groups, for signaling purposes. But a language? She didn't think so, at least [122] not like a human language with syntax and flexibility of meaning.

"I don't think they make a language," Mather said, because the displays are essentially stereotyped. And they do not comment on externals. "Perhaps they don't have anything sufficient to say."

However, she wasn't able to offer much more about it. "We know almost *nothing* about these animals," she said, in conclusion. "I learned about their complexity by doing both lab and field studies. You cannot do it only by taking them to the lab and asking simple questions."

I waited behind the line of students and professors who wanted to talk to her. I had questions. How could animals that live singular lives develop communication signals at all, never mind a complex color camouflage system when they don't see in color? And she hadn't said anything about how long squid or octopus or cuttle fish live, how long they stay with their parents, how long it takes octopuses to learn their mapping skills, their house-building behaviors or which specific disruptive signal to flash at which particular predator.

Finally, the kids moved off.

So how long do they live? I asked.

She knew what I was getting at. She laughed. "No longevity!" she said. When the young hatch, they have to scramble. "It's totally every octopus and squid for itself."

Wait, I said. You mean they are born ready to deal with an incredibly hostile environment? With predators coming at them from all directions? Does that mean they already know things when they hatch, or do they just learn incredibly fast and if so, how?

To every question I posed she said we just don't know.

10

Plantoids

Place the cursor at November, 2012, the weekend before the US election when Obama wiped the floor with Romney.

I sat with friends at our kitchen table scarfing leftover Halloween candy. They were helping me kill time until my flight to Florence. We talked about what we always talk about—politics—and we placed bets on which man/party the American electorate would choose. The awful complexities seemed to boil down to a choice between more tax breaks for the one percent, or medical insurance for the poor. A no-brainer, my friend said. Wait a minute, said her boyfriend-at-the-time. There was the usual carping about whether people *really* understand what they do when they vote, and *how* people calculate this-versus-that-versus-the-other thing in order to mark a name on a ballot. Surely proportional representation is more true to voter intentions than first-past-the-post, said one. I snickered about the micro-lifetimes of Israeli coalition governments, but that failed to convince. Then I remembered one of Francis Galton's more useful contributions to statistics.

Galton measured the estimation capacities of individuals versus groups in the context of whether universal suffrage might result in wretched political judgment or whether there might be wisdom in crowds. At an animal exhibition in Plymouth, ordinary folk were invited to pay sixpence for a chance to fill in a card with their name and address and their guess about how much the remains of a particular ox would weigh after it had been butchered and dressed. A prize was offered. Galton got the 800 entry cards and ran a statistical analysis of them. The median guess of the hoi polloi turned out to be within one percent of the true weight, whereas all the individuals were far less accurate.[123] "The result is, I think, more creditable to the trustworthiness of a democratic judgment than might have been expected," he concluded.

See? What matters is that we *all* vote, I said. (Yes, self-righteously.) The greater the number involved, the smarter the result. That's how the best choices emerge.

Emerges, emergent. Since the Summer School in Montreal, I had been tacking those words to the definitions of intelligence I was collecting like a kid pinning a tail on the donkey. It's an emergent property, I said to myself, just like that guessing game over the weight of the ox: no individual got it right, but the crowd did. To be fair to myself (and if not me, then who) I was imitating a lot of credentialed people who salted their papers and conversations with these 'e'words. The phrase 'adaptive behavior' usually emerged (sorry) alongside. Whenever anybody asked what I was working on, I'd say intelligence, have you heard about the slime mold that computes the best way through a maze with its body? It's got no brain but it's smart anyway. Intelligence is emergent adaptive behavior. Oh, how *fascinating*, they'd cry, can't wait to *read* it (meaning: they wouldn't even crack open a *précis*).

But then I began to ask myself: so how are these phrases better than the ones you started with? What do they actually mean, and more to the point, what do they imply? If intelligence is just adaptive behavior, the pop- up product of any experience, then surely *all* living things must be sentient and intelligent, not just big-brained animals like humans, chimps, orangutans, whales, and my clever dogs? If everything alive is intelligent, and sentient, what then?

For one thing, my calm morning walk would become fraught with moral choices. (To crush or to step around the ants and worms, that would be the question.)

I was counting on Stefano Mancuso, a leading plant scientist and co-founder of a new group studying plant intelligence to help me think this through. In a TED talk he gave (shown at the Summer School in Montreal in his absence due to an illness in his family), Mancuso had raised the horticulturalist's version of Mather's question. She'd asked: how come everybody ignores invertebrates when they make up ninety-eight percent of all animals? He'd displayed a lovely old woodprint of Noah's ark with the animals marching on board two by two. "What's wrong with this picture?" he'd asked. And then he answered with another question: "Where are the plants?"

He'd then outlined cognitive science's long history of leaving plants out of their studies. The exclusion began with Aristotle who placed plants near the bottom of his scale because they don't move. Plain wrong, Mancuso said. Plants do move. And they also have incredibly sophisticated sensory capacities: some can monitor fifteen different chemicals at fifteen meters. The biggest living thing on earth is *not* the blue whale, it is the *Sequoiadendron giganteum* which weighs in at about 2000 tons and lives for almost 800 years. We forget about plants, Mancuso said, because we humans operate at a very different time scale, yet without them we would not be here. Without plants, we would have no oxygen atmosphere; without plants there would be no food for the animals that eat plants and therefore the animals that eat the plant eaters.

He showed a video of a Venus flytrap mercilessly clamping shut on a bug. "This was refused for centuries," he said, "that they can kill animals." He'd followed with a slow motion film of young sunflowers in small pots turning their heads in unison, like dancers. "They are training themselves, like young animals, to track the sun," he said. "They respond to gravity, but they also sleep. They curl their leaves, they reduce movement. This sleeping mechanism is the same in plants, insects, animals."

On he went, itemizing plants' intelligent behaviors:

- able to communicate with other plants at a distance of kilometers through the molecules they make and release;
- able to communicate with animals in order to use them as vectors for sex, or to entrap them as food,

or to use them as guards to ward off other predatory animals;
- able to predict environmental changes far enough in advance to *grow* in the best direction.

He showed an image of a dodder plant, a parasitic vine that lives off its neighbors so well that it barely bothers with photosynthesis to feed itself. It grows and twines seemingly blindly until it encounters a shrub or tree. Then it sinks a few of its haustoria into the victim and tests it for available nutrients: if there aren't enough, it unhooks itself and moves on, like Dracula disdaining a bloodless neck. In other words, this plant explores, investigates, calculates, and predicts.

I found most of what Mancuso said interesting, and yet also unnerving. For one thing, I like to garden. Who wants to think of plants as intelligent, sentient, communicative beings while pulling them up by their roots (do they scream?) and chopping up their bodies for compost or salad? He made it clear that if anyone doubted him, his ideas are rooted (sorry, could not resist) in the earlier works of the giants of science, especially the studies of plant movement by the Darwins, Charles and his son Francis, in the nineteenth century.

He didn't mention Alan Turing. As it turned out, he should have.

Mancuso had recommended a hotel in a former monastery in the heart of ancient Florence, though his lab is in the Polo Scientifico, a second campus of the University of Florence in the suburb of Sesto Fiorentino. While you're here you must enjoy yourself, he'd said. In the thirteenth century Piazza della Santissima Annunziata, a few short blocks from the Duomo, his suggested hotel presented an austere face. Inside it was another story: its ceilings were high and grand, yet its rooms remained intimate. The wooden floors of its upper lounge were so elegantly aged they sounded like crackling peanut brittle as I tiptoed across them. The night bar man complained of the stupidity of the University moving some science faculties to the suburbs while failing to provide students with free transport. He was still complaining when I went up to bed. I climbed the foot-worn stone steps, my legs rubbery from too many hours on planes.

But I was still on Toronto time, and couldn't sleep. So I rooted through Mancuso's and his colleagues' key papers again. He'd published most of them with his students and with Frantisek Baluska of the University of Bonn. Sometimes their works referenced Anthony Trewavas of University of Edinburgh. While checking dates on papers, I realized it was Anthony Trewavas who first dared to broach the subject of smart plants in print in our time. He'd published a thoughtful review article in 1999[124]. The article focused on the calcium wave that kicks off the development of human embryos, and also plays a major role in plant development and behavior. He considered it to be like a chemical version of a computer. He wrote: "the true significance of the calcium wave may be in constructing a cellular intelligence." In 2002 he outlined his views in *Nature1*[25] and wrote a blockbuster paper in 2003.[126] Mancuso and Baluska's plant intelligence papers didn't appear until later.

Trewavas's *Nature* piece had an editors' please-don't-blame-us-for-this sidebar. Trewavas had an FRS after his name, meaning he is a Fellow of the Royal Society. If you want to be humbled, look at the other names on the Society's list. A person who has earned an FRS cannot be ignored even if he is saying seemingly whacky things. The sidebar read:

> "Plant Intelligence
>
> Traditional definitions of intelligence use movement as a criterion. But are the adaptive behaviours shown by individual plants 'intelligent'?"

Trewavas said they are.

He had sought to explain how plants assess their environments, grow their way to new resources, dump compromised leaves and branches, fight off pests and parasites, all without any obvious brain-like organ for processing information. It was utterly obvious to him that plants *must* process information in very complex ways. The piece made it clear that while he had been studying plant chemistry and behavior for forty years, his ideas about their intelligence had been bolstered by his very wide reading beyond the confines of the disciplines of plant science. He used concepts borrowed from cybernetics, from information theory, systems theory, and computer science. Unlike most of his colleagues who study the interactions of plant proteins, he was interested in the behavior of plants as *whole systems.* He thought about how different cells in different parts of these systems communicate. He thought about the routes important electrical and chemical signals take; in other words, the ways and means of transferring information within and among living plants.

But how is information communicated and stored in a plant if there are no neurons to do the job?

Trewavas declared there is a formal equivalence between the calcium wave, which had long since been demonstrated (by him and others) to be a major signaling system in plants, and the neural networks in animals' brains as well as the neural net architectures used in modern computers. He argued that the calcium wave in plants amounts to computation capacity. Neural nets learn and remember by increasing the number of physical connections (called dendrites) between active neurons sharing signals, or, by increasing the strength of existing connections, and therefore changing the interplay between signals and responses over time. Trewavas thought chemical interactions involving calcium could do the same job of learning and remembering, but in a different way.

Calcium plays a role in opening the pores in plants' cell membranes which permit other molecules to enter and leave. Molecules like proteins and hormones do the work of plant cells. A change in the way proteins fold makes them interact with other molecules in different ways: changed shape changes behavior. The chemical interaction in which one protein locks onto and changes another is called transduction. The signal—the information—*is* the re-shaped molecule which changes the nature of the work done. "When plants are signaled," he wrote, "they increase the cellular constituents concerned with calcium signaling....However else these data might be regarded, this is a clear example of cellular learning....So far as I am aware, this property of increasing the cellular content of signal components is unique to plants."

He produced drawings showing that the information flow through a "calcium cloud" and through a brain's neural network look very similar. He quoted Darwin on how light seems to act on both plants and animals in the same way, comparing plant responses to that of animals via the nervous system. He clearly saw plants as being constantly forced by exigencies— such as competition for light, limited nutrients, their inability to run away—to change themselves intelligently to meet circumstances.[127]

> "In responding intelligently to this multiplicity of signals, plants have become masters of phenotypic and physiological plasticity which allows them to cope with the variable circumstances that surround them. Behavioral plasticity surely demands a cellular system of considerable computing power if plants are to survive the signal morass in which they find themselves. The ubiquity of calcium involvement in plant-cell signal transduction suggests that calcium forms the basis of the intelligent system controlling plasticity."

At the end, he described how molecules, cells, tissues, whole plants, populations, ecosystems fit into an interconnected hierarchy in which interactions between each level[128]

> "...generate the level above. The 50,000 or so cellular molecular species represent the lowest level interactions between these molecules [and] construct a complex network perhaps best described as analogous to a very large, badly woven fishing net. Like all networks, the connections form molecular collectives that generate new emergent properties.

> Cellular behavior is emergent behavior as are cytoskeletal dynamics, cell division, cell growth, and cell development. Emergent properties are not predictable from the most detailed analysis of the constituents because by definition it is the interactions between the molecules that are critical. Emergent properties can be observed, but understanding is another matter."

Plants, which first appeared in the oceans, Trewavas argued in *Nature,* originally just used chemical photosynthesis to generate energy. But as plants moved out of the oceans to colonize land, they evolved new means to get nutrients and make energy, and new body parts with which to forage—specifically, roots. A plant can consist of a single root, leaf, bud, or, a plant can be an assembly of millions of these parts. "We simply do not understand this organizational plasticity," he said, "but plasticity is foresight."[129]

His 2003 article was even more pointed. He had been invited to give a major address at a plant society meeting in 2000. As part of the deal (his travel and hotel expenses were paid by the society) he promised to produce a paper for their journal after the meeting was over. When he finally sat down to do it, he wrote that plants "can best be viewed as more like a democratic confederation in their control structure, rather than an autocracy as occurs in animals, controlled by an all-embracing nervous system." Because plants live on a different, slower time scale not observable by humans, even by scientists studying their behavior, much has been missed, but when time scales are taken into account, there is little to distinguish the behavior of plants from animals. He described plants as individuals with internal representations of their realities, with memories that specify optimum situations for maximum fitness which, when compared with circumstance, specify the extent of plasticity. They have the capacity for self-recognition, foresight, and the ability to predict (growing to where conditions would be best). When he finally sent the paper off, two years later, the editors were shocked by what he had to say, but they published it.[130]

He closed a paper in 2005 saying that while primate intelligence is currently considered to be much more advanced than that of plants, when the capacity of plants as species to dominate and exert power over others is considered, then "this conclusion may well need reassessment."[131]

Mancuso's papers' also distinguished between operations at the cellular level of a plant, and the operations of the larger plant system. He seemed to say that plants' intelligent behaviors emerge out of the cooperation and coordination between groups of cells. He focused on electrical signals in trying to understand plants' information systems, circling around brain/neuron analogies. When he wasn't comparing the complexity of plant root systems to the layout of the Internet, and comparing the Internet to a brain laid out flat instead of bunched and folded, he focused on the "dynamic electrochemcial activity" of the root apex cells.

Mancuso and his colleagues[132] argued that dynamic electrical activity in plants could be a means of gathering information, transferring, and integrating it. They proposed that the cells at the tip of each always-growing root "continuously integrate internal and external signaling for developmental adaptation in a changing environment."

Which is a mouthful. But taken together, all the papers amounted to a manifesto: plants feel, plants think, plants remember, plants have something like consciousness as they interact with their very challenging environments, and, as Trewavas pointed out, their intelligent, adaptive behavior will never be captured fully in a lab where conditions are controlled. It is their capacity to adapt to extreme change that defines their intelligence. Mancuso went one step further than Trewavas: not only are plants intelligent, he said, but their particular forms of intelligence could be usefully mimicked to create intelligent machines.

All of them wrote of plants as intelligent living computers, which brought Turing to mind.

After reading these papers the first time, I had gone for a walk in my neighborhood. I found myself reconsidering the elms and ash and maples that bower over my street. I had always thought of branches, leaves, and flowers as trees' defining features, almost like their faces. We humans know each other by the shapes of our noses, mouths, the color of our hair (or lack of it) the overall shape of our bodies, the ways in which we move, and if we can't see each other, the sounds we make walking, breathing, and even the smell of our skin. But Mancuso's papers made me see that my face metaphor was wrong. Face implies a uni-polar entity, something like us with thinking heads on top and obeying muscle and bone below. Just as we have no equivalent to the elephant's trunk—the hand that smells as Kiley-Worthington described it—we have no body parts to compare to the parts of plants.

Plants are bipolar: they orient in two directions, down toward the center of the earth as well as up toward the light.

If Mancuso is right, plants recognize danger, think, organize and plan. But what body part does this work? Their roots forage for water, suss out the best soil patches, suck in nutrients, maneuver and elbow for the best position like law students searching for the same book in a library. Their branches and leaves, their flowers and fruits, are their sexual and excretory organs as well as their chemical factories where light is transformed to energy and all kinds of molecules are made.

Mancuso argued that roots do the work of integrating information for the whole plant.

Which made that old Yiddish curse go round and round in my head: may you grow like an onion with your head in the ground.

In 2005, Mancuso and his colleague Frantisek Baluska invited about 150 leading plant scientists from the US, Britain, Germany, Israel and Canada to a conference in Florence. Trewavas was also part of the group. Together, they decided that thinking about plants as organisms with a neurobiology would lead to useful experimental questions. So they created the Society for Plant Neurobiology and a journal of the same name. Mancuso also opened a new lab called The International Laboratory of Plant Neurobiology.

The application of the word 'neurobiology' to plants which have none was very unorthodox. Yet it wasn't until 2007 that the pushback began: thirty-six plant specialists signed their names to a letter objecting to the word. They sent it off to

Trends in Plant Science, a major journal, which published it. The letter was titled "Plant neurobiology: no brain no gain?"[133] This letter is referred to as Alpi et al. after its first author, Amedeo Alpi. Though they conceded that there are similarities between plant cells and all other biological cells (including action potentials, voltage-gated ion channels in membranes, neurotransmitter-like substances, and signal transduction and transmission over distance) they said "this does not imply *a priori* that comparable structures for signal propagation exist at the cellular, tissue and organ levels…" The signatories asked Mancuso and colleagues to stop using the word 'neurobiology' on the ground that plants don't have neurons, and urged that they come up with a better concept.

Mancuso and colleagues replied with a letter of their own, basically a history lesson about who else had had such ideas and how they too had been disparaged in their day. Then, in 2009 Mancuso's lab, along with Baluska, published an article in *PNAS (Proceedings of the National Academy of Sciences)*,[134] a highly regarded American general science journal. It had an innocuous title "Spatiotemporal dynamics of the electrical network activity in the root apex," but it clearly described plant roots as propagating their own signals—acting, in other words, like pseudo-brains. Since it was also edited by Emilio Bizzi of the Department of Brain and Cognitive Science at MIT, it had major credibility. Bizzi explores how systems of cells in animals combine in larger structures, and the how and why of muscle movements.

The lead author of the *PNAS* paper, Elisa Masi, described how Mancuso's lab used a sixty channel multi-electrode array to measure the electrical network signals of root tips. They observed both intense spontaneous electrical activity, and stimulation-elicited bursts of local signals. As well, the "data reveal synchronous electric activities of root cells emerging in a specific root apex region...this appears to be the first observation of synchronized electrical activity in plants."

And then came these fighting words:

> "Electrical events in plants are normally associated with an external stimulation. In the present study we demonstrate that a great number of spontaneously generated electrical signals exist in cells of the root apex. The duration (...forty milliseconds) and velocity of conduction...appear to be...not far from the conduction velocities detected in nerve cells of invertebrates such as hydromedusae or siphonophores...or hemichordate worms."

They called this electrical signal transmission "neuroid." They also suggested that synchronized oscillations in the transition zone of root apices suggested "some kind of sensory zone specialized for integration of diverse sensory input information that enable the growing apex to continuously monitor diverse environmental parameters and to mount appropriate adaptive output responses."

That too was a mouthful. It boiled down to this: roots act like neurons.

In March 2010, in a non peer-reviewed online journal called *Lab Times,* two senior researchers from Germany hacked at this *PNAS* paper with a will. Their article was entitled: "Intelligent Plants or Stupid Studies[135]" It was quite clear how the authors answered that question. Herbert Rehm and Deitrich Gradmann said flat out that Mancuso's lab's measurements had to be wrong and their data must simply be artifacts caused by slippage of the recording device they used. They also reported that one of them—Gradmann—had written to *PNAS* and asked for the article to be withdrawn and had been told to contact the authors. They claimed the authors had merely asked them to supply contrary data, which they claimed they were not obliged to do as they were making no claims. The shoe was on the other foot.

The tone was insulting. Allegations of incompetence and even of undue influence in the editorial process lay like dangerous snags just below the surface, but on the other hand, what else could be expected? No dogma is overturned without a fight. Yet it looked as if Mancuso's cadre had decided that they were not quite ready to assault the Bastille. That summer, their Society and Journal changed their names to the more innocuous *Plant Signaling & Behavior*. Only Mancuso's lab still proudly carried the neurobiology banner.

The taxi inch-wormed and honk-honked its way through Florence's grim Monday morning traffic. The driver actually shouted "Mamma mia" as he shook his fist and cursed his enemies. We caromed through a roundabout and on past Florence's huge graffitied fort into a bland district of low-rise apartments. Finally, we sped through a scatter of industrial parks interspersed with green fields. We even left the airport behind.

I'd missed this twenty-first century Florence the first time I visited in 2008, just as the world's financial system shuddered to a stop. Stephen and I had stayed in a renovated farmhouse near Fiesole, a mountain top town about ten kilometers away. We did all the usual tourist things but had been utterly entranced by a Leonardo da Vinci exhibit in one of the former Medici mansions. It featured all the machines he'd invented presented as models: they were marvels. (If you do a Google search about da Vinci's IQ you will find it guestimated at 220, higher even than Terman's guess about Galton.) If da Vinci had been backed with the capital to build these machines, the western world would have made it to the industrial age and World War I in half the time. The museum plainly showed that it's not just great ideas that shape the future, it's the wherewithal to convince others of their value, and to effect them, something Stefano Mancuso understood very well.

The taxi driver finally found the horticultural sciences building on a half-empty road called The Way of the Idea. Across the street, an open field of unmowed grass hid rusting cars and dead equipment. At the end of the street a small Roman-temple-styled church stood guard for lost souls.

Mancuso's floor to ceiling windows gave a view of fields and silos. There were conference posters tacked on the walls, a desk buried a foot deep in papers, a big computer screen, dark, thick-leaved green plants in china pots soaking up the sun. Mancuso himself looked like a middle-aged Harry Potter: his thinning grey-laced brown hair stuck up in cowlicks, his lustrous brown eyes peered at me from the glasses sliding down his nose. Have I mentioned the gray beard trimmed so carefully I kept forgetting he had one? He wore a jacket with leather patches at the elbow, good oxfords, a striped shirt. The overall effect was professorial sweetness. But sparks jumped off steel from time to time.

Behind him on the bulletin board he'd hung a sketch of Darwin, and the perpendicular wall bore a poster bearing another image of Darwin, the famous one with that beard and that bowler, looking sad and tired.

I slid into a chair across from him, pulled out my notebook and asked, what is intelligence and who displays it?

"My point of view is that intelligence is a principle of life," he said, firmly, using the syntax of a person who thinks in Italian but speaks fluent English. This stemmed from his understanding of Darwin's theories and observations, he explained. It also came from years of work done by many others across many fields in biology which showed that the so-called recent specializations of the nervous system "represent rather ancient and fundamental cell survival processes."[136]

"Every organism, it is all the time obliged to solve problems," he continued. "To find food, fight stress, fight enemies. It cannot be done without intelligence. It can't be done by programming....You can't exclude that a human is commanded by a program...we cannot exclude. But I disagree completely with this vision of life. Life is much more elastic, without boundaries, without programs, destiny..."

You mean *telos*, the Aristotelian notion of an end point development aims at?

"As in *telos*," he agreed. "But linking to intelligence, we cannot imagine an organism—the simplest you can think about—without having to solve problems. So intelligence, it should be added to the requirements to define life, you know, able to reproduce, etc. Intelligence should be there."

I was startled by his forthrightness, but not surprised. His papers are like that too: careful, yet bold.

Start at the beginning, I said to myself. Get him to tell his story. He seemed surprised that I asked, but he obliged.

"I was born in the south of Italy," he began, "in the district of Catanzaro, Calabria, in the toe of the boot. In a small town."

The year was 1965. His father was a General in the Army, his mother a teacher (both have since retired). He is the eldest of three boys, one disabled, the other a judge who deals with criminal cases. The disabled brother lives with Mancuso and his family. He came to Florence to go to university at the age of eighteen. He followed a girl, a girl who came from Florence who became his wife.

He was not interested in plants as a boy, though he was interested in nature, often going to the forest to hike. He wanted to study something in science: especially physics. But he became an agricultural scientist instead out of a futile attempt to predict Italy's economic future and position himself to profit.

"I was thinking there would be more chances to find a job in this field. I was *completely* wrong.... But during my training I started to like plants. I had never thought of a plant as actually a *living being*," he said. He started to laugh. "I had a conversion..."

In the 1980s, science students in Italian universities studied for five years, earned the equivalent of a bachelor's degree, and then produced an experimental thesis that took from six months to a year to complete. Then they went directly to a PhD. His thesis changed everything for him. "I realized plants are not just living, but could solve problems actively. This struck me heavily. It changed all my interests in research."

The subject of his thesis was the growth of roots and how they detect and find food. He had not yet read Darwin on plant movements–that happened later, when he was working on his doctorate\ "I am a booklover. I collect old editions, it's a hobby, you know, going to small shops, looking for old editions of scientific books. Today I have a big library on science subjects, mostly on plants. I started to read the original research of Darwin and Romanes, of Jacques Loeb and Herbert Spencer Jennings."

Who are Loeb and Jennings? I asked.

"They wrote at the end of the nineteenth and beginning of the twentieth century. Jennings [a leading geneticist who fought against the legal and regulatory overreach of the American eugenics movement] wrote a wonderful book on the intelligence of invertebrates but also single-celled organisms. He said we are not rating paramecium [amoeba] as intelligent because they are so small but if they were as big as a dog we'd have no doubt. It fights and forages like a big animal, like a whale."

Loeb, on the other hand, was a Pavlovian, said Mancuso.

"You know, operant conditioning, Pavlov got a Nobel for conditioning a dog with a bell....Loeb went much farther to say almost everything in life is programmed this way."[137]

Loeb, who was educated in Germany, came to the US to teach at University of Chicago and Rockeller University. Jennings was born and educated in the US. The two of them had a famous debate: stimulus- response versus adaptive behavior. But then the whole debate just "disappeared from the literature of science," said Mancuso.

"This field about intelligence completely disappeared from the history of science," continued Mancuso. "Until the 1930s, there were people working on it, for example, Francis Darwin, Charles' [third] son, who was a great scientist, a plant scientist." Francis Darwin was one of the first plant physiologists, appointed as a Lecturer in Botany at Cambridge.

In September 1908, Francis Darwin gave the President's Address to a meeting of the British Association for the Advancement of Science held in Dublin. He explained to his colleagues that plants are intelligent, said Mancuso. [Darwin said they remember a stimulus and somehow control their development and shape according to external and internal conditions.][138] Mancuso plunked at his computer keyboard to pull up the correct reference. When he found it, he turned the screen to show me: there it was, an image of a story about Darwin's speech in the *New York Times.* "I'm happy to have in my field Darwin and his son."

Back to your thesis, I said.

"I was studying roots....I saw they were growing and growing, actually they were *moving.* This was the first demonstration for me that they were not fixed organisms, but move, in a different way from animals, by *growing.* I started to think how to visualize this. There was no digital camera then. I had to study with photography, doing small time lapse photos of the roots. A completely new world appeared, it was like inventing the microscope. To me, to that moment, an organism fixed, sessile, showed a lot of movement. I could see the behavior....The movement was amazing. I saw directly the behavior of the plant and could play with it."

He put obstacles in the way of the roots, building mazes for them. He put two plants close together to see how they responded to each other. "They... change behavior according to the neighbor....They are able to do kin recognition. Any biologist will say that is only found at the highest level of animals. All plants recognize kin, well not all, I can't say all.... Two plants in the same environment compete—they invest more energy in the root apparatus growth instead of the shoots. They grow down and make a bigger root apparatus to colonize the biggest patch of soil they can. The way nations compete. When two plants are kin, part of the same tribe, not just the same species...there's no more competition. It's a co-operation. We don't know regarding the upper part of the plant. Shade avoidance, this is an automatic program response. But even here there is a small mystery. They will start first to elongate at a faster velocity to prevail. In a single tree, many parts are self shaded.... The tree can understand whether the shade is from another, or itself. Again, this should be studied as intelligence. And no one is studying it."

So where did you go with this thesis?

"At that time," he said, "I just thought it was fascinating and plants were way more active than others had realized. That was the beginning of my real interest in plants. I started my PhD at the University of Pisa. I studied with others, not plants, biophysics mainly. I was there for three years."

His project at Pisa was to develop new instrumentation "to monitor changes in cells using electric tools….I started to be interested in electrophysiology—the electrical signals produced by plants. I discovered what no one taught me, that they produce a lot of signals all around the plant, it's the same kind of signal as animals, but a different frequency or amplitude. Much lesser. Slower. But… it's just a quantitative difference, not qualitative."

When he went and looked for it, he found a lot had been written about plant electrical signals in the old literature that he liked to collect. All kinds of experiments had been done showing that plants produce action potentials, for example, the kind of voltage changes that occur in neurons transferring a signal. This work had long since been forgotten.

"From the beginning of electrophysiology, action potentials were detected by Sir John Scott Burdon Sanderson, a close friend of Darwin's," he said. [Darwin wrote to Burdon Sanderson, a leading physiologist, and asked him to investigate the question of whether the movements of plants were accompanied by electrical signals as in animals.][139] "From the end of the nineteenth century [actually 1873] we know there are action potentials."

First they were found in the Venus Fly Trap, or *Dionaea,* but then also in every plant looked at.

"Jagadish Chandra Bose, a Bengali, and a great physicist, at the beginning of the twentieth century, he was awarded a baronetcy, quite difficult to obtain for an Indian guy....He studied waves, wireless signaling, all kinds of electromagnetic waves....The second part of his life he studied just plants. He wrote five or six books on the neuron systems of plants. I have all these books at home."

Most of Bose's plant books, published between 1907 and the late 1920s after he returned to India, described parallelisms between plant and animal tissues and pointed to what he called feelings in plants. While many study plant electrical signals now, said Mancuso, it is a sideline in plant science, though "it is of paramount importance."

Why?

"Plants can transport information in the same way animals do, but you don't need neurons to do action potentials. Today, the neuron doctrine is everything, it is the action potential, the message of the brain bringing the information signal and traveling from one neuron to another through a synapse, the gap between the neurons. Some years ago, an important neurobiologist wrote a book called *You Are Your Synapses.* Well, in plants, over all in roots, almost all the cells produce action potentials that propagate. This is not a supposition. This is a *fact* demonstrated hundreds and hundreds of times. Despite this phenomenon, there are just a few groups studying it. Plant electro-physiologists are such a small community I know almost everyone in the world who does it. Yet there are hundreds of thousands who study this regarding animals."

For his doctorate, he'd asked how it is that plant electrical signals are so similar to those of animals. It had become dogma by then that the only excitable cells in nature are neurons, cardiac muscle cells, and glands.

Look on the Net, Mancuso said. "You'll see no mention that plants are full of excitable cells.[140] From a biological point of view, this is complete nonsense. If an alien came to this planet it would say, life on earth is *plants*."

Then he veered off into language that political philosophers such as Machiavelli and Hegel would have found familiar. "In every kind of relationship there is a master that is in charge, and a slave....in our relationship with plants, plants are the master, we are the slaves....It's the most stupid thing, to ignore everything, the organisms on which we depend....Outsiders think science is a world of ideas where logic is running everything. Not true. Relationships, power, politics are almost more important than the logic of the idea."

It was these realities of academic politics that explained why he continued on for the next fifteen years without writing a word about the interesting behavior he had seen in plants as a student. First he found a position here at the University of Florence. Then he worked his way up the academic ladder.

"In 2001 I became a professor here. Not because of my idea, but thanks to my classical papers, my molecular biology papers."

Then, and only then, did he tackle his real interest, plants as intelligent systems. First he went out and got money to set up a lab.

So where did you get it, I asked.

"The money…came from a bank foundation so kind as to trust my idea and fund my lab," he said, smugly

How did that happen? I asked.

"They gave a big amount of money for three years."

"How much?"

He looked a little bashful, as if he didn't want to crow. "Two million Euros, between 2005 and 2008," he said.

That was the equivalent of three million Canadian dollars, a huge amount of money for a new lab investigating something so *outré* as plant intelligence. That kind of money meant he wasn't just a leading scholar with a good idea, he'd become a science prince, the kind of rainmaker students will flock to. As de Waal had explained, doing research in an institute is all very nice, but without students to generate new ideas and produce new papers, it will soon peter out. To attract students, you need a pile of grant money to sustain them. This became even more important after the 2008 global financial meltdown. The competition for research money in Italy became more than fierce. There are only two private universities in Italy, the rest are public and all the salaries at public universities are paid by the government, which had begun whacking at their budgets.

"In the last four years there have been no new positions at all. In 2008, the University of Florence had 1,300 researchers," he said. "Today we are just 800. Retirees are not replaced. So the people who work here are paid out of research funding. We need to compete for funds. Most of the money today, 300,000

Euros a year, comes from a grant of the European Union that we won, also from the Italian Ministry of Research. Also we have some money previously from Defense."

What are they interested in? I asked.

Robots, he said. He had been working on what he called soft robots, robots modeled on the flexible movement of plants.

So let me be sure I understand this, I said. You had been interested in this idea about investigating plant intelligence since your doctorate but you didn't dare do anything about it until after you were appointed a full professor?

"If I had been controversial," he explained, "no way I'd *become* a professor....The idea was in my head all along, but not published. As a professor, no one can fire you anymore. So I can do what I want to publish, my idea."

So when did you first declare yourself?

"First publication when I said plants are intelligent organisms was in 2004."[141]

And yes, he had read Anthony Trewavas' papers first. Trewavas, being a Fellow of the Royal Society, "could say this." However, he'd paid a price. "He's now known as Crazy Tony because of this idea. He is a wonderful scientist...you cannot find any weakness in his papers, so they just say he's crazy."

He had met his colleague, Frantisek Baluska, the lead author on that first paper and an associate professor at University of Bonn, after Baluska wrote him an email to ask for more information about another paper he'd written. It was a "classical" paper, meaning it said nothing about plants being smart. One email led to another and he invited Baluska to give a seminar in Florence. When they met, they realized they were both interested in how plants process information. "So we work together till now," he said.

When he and Baluska attended a large meeting of the Society for Experimental Biology in Edinburgh in 2004, they realized that almost all the papers dealing with neurobiology were about humans. There was nothing on plant cells. "So we started to joke over a beer that we need to organize a meeting on plant neurobiology, since it's almost the same. If I'd known about the furore this name would causeI'd have chosen another name. *Of course* we know there are no neurons in plants. Except for neurons, all the rest is the same."

Maybe there's another problem with your papers, I ventured. It's as if you're saying two things: you say plant root apices act like neurons and so the roots are like a brain, but then you also say that neither neurons nor a brain are necessary to process information.

"Look," he said, "the machines of the brain are almost there. There are action potentials, neuro-active substances..."

Like?

"Glutamine, serotonin, melatonin, acetylcholine, adrenalin, dopamine, all these substances known in animals are also in plants, doing the same thing, or very similar. It's a strong analogy. From the point of view of science, to look at the plant from a new discipline called neurobiology will have an impact on the way we study plants. Look at the huge infrastructure for studying the brain. We can use these procedures and translate to plants."

It was after that Edinburgh meeting that he got serious about finding money. In Italy a law passed in 1992, regulations established in 2000, requires banks to create foundations to reinvest some of their profits in something of benefit to the public. A private bank started in 1829 by 100 citizens of Florence, later called Ente Cassa di Risparmio di Firenze, created such a foundation in 2000.[142]

"In Florence, usually this sort of money goes to restoration of palaces, museums, the arts in general. I had a chance to meet one of the committee deciding on the destination of this money, a physics professor in Florence," he said. "I asked if some could be used for science. He agreed and started to fight on this committee for a certain amount to go to science."

Relations were set up with granting agencies and universities. A competition ensued, which Mancuso entered.

He sat back a little in his chair to announce the obvious. "Our project was well received, to build the first plant neurobiology laboratory in the world."

With his grant in hand, "me and Frantisek invited more than 150 plant scientists here, the most respected were here. No one had anything to say about plant neurobiology at the meeting. We've had five meetings year after year. After the fifth we changed our name to the Society for Plant Signaling and Behavior."

So people came to the meetings but didn't get excited over that word? I asked.

No one was bothered by the neurobiology word, he said.

But after the 2005 meeting, a colleague called Mancuso to say he had been in Paris at a conference and a petition had been circulating there asking people to write a paper objecting to the name neurobiology and its implications. The caller had been asked to sign it but refused.

I asked, this is Alpi et al?

"Yes," he said. "Many of them, in the following years, told me they didn't read the paper. They also didn't read ours. I know Alpi. He is also retired. Many of these guys are very old. I have a good relationship with Amedeo Alpi. Before the paper, he never spoke to me about it. He's only eighty kilometers away. Normal would be to meet and speak about your crazy idea. But no one did it."

So why wait two years, and then attack? I asked.

"These guys didn't imagine the success plant neurobiology would have....The success was public. We had a network of scientists interested in a new/old aspect of plant biology forgotten for sixty years. They came out of the catacombs. We organized. We funded ...our Society and very important scientists decided to participate. Societies are the equivalent of parties in politics," he said.

How do you mean?

"They are the centers of interests. We know each other and can participate on competitive grants together, help each other to pursue a goal in science...."

This reminded me of the political doctrine of former Canadian Prime Minister Jean Chretien. He had opened his memoir with this definition of politics: "to be frank, politics is about wanting power, getting it, exercising it and keeping it."[143] Only later did he mention making the world a better place.

"It's the same in science," said Mancuso. "Without the use of power, you cannot have your idea accepted in the mainstream of science."

To push forward his radical ideas, he believed he had to be personally unassailable, have tenure, and money. Then he and Trewavas and Baluska had to build a Society— a party or a faction of shared interest. Yet he made a simple miscalculation that the wily chimp Yeroen would have avoided.

"It would have been more strategic," he admitted "to involve in funding of the Society and in the meeting in Florence more important people. One important professor wrote me a strong letter complaining he was not involved in organizing the meeting and his Society was not involved. This Society dealt with plant physiology....I'd asked to be a member and they didn't admit me because I am not a physiologist. He was so upset he wrote to the director of my department saying how nasty I was in organizing this meeting. This director was on my side. This is the way in science, it's not just Italy. Frantisek is in a much worse position in Germany."

Their 2009 paper in *PNAS* made things much worse. Not only had Mancuso managed to corral a pile of money, but he and his colleagues were being published in major journals.

"Anna Moroni is a respected plant electro-physiologist at the University of Milan and the wife of Gerhard Thiel of the Technical University-Darmstadt in Germany. After the *PNAS* paper, their society invited me to give an invited lecture," said Mancuso. "At first glance it was a big honor."

The invitation had come from a British scientist at Heidelberg so he never suspected that his opponents in Germany might be behind it. He had not suspected anything. He gave his talk, which lasted about forty minutes. In the normal course of things, there would have been a few tough questions after his speech and the chairperson of the panel would have cut them off after about ten minutes. Instead, all kinds of people went after him from the floor.

"There were some 200 people in the audience asking stupid questions, not waiting for answers, it went on for an hour and a half...without people having read the paper, like, 'you say plants have brains,' 'no, I never said that,' it was really a trap...even the chair is looking at me with hostility....Today, science is just a competition for resources, fame, power. On the contrary of other fields like sports or politics, where there are rules..."

I laughed out loud at the idea of rules in politics, especially in Italy.

...A kind of *tradition,*" he corrected, but "in science there is *nothing*.... The more rude you can be, the more you are looked up to. And people are scared..."

Mancuso and his colleagues had wanted to do something better with their Society. "We wanted to cooperate, not compete, but even cooperation is a nightmare, even inside the team there is a fight for power, first name or last name on the paper....Today, scientific meetings are a nightmare.... You never will hear something new if research is not yet published, you will never speak of it in a meeting in case someone copies you....This is the worst part."

Which brought me to the *Lab Times* paper. I broached it carefully. Though steam did not blow from his ears, it was soon obvious he was still furious about it.

"Rubbish," is how he characterized it. "....We sent our paper to *PNAS*. They sent to three reviewers. All were happy with it. We published. The normal way is, if it is possible there is an error, you produce another paper saying error, it is sent to three colleagues to know if it is true or not."

Lab Times is not peer-reviewed. Yet there was a suggestion in the *Lab Times* paper that peer review at *PNAS* had been manipulated or dispensed with altogether due to the influential Bizzi who edited the paper. The implication behind this was that Bizzi is Italian and Mancuso is Italian, and Italians stick together.

So did Bizzi get it in *PNAS* for you? I asked.

"It's completely false," he shouted. "I asked the director of this journal to answer. They never responded, not even to say no.... It is defamatory. It says we can't measure. It says Bizzi, because he is Italian...is my friend. He is a neurobiology professor at MIT,[144] and no, I never met him. I know him by his fame. It's unbelievable. I wrote *PNAS* to ask if they want to take action. They say it's common. Many colleagues said take no action. It would give him importance."

The lead author, I asked. Do you know him?

Of course he did. "This guy is a retired, respected German plant physiologist....In Germany," said Mancuso, "there is big opposition to this idea. It's a faction. They try in every way to show this opposition."

Regardless of his anger, Mancuso had expected this kind of behavior.

"Science is made by controversy," he said. "But I would be happy to have good critics. That is the way to make your science better. I never found a single word against my idea to make me think a little bit. 'They have no brain!' Okay, I know..."

He showed me a small circular plastic maze with a small metal ball inside, the kind of party favor that used to come in boxes of Cracker Jacks when I was a child. There was a picture of a boxwood maze printed beneath a sequence of plastic channels. The point of the game is to shake and rattle the plastic box so as to move the ball to the maze's exit, and out.

"Look, I bought this three days ago," he said. "I bought 150 of them. It cost me seventy Euros. Very nice. You take out the balls and the background and you use it for plants to see how much time it takes the roots to solve the maze, to arrive at the center where the reward of nitrate is.... If you were inside this, you would have to, on average, make five or six wrong turns before finding the solution. A maze like this is used for mice to measure their intelligence. The plant roots are *genius.... They never make a wrong turn.* They spend no energy making the wrong decision. Another feature of intelligence is the efficiency of your behavior."

I stared at his plastic maze and thought immediately of Andrew Adamatzky's group putting slime molds through mazes in the UK. I just assumed Mancuso had never heard of Adamatzky, that he had found his own way to this kind of experiment, a science version of parallel evolution. After all, why would a plant scientist know about unconventional computation going on in another country?

Hey, I said, did you know they make slime molds run mazes in the UK?

"I am *collaborating* with Adamatzky," he said as if only a halfwit could think he was unaware of someone so obviously important to his work. "I am pretty sure we will be able to demonstrate you don't need neurons to have intelligent behavior."

He said he had to go to a meeting, but would return. I went off to find a place to eat lunch.

The wind was blowing hard when I walked back to Mancuso's office. He was waiting for me.

The first thing I asked was, "How did you get involved with Adamatzky's lab?"

Mancuso said he Baluska had made a contact with Adamatzky through Daniel Robert at University of Bristol who studies how insects detect sounds without anything resembling ears. Mancuso, Baluska, and Monica Gagliano, of the University of Western Australia, had contacted Robert because they are interested in how plants detect sounds. One contact led to another until, along with Adamatzky's group, Mancuso's lab had made an application to the European Commission's committee on unconventional computing and won a grant from its Marie Curie Training Network.

Wait a minute, I said. You're interested in how plants detect sounds? What about plants and sounds?

I thought for an awful moment that he was veering into *The Secret Life Of Plants* territory, a wildly popular book published more than thirty years earlier which purported to show that plants have emotional reactions to sounds, music, speech, and even what the humans around them are thinking. The book, by Peter Tompkins and Christopher Bird, started with the story of Clive Baxter. One night in 1966, Baxter, who used lie detectors in his business, impulsively attached his galvantometer's electrodes to the leaves of his office plant, and got a response. He believed the plant reacted to the thoughts he'd been having of doing damage to it.

Oh boy, I thought. Maybe this is all nuts after all.

Mancuso explained that he had been doing experiments to demonstrate that plants "detect and grow to or away from the source of sound depending on the frequency.…Plants not only can react to sound but produce it during growth," he explained, "as a byproduct of growth, there is a crackling sound. Probably all the roots can use this sound regarding position and velocity of growth."

You're not talking about plants communicating with sound? I asked, very carefully.

"A real message is a communication, but a byproduct can be used. It is still *information,* but not produced *for* me," he said. He seemed to mean that a deer walking in the forest might make sounds that a hunter could make use of, but they weren't a message meant for that hunter.

How did the sound experiments start? I asked.

A winemaker came to him, the owner of a great winery which produces wines that sell for 300 Euros a bottle. The vintner had been using classical music in his vineyard for years obtaining wonderful results, he said.

Mancuso made a face, as if there was a certain smell.

"I said 'I get every day mail from really crazy people…and from vegans. They hate me even more when I say, joking, 'you don't eat animals because you hate plants!'...The vintner asked to talk to me. I was reluctant but in the end he came. He's not crazy at all. He showed growth results [of playing music to his grapes] that were quite well recorded. I said, okay, we will have measurements in your vineyard. Of course plants have *nothing* to do with *music*...but sounds affect the growth of plants and we can do experiments."

Bose Europe (the speaker/stereo manufacturer) agreed to fund the experiments to the tune of about 100,000 Euros over two years. The big problem became control: how could he absolutely control the frequency, volume, and position of sound in a field? How could he get the same conditions, with sound and without sound, to compare the growth of different plants? He had finally gotten decent results after three years of futzing with the method.

So what do the speakers play? I asked.

Mozart, he said. Because the name of the vineyard is the Mozart Vineyard and since the music would be played twenty-four hours a day, it had to be something the neighbors would find acceptable. But he didn't just put Mozart on the soundtrack.

"We arranged a trick to enrich the Mozart with the frequencies we wanted to study," he said.

Earlier lab studies done by Gagliano, Mancuso, and Robert, published in *Trends in Plant Science* in 2012[145] had shown that a sound wave of between 100 to 600 Hertz "is most effective in driving the physiology of plants." Hidden in the Mozart, they put out a 200 Hz tone to see what happened to the plants.

So what happened?

It's not yet published, he said. The grapes were only picked a month ago. He was waiting for the analysis. But then he added: "The results are untouchable and demonstrate that plants growing with sound have bigger sugar content in fluids, more polyphenols which is important for wine quality, and these get ripe two weeks before normal. That is a huge help in viticulture. The last weeks are critical regarding fungi, moisture."

It was time to find out more about the plant robots he'd mentioned earlier, I thought. Plantoids is what he'd called them.

Tell me about the robots you work on, I said.

"It's a nice part of the story," he replied. Every other kind of living thing was being looked at by those wanting to make more efficient machines, everything from snakes, to insects, to humans, to whales, but not plants.

"I proposed it first in a story in *Wired.* After that there was a lot of laughter about the story but I am accustomed to this reaction now." Besides, he got a grant from the European Space Agency to study the question.

"You look at natural organisms to see what they do better... what can plants do better? They can colonize better than other organisms. They can plow the soil, the ground, with efficiency in an incredible way. I proposed a plantoid for the exploration of space...They send to Mars a robot as big as a jeep taking measurements. I said send to Mars something inspired by plants, a big bag full of robotic seeds, you open it in the atmosphere to spread thousands of small robots using roots to explore a small area of soils, a completely different strategy, a lot of small, cheap plant-inspired robots. This study was done in 2008. We had last year a European Union project, a two-million Euro project."

I was about to ask him if he knew anything about evolutionary robotics when he explained that his group cooperates with two others, a robotics group at University of Pisa "and Dario Floreano in Switzerland," he said.

I'm sure my mouth fell open. Dario Floreano was the man who had given a talk in Montreal about evolving intelligent behavior in robots. It seemed shocking that a plant specialist in Florence would team up with a robotics group in Switzerland while also working with an unconventional computing lab in Bristol studying slime molds. It was as if everybody, everywhere had heard the zeitgeist sing and grabbed on to its skirts at once.

But how do you know Floreano? I asked.

"He's one of my closest friends," he said. "We are trying to produce a working prototype of a plantoid. With Dario, we want to get information regarding plant communication back from robots. Biology can give us ideas regarding robots, but robots can give ideas back regarding biology."

He must have read Floreano's report on how robots with no programmed capacity to communicate nevertheless evolved one, I thought. I was almost right.

"The specific task with Dario," he said, "is to understand through simulation and actual robots how each root communicates and behaves with other roots. Each plant has hundreds of millions of single roots."

Mancuso's group had recently published an article describing how corn roots coordinate their growth movements like animals in a swarm.[146] Mancuso's group argued that "roots can be influenced by their neighbors to induce a tendency to align the direction of their growth. In the apparently noisy or disorganized patterns formed by growing roots, episodic alignments are observed as the roots grow close to each other. "These events are incompatible with the statistics of purely random growth," they'd reported.

The question is: *how* do they coordinate their behavior? By chemical signaling? By electrical repulsion, or attraction? By sound?

"We produce a paper," said Mancuso, "....Me and the lab showed that plant roots behave like a swarm, but not all the time. Dario works with swarm robots. We want to apply that to the way plant roots behave to extract an idea about this kind of diffuse intelligence you have in roots."

What do you mean by diffuse?

"Every single root is a few hundred of computing cells, but altogether, a root apparatus is like the Internet. We need to find out how they work, which will also help us with network studies."

So you're connected to Adamatzky, and to Floreano, I said. Who else?

He suggested I should get to know the works of a man named Michael Levin, a professor at Tufts in its Center for Regenerative and Developmental Biology. They had never actually met, but were cooperating by email.

What does he do? I asked.

"He—we— are studying how memory and learning could happen without neurons, in just normal cells. We are studying bioelectricity, normally produced by cells as a leading signal for pattern formation. Today it is part of the forgotten side of science.... This electrical current, with probability, is used in pattern development. How to build a body is not just genetic."

He handed me two of Levin's papers on bioelectrical signals and pattern formation. I could see how they fit with Mancuso's interests. Levin had shown that by changing the electrical activity in a frog, he could make an eye grow on a different part of the body altogether. "Bioelectricity is driving the pattern formation of the body. And it also controls growth," said Mancuso.

According to Levin's website, he had started his career in artificial intelligence, then moved to cell biology. In two papers he dealt with how changes in voltage can guide the developmental patterns in the cells of embryos. In one, he argued that flows of ions somehow carry information, that a voltage change can induce a change in the when and the where of gene expressions: he characterized this as "a true epigenetic layer of control."[147],[148] In the second paper he raised the big issue: where does body shape come from? Is an individual's body plan an emergent property of individual cells just following simple rules, or, is there a larger map or pattern stored somewhere?

Frankenstein's monster came to mind, slabs of flesh zapped into motion by the right jolt of electricity. The light outside said the afternoon sun was heading down.

What are you working on now, I asked Mancuso.

He didn't want to show me his last paper because it hadn't found a publisher yet, though he called a draft up on his computer. "Plants learn faster and forget slower in environments where it matters," he said. "I want to put it in *Nature* or *Science.*"

He and his colleagues, lead by Monica Gagliano at University of Western Australia, had run a series of experiments with the plant *Mimosa pudica.* If you touch it, the leaf closes almost instantly. Why? "To scare insects," Mancuso explained. "But it's energetically costly. We thought we could teach it that a certain stimulus is not dangerous and so it would learn not to close."

This too was old terrain, revisited.

Stevan Harnad's mentor at Princeton, Julian Jaynes, published a book called *The Origin of Conciousness in the Breakdown of the Bicameral Mind* back in 1976. In the introduction, Jaynes marshaled a history of ideas about consciousness, including the notion that it is an *emergent* property—something new that arises when two separate things (like sodium and chloride to make salt) are coupled together. Jaynes described his attempts as a student to investigate the origin of consciousness. He'd dreamed up the idea of trying to teach a plant to learn something, and then to work his way *up*, as he put it, from plants to paramecia, to flatworms, earthworms, fish, reptiles, etc. He thought he would be "chronicling the grand evolution of consciousness" as he climbed a complexity ladder.

He started his experiments by trying to teach a mimosa plant.

"The signal was an intense light; the response was the drooping of a leaf to a carefully calibrated tactile stimulus where it joined the stem. After over a thousand pairings of the light and the tactile stimulus, my patient plant was as green as ever. It was not conscious."[149]

Mancuso's group by contrast had managed to demonstrate that mimosa can learn and remember. They grew mimosa plants in pots. They dropped these pots a distance of five centimeters onto a table, which didn't hurt them. They repeated these drops many times. At first, the mimosa curled their leaves each time. On the sixth or seventh drop, each mimosa stopped curling its leaves, having learned that the drop was not threatening. Yet "if you change to a touch with a pen, it closes immediately," Mancuso said, meaning the plants clearly distinguished being harmlessly dropped from being touched.

"We thought this response was energetically linked, that plants grown in poor low light versus light rich plants would learn with a different velocity," he said. "In low light, they learned that the stimulus was not dangerous at a much faster rate."

Why?

"No energy to waste," he said. "If I remember, (he checked his screen), they were forty percent faster in learning."

And not only did they learn, they remembered. "After forty days, these plants remembered the stimulus. They could still discriminate forty days later. In just one paper we demonstrate they memorize, learn and store the memory for at least forty days. Many of my students have less memory," he said. And laughed.

They had sent this paper out to colleagues to review and had received no objections. But he was under no illusions about the paper's publication fate. *Nature* he was sure would not run it. They would likely not even send it out for review since they only review those papers they see as important. "You cannot *imagine* the time we lose to have our research published....In the old days scientists published the book and put in everything there...."

Can you show me around the lab? I asked.

We went down to the main floor. There were several rooms with very serious equipment suitable for the measurements of plant behavior under controlled conditions of light, temperature, humidity, pressure. There weren't many students around. We went into the greenhouse which was divided into separate growing regions. There were several pots pinned to each other, the divided roots of a single plant growing inside two pots. Their question: will the roots recognize each other when plunged into the same pot later? There were some pepper plants planted in pails.

He stroked the pepper plant's dark leaves. "Pepper produces a toxin around its fruit," he said. "Most animals don't like. Only a few birds do."

This made pepper an odd sort of plant because plants mainly use chemistry to attract insects, not to kill them.

"They insects are needed to transport the pollen. They set up a market. The plants are the sellers. The buyers are the insects. The publicity is the flower. It's a serious business. When they give pollen to be spread by insects, they give in exchange nectar. It's just an energy cost to the plant. Insects like the sugar."

There are plants that are honest and pay for the job of spreading their pollen with good nectar, and then there are plants that are dishonest, such as orchids. "They cheat the insect. They give nothing in exchange, but they have a wonderful trick of producing a flower like the female of the insect which is even producing the same *pheromone* and *texture.* No way the male can detect any difference from the real female..." So the insect crawls into the orchid, has its way with it, and leaves... "and they transport the pollen."

This brought him back to the meaning of the bigger story behind all his research, that Hegelian master-slave relationship between plants and animals.

"The plant is the master," he reiterated.

Over the long, long course of their evolutionary history from the first algae in the ocean that learned how to create energy from light, plants have developed brilliant tricks of camouflage and attraction "because they cannot move," he said. That is why they must be acutely sensitive to what is going on around them and to any change coming from their environments. An animal "can run away." No plant can. "They need to sense *with anticipation* to modify their metabolism to survive much faster than an animal can sense.... They need animals in key moments of life—to carry their pollen, for seed transport or defense, to defend against a predator. They must convince them to do a job for them."

It was the science of this master-slave dialectic that was new.

"Michael Pollan [author of *The Botany of Desire*] says they even manipulate humans. There is the paradox of the drug reward. The paradox is that almost the totality of the drugs used by humans for pleasure are produced by plants. Caffeine, nicotine, heroin, cocaine, morphine, even scientific drugs, normally we copy the plant molecule. Why should a plant create a drug dependence in an animal? The textbook thesis is they are toxins to keep animals away. But that is not the case."

The substances that plants make that create dependence in humans act directly on what Mancuso calls reward neurons.

"These things are activated in our brains when we do something good for us, when we eat something we like, or have an appointment with the girl of our dreams....These drugs activate the same groups of neurons. This is, from an evolutionary point of view, very important because this group of neurons is there to say we are doing something good for life. Yet plants can chemically manipulate these neurons and create a dependency. Their aim is to make the action be repeated."

In other words, plants want the human or any other interested animal to take the pollen, eat the fruit, spread the seeds again, and again, and again. "They make us repeat actions that are good for survival. Plants manipulate these neurons."

"We need to consider the plants we consume from a biological point of view," he said. "The success of a species is in its number and distribution on the planet. Now think. You are a plant like wheat. The goal is to spread across the planet at the fastest velocity and conquer available land. A wonderful strategy is to enter a relationship with humans. You produce fruit, seeds, everything we consume ..."

You mean they *decided* to work with us? I asked. Ridiculous idea, I was thinking, he can't mean that. But he did.

It really riled him that we don't want to consider the idea that a plant can make a decision to be in a partnership with us, when we find it totally believable of dogs or horses. "What you say about the wolf is that it decided that if it lives in a band with humans it's a great help. In exchange, the dog makes a job. The domestication of animals was good for both of us....Same for plants. We eat about ninety percent of our calories from five or six species of plants. They take this *decision,* an evolutionary decision, but it was a decision," he said.

And, he continued, of the million compounds we know, almost all are made by plants. Why? "The answer is they need all these compounds to communicate, to manipulate, to shape the way the *animal* behaves."

The first life forms on land were plants. But, he pointed out, the first flowering plants didn't appear until much, much later, long after the appearance of mammals. "They are very modern," he said. And he believes they evolved specifically to take advantage of the existence of animals.

So, he asked. How is it that plants represent ninety-nine-point-eight percent of all living things on the planet by mass? "By being stupid, fixed, low level organisms? It is absolutely *impossible.* They are intelligent diffusioners, but more able than us to sense and communicate."

So you are saying they are smart enough to shape our behavior, I asked.

"Sure. Without plants, if we are living in towns with no garden or parks there will be a lot of serious pathology. More suicides, higher levels of depression and social issues, higher numbers of violent acts and this is studied a lot."

He was right about that. I was familiar with studies that show patients heal faster in hospitals if they can see a tree outside their window, and students with attention deficit disorder do better on tests given in a garden than indoors.

So getting back to intelligence, then, I said. What do you think it is: a property that just emerges from life?

"In principle, I would say yes, it's an emergent property of complexity. But immediately after, I think…"

He paused for a long time.

"Intelligence is also linked to evolution. It's not limited to living organisms."

He had arrived at the scary place I had been trying not to think about since Montreal, the idea of non-organic intelligence, intelligent machines. Turing had had no doubt that a machine could be—would be—considered intelligent if it could learn, and that learning could happen in a non-organic system. Turing wasn't sure whether it would be best to teach a machine by feeding it information through a program, or by giving it *experience,* but both, he'd said, should be tried.

"No one has a problem speaking of artificial intelligence," Mancuso said. "We have smart phones, intelligent houses, everything is intelligent everywhere."

This is why it aggravated him that so many people won't accept that plants *must be* intelligent systems. But then he circled back to the paramecium, the foraging, fighting single cell: if one found that intelligence is an emergent property of complexity, how to explain a single cell's ability to solve problems? Surely that meant a single cell is also extremely complex?

"We know *nothing* of what life is," he said emphatically. Plants "digest with no stomach, and they move by growing. And they think without a brain. They have everything distributed...All cells respire, digest...all are able to process information. For me, the definition of intelligence is the ability to solve problems. They are at least as intelligent as other animals....Each single plant is a colony of millions of radiating modules and each tree is part of a bigger community. The biggest living organisms are aspen forests. They are a single large organism....all the plants are interconnected. That's the second level—a plant is a big colony. Then all the plants together are connected to each other...."

From a single foraging cell, to a group of cells acting as an organism, to groups of organisms acting as a colony, to the interaction of all such colonies in what Lovelock called Gaia, did each have its own form of intelligence, a different kind emerging at each rung of the complexity ladder? Or, I wondered, is all intelligence of a piece?

That question would have to wait for another day.

11

Say Hello, Robots

Murphy's Law promises that what can go wrong, will, yet I'd placed myself at the mercy of a cat's cradle of transport systems built by generations of human engineers. I had to get myself from Florence to Geneva, Geneva to Lausanne, Lausanne back to Geneva, Geneva to the UK and London, London to a town called Lewes, in less than thirty-six hours via taxis, shuttles, trains, planes, and subways. One missed connection and my careful arrangements would disintegrate like a feather fascinator in a stiff wind. In fact, one part of my plan had collapsed before I'd even left Toronto.

I had arranged to visit the renowned independent scientist James Lovelock at his home near Abbotsbury, in West Dorset. He had argued some years earlier that global warming is Gaia's adaptation to our industrial assaults upon her. By the end of this century, he said, we humans will be reduced to a few breeding pairs huddled in the Arctic as the rest of the planet gets too hot for as long as 100,000 years. But then he'd recalculated and recanted. His next book would contend that we humans will help Gaia adapt to the conditions we love to create, in a way that we can live with, and that a new era called the Anthropocene is upon us.

I had an urgent question for Lovelock: if adaptive behavior is the defining feature of intelligence, if it can emerge when any two cell interact, did that also mean that smart machines could become the moral equivalent of living things, also part of Gaia's smart, self-regulation system?

But at age ninety-four, Lovelock's heart had urgently needed a smart machine called a pacemaker. He'd cancelled.

Round one to Murphy.

So instead of going to see Lovelock, I was flying to Switzerland to see Dario Floreano, a man who uses evolutionary theory as a design tool. I had been booked on an airline called Darwin Air. The in-flight magazine's title? *Evolution.* The zeitgeist was having its way with me.

I held my breath in Geneva's arrivals hall: a lost bag would signal round two to Murphy. But there it was—the very first to plop onto the carousel. I found the phone for the airport hotel and its shuttle bus appeared in minutes. The free Wi-Fi worked (in the lobby) and my room was just fine. I woke at dawn to the TV blaring Obama's America-is-more-than-the-sum-of-its-partisans victory speech, rolled my luggage to the hotel's curb, took one quick breath of the crisp Swiss air, climbed into the shuttle, and was whisked back to the airport. I found the luggage lock-up, left my bag there, walked downstairs to the train station, bought my ticket at a kiosk just as my train slid up to the platform, silent as a great white shark.

It carried me through Geneva's suburbs to Lausanne along the shore of Lac Leman. I had been here years before during a steamy summer when the lake was murky, its perimeter festooned with plastic bags and used syringes, the sky the color of cotton stuffing. But this time it was a shocking cobalt blue and Lac Leman its fairy tale mirror. Snow-crested mountains towered over everything. The landscape looked as super-real as an old fashioned postcard, the kind we bought as kids on rainy vacation days when there were no more museums to haunt.

I got off at Lausanne's grimy old train station. The Metro stood across the street beside a MacDonald's, exactly where Google said it should be. The Metro carried me out into the suburbs to the last stop, a long platform with an underground tunnel to the campus of École Polytechnique Fédérale de Lausanne (EPFL). I followed the students, climbed a steep flight of steps, crossed a bridge over a ravine to the main door of the building housing Dario Floreano's lab.

I arrived exactly on time.

Murphy would have been amazed.

Dario Floreano's Laboratory of Intelligent Systems occupied a series of rooms off an ordinary corridor in a modern brick building. There were conference posters tacked on the walls displaying past efforts, including one about soft robots made of something that looked like bubble wrap. There was a hyper-neat electrical lab where robot prototypes were wired up and tested, and another large room where prototypes were being built by students and postdocs.

Floreano had meetings and was not there yet. He's a full professor, directs this laboratory, and The Swiss National Center of Competence in Research Robotics. He has also advised the European Commission's Division of Future and Emerging Technologies. He has been a governor of the International Society for Neural Networks, and has been on the editorial boards of nine different journals. His list of expertise goes on and on: bio-inspired robotics; bio-inspired artificial intelligence; computational biology; artificial evolution; neural networks; and swarm behavior. He'd been able to free just one hour for me in the afternoon. But he had assigned one of his postdocs, Steffen Wischmann, to give me a tour of the lab and its robots. I found Wischmann staring at a large computer screen in a smallish office stuffed with desks.

An ethereally slim young biologist with soft brown hair, he had a whisper of stubble, big brown eyes, and the pallor of a person who spends too much time in labs. Since 2008, he had been a postdoc/research associate in two labs at the same time, this one, and Laurent Keller's group in the Department of Ecology and Evolution at the National University of Lausanne (its campus is right next door). Wischmann's MSc thesis involved using an evolutionary process in a computer to design the shape of a body and a control system for two-legged movement. To his surprise, he ended up with a model that moved with a kind of slither.

We went to an open area off the corridor and sank into a leather couch near the espresso machines. A flying robot prototype with see-through wings hung from the ceiling. Wischmann spoke in a voice so soft that catching his words was like grasping for fireflies.

A journalist is not supposed to confess to readers that she doesn't really want to think about that which she is reporting on. But we're friends now, so I must be honest. As I sat listening to Wischmann explain his lab's work, I could feel sweat gathering under my arms and panic rising in my throat. I struggled with what he was telling me. It isn't just that I don't speak higher math and do not enjoy a walk through the valley of the shadow of computer algorithms. There is something fundamentally frightening about the notions that Floreano, Wischmann, and other evolutionary roboticists rely on. They graft life principles onto inanimate tools that then behave as if they're alive. The idea that intelligent plants *decided* to become our daily bread is merely upsetting by comparison: at least plants *are* alive. These smart little robots of theirs made me want to run back to the welcoming arms of nineteenth century vitalism which posited that the spirit that animates all living things is utterly absent from inorganic matter. Please, I wanted to say, can't we hang on to that kind of exceptionalism just a little while longer?

As Wischmann explained it, Floreano's EPFL lab uses evolution as a double-edged tool. They employ a stripped-down version of the mechanics of evolution to design computer-simulated robots, and then build populations of real robots that display intelligent behaviors. But they also use real and virtual robots to test sub-theories of evolution as proposed by biologists observing nature. The interplay between the theoretical and the simulated, between nature and the machines, gave their work its salience and strength.

The basic principle at the root of it is this: any body, in interaction with its environment and impelled to survive and thrive, will—must— generate intelligent adaptive behavior.

Robots "situated" in an environment will become smarter through the application of the same random evolutionary forces that shape living things, that have shaped you and me.

The idea that small random changes create endless variety over time is the rock upon which Darwin built his original theory. However, the notion that random change generated my own intelligent behavior is not what I like to tell myself when I get up in the morning. I spin a different narrative about how I solve problems by dint of reason and logic and cultural sophistication and do this better than most because I have been certified smart and have a nice, wet, organic brain. I like to think that distinguishes me entirely from a brainless robot.

Of course by the time I sat down on that couch with Wischmann, I had been pretty well indoctrinated into the notion that intelligent behavior involves much more and much less than a capacity to do math and memorize Shakespeare. I was almost convinced that intelligence emerges out of specific struggles to survive and replicate in everything from the slime mold to giant colonies of aspen. But the work of this lab forced me to consider that new (yet familiar) forms of intelligence can also emerge from autonomous machines. Somehow, machines that do not eat and do not reproduce have been hoodwinked into behaving as if shaped by the contingencies of life, and they are somehow able to act in ever more intelligent ways as they accumulate complexities. It made me extremely uncomfortable.

Oddly, in the light of the seamless trip I'd just taken from Florence these ideas seemed even more upsetting than when I first encountered them in Montreal. My little voyage had been almost flawless thanks to generations of applied human reason and skill. It seemed to demonstrate that there is a Wall of China-sized boundary between the adaptive behaviors of people who build dams and airplanes and trains and metros, and those other living things whose big inventions include the use of grass to poke into termite mounds. Yet these people had made smart machines by copying the same forces that made us. It seemed obvious to me, therefore, that one day machines would become much smarter than we are, and that we could end up with the orangutans and chimps and bonobos on the wrong side of a new smarts boundary.

So I desperately wanted to deny instead of understand. Our smarts can't come just from randomness! I wanted to shout. We're special! So there must be something wrong with this work.

Instead, I tortured Wischmann with questions that arose from an article Floreano and Laurent Keller published in 2010 in the online journal *PLoS: Biology* under the title "Evolution of Adaptive Behaviour in Robots by Means of Darwinian Selection."[150]

Floreano is a psychologist, while Keller is an ant expert interested in the evolution of social behavior. Keller's lab has demonstrated, among many other things, that an ant colony that once originated in Argentina but somehow made its way to Europe, possibly hitchhiking on a cargo ship, formed two super colonies when it arrived. The largest stretches from Italy all the way to the Atlantic coast of Spain which makes it the biggest cooperative unit ever recorded.[151] The super colony members treat each other as kin though they are genetically different and widely dispersed. They treat members of the other super colony, and all other ants, as enemies.[152]

The *PLoS* article explained their method, specifically how "the process of natural selection can lead to the evolution of complex traits seen as adaptive behaviors." First they generate many computer models programmed to receive virtual 'sense' information that is fed through simple, but unique, neural networks. This process is meant to simulate an evolutionary process involving random genetic mutation, and the recombination of genes through sex, in a population of virtual robots. The neural network settings, their interconnections, are first chosen at random, but then are refined over time by the programmers. These networks are connected to virtual robot parts that can move but are constrained by the rules of physics, gravity, acceleration, etc. These simulated robots are then set a task. Their performance is judged by the programmer who throws out the worst performers, and makes random changes to the neural networks of the best performers, including re-combinations of different paired networks.

A neural network is a radically simplified computer version of the real thing. Real neurons build physical connections to neighbor neurons which link together in a network as they receive and send electrochemical signals back and forth. The more neuronal connections, and the greater the intensity of these interactions between neighbors, the stronger the information bond between them: thus, we learn. Neural network-style computation enables modern computers to learn by changing or weighting the frequency of interactions between one part of a network and another.

Floreano and Keller used the word 'genome' to express the idea that they are altering each neural network in a way similar to how evolution works on the genes of real animals which change through mutations passed on to the next generation. The programmer both defines the 'fitness' task and judges the performance, thereby acting as Darwin's selecting hand.

The most successful models produced by this kind of artificial evolution are then embodied as a real robot. They build a small population of robots each with a unique version of the computer-refined neural network. They wire up the wheels, or whatever other means of movement they employ, to motors (known as actuators) and also to various sensors that react to light or some other signal, all running through the neural networks. The coupled relationships among all these parts are comparable to the connections between an animal's sense organs, its muscles, and its brain. Some real robots will perform the fitness task better than the others, and will be the basis for the next generation, and on the evolution goes.

With these means they had achieved astonishing results. As they put it in the *PLoS* article:

> "Just a few hundred generations of selection are sufficient to allow robots to evolve collision-free movement, homing, sophisticated predator versus prey strategies, co-adaptation of brains and bodies, cooperation and even altruism. In all cases this occurred via selection in robots controlled by a simple neural network which mutated randomly."

I had spent quite a few hours with this paper, reading and re-reading, hunting for the magic trick because, come on now, how can random changes turn into intelligent behavior in machines? There had to be a flaw. At first I even thought I found one, so I hung on to it for dear life. They claimed that the field of evolutionary robotics was inspired by Alan Turing's *Mind* article of 1950, the one that describes the Turing Test:

> "He [Turing] suggested that intelligent machines capable of adaptation and learning would be too difficult to conceive by a human designer and could instead be obtained by using an evolutionary process with mutations and selective reproduction."

Aha! I thought. Not so. Turing wrote about creating a computer that could fool people into thinking it can think. A machine that could fool us, Turing argued, had to be a machine that can learn because even big teams of humans programming such a machine would take way too long to get the job done. So the machine had to be able to learn and remember things for itself. One way to make a machine learn would be to model it after how humans learn—not adult humans (too complex) but children. Turing divided this modeling problem into two parts, the child program, as he called it, and the education process. And this is where I thought Floreano and Keller had misinterpreted their master. Turing said:

> "There is an obvious connection between this process and evolution, by the identifications [:]
> Structure of the child machine=hereditary material
> Changes of the child machine=mutation
> Natural selection=judgment of the experimenter.

> One may hope, however, that this process will be more expeditious than evolution. The survival of the fittest is a slow method for measuring advantages. The experimenter, by the exercise of intelligence, should be able to speed it up. *Equally important is the fact that he is not restricted to random mutations.* [Italics mine] If he can trace a cause for some weakness, he can probably think of the kind of mutation which will improve it."

Turing had used evolution as a *metaphor* not as a prescription for a programming method. And he had said nothing about embodiment. In fact, he had proposed that his childlike learning machine would *not be* a robot. "It will not, for instance, be provided with legs," he wrote, "so that it could not be asked to go out and fill the coal scuttle," Neither was he interested in giving it senses. All he required was a means to communicate between the teacher (the programmer) and the child (the learning machine.)

See, I reassured myself. They've turned Turing's work upside down.

Yet a different part of Turing's paper had definitely shaped their thinking. It was what Turing said about randomness as a method of search. Turing's reliance on randomness was a product of his career as a code breaker searching through billions of possibilities which had to be narrowed down by insider knowledge, clever logic, and statistics. It was also a result of his limited understanding of genetics. He wrote:

> "a random element is useful when we are searching for a solution of some problem....Now the learning process may be regarded as a search for a form of behaviour which will satisfy the teacher....Since there is probably a very large number of satisfactory solutions the random method seems to be better than the systematic....
>
> "It should be noticed that [randomness] is used in the analogous process of evolution. But there the systematic method is not possible. How could one keep track of the different genetical combinations that had been tried, so as to avoid trying them again?"

In 1950, Turing couldn't know that the DNA in cells *is* a record of previous genetical combinations already tried. Neither could he know that the notion of evolutionary change *only* happening due to random mutation and sexual recombination was about to be thrown out by Barbara McClintock, Lynn Margulis, and many others.

In short, it seemed to me that Floreano and Keller had oversimplified how evolution actually works and claimed Turing as a mentor when his ideas went in another direction entirely.

Yet a little voice within said, so what? The robots do what they say they do. How do you explain that?

Floreano and Keller described how their simulated neural networks produce a variety of robot behaviors:

> "...an artificial genome can describe the strength of synaptic connections of an artificial neural network

that determines the behaviour of the robot. The input neurons of the neural network are activated by the robot's sensors, and the output neurons control the motors of the robot. With a population, each individual has a different genome describing a different neural network (i.e., different connections between neurons) thus resulting in specific individual responses to sensory-motor interactions with the environment. These behavioural differences affect the robot's fitness which is defined for example, by how fast and straight the robot moves or how frequently it collides with obstacles. At the beginning, robots have random values for their genes, leading to completely random behaviors. The process of Darwinian selection is then imitated by selectively choosing the genomes of robots with highest fitness to produce a new generation of robots. In this process, genomes are paired (to allow recombination) and random mutations (i.e. character substitution, insertion, deletion or duplication) are applied with a given probability to the new genomes. This process of evolution can be repeated over many generations until a stable behavioural strategy is established."

I thought I had them there, too. This method didn't even begin to capture the complexity of real evolutionary change over time, I said to Wischmann. Did he know about the work that shows the chromosome breakages that can lead to big behavior changes are mainly non-random, the result of physical forces? Was he aware of jumping genes, retrotransposons, controlling elements that aren't genes at all but control their expression? And what about epigenetics, how the environment and the body negotiate the future?

I could see he was getting hot under the collar. He isn't a geneticist. Though he is a biologist, he is interested in the mechanics of motion. I was being unfair.

Yet I couldn't stop. Randomness is not the only way evolution works, I ranted, as if picking at their method would somehow change the results and restore life's specialness. And where's the connection between a change in a few base pairs of DNA and changed behavior? I said. If you don't have complexity in your model, what have you got, really?

We must keep things simple to keep control and separate effects, he said. Science operates by reduction, he said.

But then you can't draw conclusions that what you've done means anything about reality, I said. What does selection really act on anyway?

There are modern biologists who argue that selection acts on genes which operate like organisms whose purpose is to have more progeny. Others argue that selection acts on groups, on whole populations.

Floreano and Keller acknowledged in their article that they didn't understand how the connection between genes and behavior works.

> "Genes do not specify behaviors directly, but rather encode molecular products that lead to the development of brains and bodies through which behaviour is expressed. An important task is therefore to understand how adaptive behaviours can evolve by the mere process of natural selection

> acting on genes that do not directly code for behaviors."

Yet there was no denying that their robots produced unexpected behaviors. One group of robots created a communications strategy (using light signals) that enabled them to point out poison (signified by a colored floor pad) versus food (a charging station with a different colored floor pad) to their fellows. Another group worked out that it is better to share a big task (moving a puck twice the size of a single robot to a charging station), than to try it alone. In other words, the robots did what animals do without being specifically programmed to do it. And more: groups with closely related neural networks evolved cooperative behavior and altruistic behavior just as related animals and plants do, working together to ensure the group's survival (access to a recharging station). This never happened between robot groups with different neural networks. Non-living things, in other words, solved problems *as groups* that they couldn't solve as individuals, just like animals and plants do, suggesting machines too will be able to develop societies, cultures and politics. And war.

I could read their words, I could understand their words, but I didn't really want to acknowledge them. There had to be a line between living and non-living groups, didn't there? That was the heart of my discomfort, that vanishing boundary.

So where is the line between living and non-living systems, I asked. Do you draw one?

Wischmann said something that is alive has to be able to reproduce.

But that could mean a genome (the entirety of the DNA in a single cell), because it replicates every time the cell does. Is a virus alive?

Well, they reproduce, Wischmann said.

But not on their own, I said. They hitch a ride.

He acknowledged that.

Okay, prions, I asked. These are wily proteins that maneuver through brain tissue, reproducing themselves at will, nobody knows how or why, and they wreak havoc.

I don't know, he said.

I found myself getting angry. I wanted there to be an impermeable boundary between the living and the-who-knows-what that is more substantial than just the ability to reproduce. It's not right to get angry because someone is saying something that is difficult to hear. I knew I was being disagreeable. So I changed the subject.

You work with Stefano Mancuso, I said. What do you think about plant intelligence?

Wischmann thought Mancuso might have pushed his plant-roots-recognize-kin argument too far: he'd heard others had failed to replicate those experiments.

Show me through the lab, please, I said.

We walked into the main room where prototypes were being built by several postdocs and grad students. Of the thirty people working in this lab, only two were biologists, the rest were engineers and computer scientists. People sat at their desks hacking up pieces of Styrofoam, shaping frames out of wires. The walls and ceilings were festooned with their efforts. I pulled out my tablet and began to shoot pictures.

Wischmann showed me the round robots which taught themselves how to signal each other and how to share the task of moving a large puck around, though not programmed to do either. They were as tiny as Tom Thumb. Some larger robots—autonomous drones—hung from the ceiling in front of a window looking like crosses between micro-helicopters and oil rigs. Some can hover. Some can fly. One, shaped like a boomerang can be flung up to see what's happening over the horizon and come right back with pictures: rights to that have been bought by a French company. Some will fly in perfect formation. Some are designed to bump into things, roll over, and right themselves because they're protected inside round cages. A young woman was trying to make a form for a new kind of wing. One of the postdocs showed me a flexible film that mimics the compound vision of the housefly: one day it will be sewn to a blind person's hat or shirt and vibrate a message about obstacles.

I took a lot of pictures. By the time Floreano appeared—late and apologetic—I was sure Wischmann was glad to see the back of me.

Floreano sat down on one side of a round table in his office. I took the chair on the other side. There were no papers, no files, nothing but a single book on his desk, a few more in the one bookshelf. The only sign that someone actually worked here was the variegated benjamina leaning in a corner by the window. His short dark hair was flecked with gray, his jaw displayed just the right amount of stubble. White pants, a blue jacket over an open necked white shirt, a flash of silver from a watch. He put one ankle over a knee, as if relaxed, yet he seemed too alert. An attractive man who cares about how he looks, I thought, yet too careful by half. I had to remind him why I'd come to see him.

I'm working on intelligence, I said, which may be the stupidest thing I've ever done in my life.

He smiled. His whole demeanor changed.

Everybody I talk to seems to be connected to the next guy, I burbled. I know you know Inman Harvey. I just came from Mancuso who works with you too. Are you working with Adamatzky as well?

"Stefano is a newcomer to our community," he said.

Was that a defensive bristle? Or just a hint that there is a hierarchy and Floreano—not Mancuso—is on top?

They'd only started working together a few months earlier, he said, "through the coordinator of the European Commission's Plantoid program, Barbara Mazzolai, she works at Pisa [at the Instituto Italiano di Tecnologia, where she is coordinator of the Center for Micro Bio-Robotics]."

There had been a joint meeting on bio-engineering at which both he and Mancuso made presentations. "There were many points in common," he said. And there he stopped.

Hadn't Mancuso said they are old friends? I was sure he had. Better get off the common links theme, I thought.

Look, I said, I really want to know how you got into evolutionary robotics.

He said it began with what he saw as the failure of psychology to define or explain intelligence. He had trained as a psychologist in cognitive science at the University of Trieste in the 1980s, specializing in visual perception, though he was really more interested in intelligence. His professor had referred him to a reference text containing about 1000 definitions of same. "The dominant interpretation of intelligence then was that it functioned like a computer with a memory, a processor with inputs and outputs. The main text was a book called *Human Information Processing*... I liked it. Computers were becoming more available..."

At first he thought this analogy to computers would lead him somewhere. But soon he realized that the new computer analogies weren't much more useful than earlier ideas about how the brain does its work. People working to generate artificial intelligence in computers had run into a brick wall. They had begun to wonder if maybe the problem was the way computers are organized. In 1985, Floreano continued, "...there was a famous book about parallel distributed processing which made neural networks popular." It was called *The Connection Machine* by Daniel Hillis, a doctoral student at MIT, who had invented a new computer architecture that would leave von Neumann's one-operation-at-a-time version in the dust. Hillis proposed a general purpose, massively parallel processor which would integrate processing and memory so that "allocation of memory to represent data automatically allocates processing for manipulation of that data."[153] Distributed tasks could be performed at the same time in small processors networked together. Soon networked parallelism became the new computer metaphor for how brains work.

Then Floreano came across a book by J. J. Gibson called *The Ecological Approach to Visual Perception.*

"That book is telling us that the body and behavior are an important part of what intelligence is," Floreano said. Yet the idea that the body shapes intelligence had not appeared in any of the 1000 definitions in the text to which his professor had referred him.

So he decided to use Hillis-style parallel processors to generate bodies-in-motion simulations in order to answer questions about what intelligence is. To do that, he needed formal training in computer science. He studied for his second Masters, in neural computation, at Stirling University in the UK, because it was the only university in Europe with a parallel processing system. It had 1064 processors all hooked up to each other. The system filled a room.

"It allowed me to create the first simulation of a robot acting in an environment with a neural network," he said.

But then life intervened. He had got married before he graduated. His wife, who was from Hungary, could not work in Italy so he had to find a job. First he got one teaching. Then he got another selling insurance. Four years went by. Eventually, his former supervisor suggested he could continue his research on neural networks while working for a pharmaceutical company in Rome, which he did. And that's when he read the work of his real mentor, not Alan Turing, but Valentino Braitenberg, then still a professor and director at the Max Planck Institute for Biological Cybernetics in Tubingen, Germany.

Braitenberg was a neurophysiologist and anatomist who asked questions about how bodies shape brains, and how intelligence emerges from the body's movements. He'd published a book in 1986 called *Vehicles: Experiments in Synthetic Psychology*. It was a thought experiment "on how intelligence could emerge out of simple mobile body with sensors and a simple cross wiring to control the motor. Put it in an environment and it will display behavior which in an animal you would call intelligent," said Floreano.

"He showed that with more complex wiring it would show fear, love, etc. It was an intriguing thing, part of a wave of thought regarding intelligence...that it needs embodiment. It was like a box of tricks that could work. I was one of the first to build his thought experiment."

By 1990 Floreano was working in Rome as a Research Fellow on a small team at the National Research Council trying to evolve small virtual 'agents' that display intelligent behavior. Inman Harvey, at University of Sussex, was on the same path. Neither group published anything until, using the ARPANET (precursor to the Internet), Sussex published a manifesto. "We had a paper, Econets, ready to publish on embodied neuro-nets....I was on the team, not the lead author."

He enrolled in a PhD program at University of Trieste. He wanted to set up experiments where many robots competed for scarce food. What would happen if they could talk to each other, he wondered. Trieste didn't have a computational system big enough to handle this work so he went back to the UK to work again on its parallel processors. He was trying to grapple with "a new definition of intelligence. In my thesis I said intelligence is any adaptive system that can allow an organism to adapt to a changing environment."

He wanted to show that one adaptive system is no better than any another, and that "the environment can shape the abilities if you have a proper adaptive process." And where could one observe such an adaptive process? In nature, it is called evolution.

"My thesis was an experiment with small robots in an increasingly challenging environment. They got more abilities by adapting. So there is a forcing of intelligence by the requirement to survive."

But after the second year of working on this project he realized that the computers were too slow to simulate all the constraints of the laws of physics that act on bodies, along with the possible behaviors of the simulated robots, along with their simulated environments. "I thought I needed a real body to investigate what intelligence is...a body that moves in an environment....I thought I need real robots."

But there were no appropriate robots out there for him to work with. "Until the 1990s, robots were big machines on factory floors."

In 1992 he was invited to a very select conference held in a castle in Trento, in the north of Italy. DARPA paid Floreano's travel expenses. About twenty people were brought together to discuss what needed to be done to solve the problems of robotics "and change the perception of intelligence," he said. Rodney Brooks from MIT was there: ten years later his company, iRobot, would launch the Roomba, a vacuum that roams autonomously through a house using a very simple neural net to find dirt and avoid obstacles. Inman Harvey from Sussex was there too.

"I met a person there from EPFL," he said, "Francesco Mondada."

Mondada, a professor at EPFL, is an engineer. He and his colleagues had built a very peculiar round robot, as small as the tip of the finger, called the Khepera.[154]

"They were laughing at him in the robotics community," said Floreano. 'What work can it do?' I said wow! It could be connected to a computer through rotating contacts. I said this is what I need, it's small enough that the laws of physics are in our favor meaning a Khepera robot wouldn't break easily if it fell off a table or banged into a wall. I said I could use them. They invited me here to EPFL for four months...and that changed our careers *extremely.*"

He and Mondado wanted to know if a neural network connected to a moving/sensing robot could evolve a place memory. The memory of place had been studied in rats which are able to hide food and find it later, as well as remember where their nests are. Physiologists had shown "this is mediated by two types of neurons in the hippocampus." One type is a population of cells that records environmental landmarks, called place cells. The other is a neuron that changes the pattern of its activity relative to where the place is. It keeps track of the direction of the animal.

"We didn't know if the robots would do this," Floreano said. "But we were not surprised when they did....You see the neural net being shaped by cues in the office, the walls, the doors, the windows of the room where the robots were located, but we didn't know how the system did it."

There were only ten connections in their first neural network. To evoke evolutionary processes, there had to be some kind of contingency, or challenge. They decided it should be the need for the robot to recharge itself to keep going. There was a charging station set out on the floor. The robot had to be able to find it.

Floreano began to vary the ten connections, one at a time. He realized some were active when the robot was in one area, another related to the front of the robot as it oriented itself to the charging station. This back and forth between the two kinds of active connections was sufficient "for the robot to time when to recharge. It's similar to a computation strategy used by animals faced by similar challenges," he said.

"I had all the experiments in my head," he said, but still, he and Mondado "worked day and night" to get them done. Then he had to write them up and defend his thesis in Trieste. "My supervisor threatened to throw me out," Floreano said. "He said I should do a proper psychology thesis, not play with robots. He thought my thesis would not make any sense."

I could see his supervisor's point. Who would have thought that tiny brainless machines with simple neural nets, sensors, and wheels, set in a box on the floor could say anything at all about the way brains function in living animals? And never mind the psychologists, the biologists didn't get it either.

"They wanted to make theoretical models to explain things, such as what are the conditions under which ants distribute themselves," Floreano said.

But Floreano believed his results validated Braitenberg's proposition that the body shapes intelligence. "We presented results at the same conference as Inman in 1993. We both... used evolution to come up with a control system to make the robot autonomous. We showed the robot could find and remember the place where they could recharge."

He began to go to conferences on the biological basis of artificial intelligence. He deliberately courted biologists like Laurent Keller. He wanted biologists to get a better grounding in his experiments. Biologists, for their part, wanted better theoretical models that would more accurately predict animals' behaviors under certain environmental conditions. They formed a society with meetings every two years, and started a journal called *Adaptive Behavior*, a much less controversial name for the study of animal/robotic intelligence than *Plant Neurobiology* was for the study of plant intelligence.

Floreano got his PhD, then a research fellowship in Rome, then another at the Sony Computer Science Laboratories in Tokyo. In 2000, he was appointed as the first Swiss National Science Foundation Fellow in bio inspired robotics at EPFL.

By then, robot kits and bio-inspired robots had become main-stream. Floreano and his group invented a flying robot, an autonomous drone, for civil applications. But more to the point, people from different disciplines began to see that bodily interaction with the environment shapes intelligence.

"There is a different understanding of what intelligence is in the artificial intelligence and modeling community," Floreano said. "Psychologists finally understand the body is important...but the idea of embodiment is still not accepted in neuroscience....Neuroscience meetings have 10,000 participants, it's where the money is...we are obsessed with the brain. But without the hormone system, the brain is nothing. They should learn what robots can teach...."

In 2005, the same year Mancuso got his laboratory up and running in Florence with the bank's money, Floreano set up the Laboratory of Intelligent Systems. In 2011 he became a full professor and Director of the Swiss National Center of Robotics.

So where does your money come from and how much do you get? I asked.

He, like Mancuso, seemed nonplussed, as if this was an unusual and possibly rude question, but he answered. He gets funding from EPFL; he gets money from the European Commission for Research on Robotics; from the Swiss National Science Foundation; from the Swiss Army, and from some companies.

Which ones?

From Sony and Toyota, he said. "They are very open- minded. Sony is interested in how robots and mobile devices could spontaneously adapt to human needs. Toyota is interested in a neural network to detect impending collisions."

So what's the annual budget here?

He shifted uneasily in his chair. "I don't know for the year. I have thirty people in the lab and I need more so about two million Swiss Francs per year....Every year I have to raise new funds for new projects, about 200,000 Swiss Francs for post-docs, PhDs, computers and software and travel per project."

In fact, he spends about fifty percent of his time on fundraising and the management of projects, on overseeing reports as required by grant oversight rules, getting sign-offs from various departments.... "It's a nightmare. I have four in the lab to do that."

Still, the largest pile of money under his control is due to his role as director of the Swiss National Center for Robotics. It has made a twelve year funding commitment of sixty million Swiss Francs, doled out in packets of twenty million every four years.

And yet: even as doors opened and his ideas were accepted and the disciplines came together and the money rolled in, evolutionary robotics began to shrink to a smaller part of his activities. He'd hinted at problems in Montreal. It's not fast enough, he'd said then. When I'd looked for current papers in this area by him or by Inman Harvey there were few.

Why so few new papers? I asked.

Sometimes it just doesn't work, he said now.

I was relieved to hear that.

"We did a lot of experiments in a few years and then we hit a complexity wall," he said.

The robots just would not come up with new "surprising capabilities." That meant they couldn't use the method as they'd hoped to develop control systems for autonomous robots with many sensors.

Is it because you've found it difficult to mimic the way nature really works? I asked him this question more politely than I had asked Wischmann, probably because he'd said it wasn't working, which made me feel better. I didn't have to worry about robot groups forming complex cultures immediately.

"We define the fitness function so of course it's not as open-ended as nature," he said. "The fitness function is unnatural. The other thing could be the genetic code we use does not help us find new structures as we get in brains."

Unlike Wischmann, he didn't defend his ideas by reference to the scientific need to reduce things to their simplest forms. "We are for sure getting something wrong," he said.

Maybe because evolution does not occur only by random mutation and recombination, I muttered.

He thought it could be something to do with the method of embodiment. The bodies of the first robots they used were not changeable, not adaptable to experience. The neural nets could change, but not the bodies, or the sensors.

"In nature there is a co-evolution of the body and the brain," he said. The brain becomes adaptable as "the body does the computation," was how he put it. "We need new ways to think of robots that allow for this."

He, like Mancuso, had been re-reading Darwin. According to Floreano, Darwin said "there takes place not only selection of the fittest but also copies of themselves. A tree in a field has no competition and can make copies. If viable, they make copies. So perhaps a new algorithm does not require a fitness function."

The more he talked, the more I heard echoes of Mancuso's insights about plants which develop over time according to an encoded narrative, and also copy themselves to do it all over again. Floreano said he was looking at ways to capture how genes encode for whole structures that grow over time. He thought he was on a good track. And he was also tackling the unchangeable robot body problem. They were trying to develop embodiments that "allow you to recombine parts." They were looking at the way living cells will sometimes forage alone and sometimes come together to cooperate, which sounded suspiciously like the behavior of slime mold.

"Robots are usually made of screws, things that are rigid, and most important are the connectors that fold and latch. The more blocks you add, the more heavy you get. So, I am looking at soft components you can deform and squeeze as a potential solution to systems that can reassemble."

He was working at a small scale—five to ten centimeters. How to connect and disconnect without cumbersome latches and connectors? "Cells have the same problem—they have to connect and not connect," he said.

So he was developing robot cells that will come apart, allowing them to be swallowed and then to reassemble. Such a system of cells could enclose a tumor and suffocate it, or carry in and release a drug in the correct place and then disassemble so it can be expelled. The main problem he had designing this sort of robot by hand, not using the evolutionary method, was teaching it when and how it should disassemble.

He, his postdocs, and Nicolas Franceschini of the Aix-Marseille Université, had also been trying to mimic the compound visual system of the fly, useful for wide field detection such as collision-free navigation of vehicles on the ground or in space, and also to test insect vision theories. They were working out ways to get their system onto a substrate so soft and thin it could adapt itself to a human body. [Two years later they would explain their achievement in an article titled "Miniature curved artificial compound eyes," in *PNAS*.]

I'll show you, he said.

We walked down the hall to a small room where some prototypes were displayed on a shelf. He introduced the postdoc doing the work. While Wischmann had mentioned that this system could translate visual information into a vibration, it seemed to me it could be aimed at something much more complex, a machine interface with the optic nerve of the person wearing the system. I asked about that. They shared looks.

It's multifaceted vision, Floreano said, not like ours.

But Wishmann had also mentioned the speed with which humans can adapt to visual information presented upside down: we learn to see it as if it was right side up. I pointed that out. The postdoc pointedly said nothing. Floreano pointedly diverted my attention to other developments.

Such as: autonomous drones flying in cages that can bump into walls, right themselves, and keep on going. These bumble-drones can exploit crowded conditions. They can "fall on the ground, get up, stabilise. And we can then make an inference about the shape of the room they're in."

Out of these physical interactions they had developed an algorithm that computes the vertical structure of a dark, enclosed space through the movements of the bumble-drone. "We got it from nature, and we can put it into mechanical principles....We can use it for exploration of caves and dark systems."

In other words, new smart behaviors were still emerging from these brainless robots after all, simply by putting things together in new ways.

"These things come together," he said. "We merged jumping robots with flying robots. Put that together with vision. We use evolutionary robotics for understanding real gene regulatory networks. At the end of the day, it all connects."

Yes, I thought. Exactly. That's the scary part.

At the end of 2014, Floreano's lab offered the world RoboGen, a new software/hardware platform for the practice of evolutionary robotics, involving off-the-shelf components, a 3-D printer, and complex neural networks run on a microcontroller. Now, anyone who wants to try their hand at creating intelligent behavior in any kind of robot they can imagine, from students, to researchers, to "hobbyists," is free to give it a go.

12

In Lewes

Inman Harvey, who does philosophy of mind with a screwdriver, had invited me to his house, not an office at the University of Sussex's Centre for Computational Neuroscience and Robotics. You can come any day but November fifth, he'd said. He didn't say why that date would be a problem. At first, I assumed he had appointments. But then, somewhere in the dim recesses of memory, firecrackers and bonfires crackled. Hah! November fifth is Guy Fawkes Day, last celebrated in my prairie town when I was about five. Utterly thrilling it was, too, with rowdy trash fires burning holes in the black sky.

I retraced my trip on the Lausanne LRT from EPFL to the train station, arrived at the Geneva airport, flew to Gatwick, took the Express train to London's Victoria Station, took a taxi to a B&B and a shoebox-sized room, and walked the next morning back to Victoria. The train for Lewes left on time. This brought the final score to Murphy 1—Dewar-an-exhausted-but-victorious 1000.

All the way, I kept thinking about the lessons Floreano and Wischmann taught on the nature of intelligence: how moving, signaling, challenged robot bodies produce intelligent behaviors as they link or communicate with others. Dynamic systems, that's what their robots are. And even as I was carried along by a network of human-made dynamic systems, I found myself thinking: maybe there is *no* boundary between life and machines. Maybe what you fear about machines taking over makes no sense; we, the living, are just wet, messy dynamic systems made of molecules which fold and twist and sidle along to make cells, which mesh with other cells, which add up to tissues, which work together to make individuals, who form groups, families, tribes, nations and so on.

Consider yourself as a smart machine of a community of machines, I said to myself as I found my seat on the train. But that brought me back to Stevan Harnad's question: where does consciousness come in? How do we get from relationships and movements to organisms that know themselves? Do smart machines get there too? Or do they lack something vital?

The train ploughed through London's shoulder-to-shoulder neighborhoods and then turned south into the countryside. Dead flat eventually gave way to hill after hill rolling down toward the coast. The sky was a Constable in progress: first, it was pewter; then watery blue; then milky wisps of vapor slipped in from nowhere; then darkening clouds billowed everywhere; then watery blue returned once more. Sheep wandered the hilltops, ivy coiled and spread along the low points beside the rail bed and through the hedgerows and spreading oaks (at least I think they were oaks).

The woman sitting across from me was also going to Lewes, to help her brother. And you, she asked. I gave her a raggedy summary of my quest: how I was learning that intelligence seems to arise from bodies enmeshed in relationships and challenged by change. I wanted to say: do you know the Zoroastrian idea about evil and goodness locked in struggle, or the Taoist yin yang symbol? But I bit my tongue. Why dredge up old, spirit-based notions to explain the unseen and unwritten patterns shaping our lives? They were interesting encapsulations, but they explained nothing.

"Fascinating," she said, visibly willing her phone to ring and save her.

Yet I hadn't even described Floreano's clever little pucks on wheels animated with a zap of Darwin, or the drones that compute the shape of a dark room as they bumble through it. How the hell could I summarize that? And after spilling out my latest version of what smartness is made of, I had to ask myself: did I really believe that intelligence—hers, yours, mine—is made of nothing more than conjoined bodies moving through a balky environment infused with the need to continue?

It seemed to me that mysteries remained and that describing intelligent behavior as a product of dynamic systems might be another nice observation, but no explanation at all.

Inman Harvey had mentioned in a paper that evolutionary robotics, the field he helped establish, has been called the new behaviorism.[155] (The new barbarism is what the religious would call it.) Its basic insight, that dynamic relationships in a challenging environment *force* adaptive (aka intelligent) behaviors, suggests that the American behaviorist psychologists got something right. Behaviorists argued that intelligent behavior has nothing to do with a hidden mind. They wasted no time pursuing how internal representations of the world are made, or trying to locate the calculating homunculus within. There is no there, there, they said, just sensory reactions. Evolutionary robotics also replaced the mysterious mind with interconnected systems flailing away.

Harvey's papers were both philosophical and practical. Like Turing, he used mathematics to trap larger meanings, particularly the mathematical nature of dynamic systems. In one article he wrote:

> "A dynamical system is formally any system with a finite number of state variables that can change over time: the rate of change of any one such variable depends on the current values of any or all of the variables in a regular fashion. These regularities are generally summed up in a set of differential equations. A Watt governor for a steam engine is a paradigmatic dynamical system ...and we can treat the nervous system plus body of a creature (or robot) as one also. ...One thing that is very rapidly learnt from hands on experience is that two such independent dynamical systems, when coupled together into (e.g.) steam-engine-plus-governor treated now as a single dynamical system, often behave in a counterintuitive fashion not obviously related to the uncoupled behaviors.

Treating an agent—creature, human or robot—as a dynamical system coupled with its environment

> through sensors and motors, inputs and outputs, leads to a metaphor of agents being *perturbed* in their dynamics through this coupling, in contrast to the former picture of such agents *computing* appropriate outputs from their inputs. The view of cognition entailed by this attitude fits with Varela's characterization of cognition as *embodied action*.[156]

This view of intelligent behavior brought to mind a David Cronenberg nightmare of desperate bodies swimming in a jelly with a life of its own.

Harvey had also published on Lovelock's well known dynamic system, Gaia. As Lovelock had long ago pointed out, the Earth's sun is twenty-five percent hotter now than when this planet was born. Yet somehow, the coupled interactions between living things, atmosphere, sea, and rock, have held average planetary temperatures in a range amenable to life for over 3.8 billion years. He asked, why hadn't the byproducts of life—oxygen, carbon dioxide, methane, etc.—turned the earth's atmosphere into such a hot greenhouse that everything just cooked? Why does this temperature range persist?

In the early 1980s, Lovelock had proposed an explanation. Alone, and then with his colleague Andrew T. Watson, he built a simple dynamic computer model called Daisyworld. It is made up of black daisies that absorb heat and white daisies that reflect it. The model appeared to show that these opposing tendencies tip the temperature first one way, giving rise to more of the other kind of daisy, which tips it the other way, favoring the multiplication of the opposite kind, and so on, back and forth, up and down like a see-saw, but therefore always keeping the average temperature within a range conducive to the lives of black and white daisies.[157] This coupled relationship with temperature resulted in the self-regulation of the system as a whole, a phenomenon called homeostasis.

In 2002, the Gaia theorist Timothy Lenton and his colleague Marcel van Oijen built another model of Gaia and claimed that with mutation added in, their model "exhibits self-organizing adaptive behavior in its response to forcing."[158]

A few years later, Harvey and colleagues[159] showed that the self-regulation described by Lovelock's model could actually be achieved with even fewer built-in assumptions than Lovelock had used. In 2004, Harvey demonstrated that all that is needed to keep a Daisyworld model operating within a viable temperature range is a mathematical hat function tied to some form of a feedback loop. A hat function is an expression that looks like a witch's hat when it is laid out on a graph. It starts at the bottom of the graph, climbs up to a peak, and then goes back down to the bottom again. This kind of function "with any feedback, positive or negative, can lead to homeostasis through rein control,"[160] Harvey wrote. Such a system did not need to evolve, though "it was compatible with" evolution.

Now I wanted to know if Harvey believed that the models describe reality. Do the unimaginably complex couplings of the world's systems force adaptive, intelligent behavior out of Gaia? Is this planet smart?

Lewes has a Victorian-era train station adorned with lots of white wooden gingerbread. In the parking lot, Harvey stood by his van trying to figure out which of the passengers streaming past him might be me. He's a big man with a large, balding white-bearded head reminiscent of Darwin's. Dark beetling eyebrows said he had been dark-haired once.

We drove a few blocks along the town's main street, quaint with Tudor overhangs, cobbled walks, painted signs, flower boxes. All very English and polite it looked, though rivers of blood have soaked its soil. Harvey was talking about history. First, he described the generally pleasant chaos of Lewes on Guy Fawkes Day when 70,000 people descend on a town of 15,000 to watch the fireworks. Guy Fawkes Day has been celebrated in Lewes without fail ever since a group of Protestants—seventeen of them—were burned at the stake in the Lewes square during the reign of the Catholic Queen Mary. The castle above the town had been built on the site of a previous Saxon castle during the rule of William the Conqueror, who triumphed at Hastings in 1066. It was used to keep the Saxon natives in their place. Tom Paine once lived in a house over there, he said with a wave of his hand. And we have terrorists too, he continued as we crossed a stone bridge. We'll park here, he said, pulling into the curb in front of a policewoman. She stood on the sidewalk, her hands clasped behind her back, on duty.

They don't like paying for parking here you see, Harvey said, so when the parking meters were put in, they blew them up. On Guy Fawkes' Day.

I thought he was joking. But when I got out of the van, I could see that the policewoman was guarding a decapitated meter.

Don't let it spoil Lewes for you, she said to me.

Harvey led the way on foot up a cobbled street of terraced houses that wound up a hill, the way growing so steep I was huffing before we went more than twenty meters. He stopped at a big door with a round knob in the middle. There was a plaque beside it that said The Old Inn, Beadle's House of St. Michaels circa 1690. We stepped over the threshold into a room built for people of a different era—smaller people. The living rooms (there were two) had low ceilings and low doorframes. In the dining room, a table had been pushed against one wall to leave space for an armchair and footstool. Pale herringbone wooden floors said this house had been carefully restored. There was a view of a rock garden with potted hydrangea, arbors, plus a rambling stone wall that climbed the hill. That wall is about 900 years old, Harvey said. And by the way, Tom Paine lived in that house right there.

The American rabble-rouser and pamphleteer Tom Paine, I asked.

That Tom Paine, he said.

So is Lewes a haven for revolutionaries then?

He made coffee and plopped into the armchair, his legs stretched out. His head dropped down toward his chest. He spoke slowly, with time for consideration. I was curious about what kind of life he'd led. On the one hand, this house announced that a senior British academic lives here. And yet, there was something about him that didn't fit that box, something counterintuitive about this system. He said he is an independent scientist, though a Sussex Fellow, and still takes on PhD students. He only attends the kind of academic meetings he prefers. Upper class like Turing, I thought, but that turned out to be not quite right either. He said he came from Bristol, from a "middling rich," merchant family. His father worked in a beer, wine, and spirits company and so was wealthy enough to send him to a public (aka private) school. And then it was off to Cambridge, on a mathematics scholarship to Trinity College back in 1966.

Whoa, I thought. That is like having a sign permanently affixed to one's forehead saying this person is so clever Francis Galton would have liked to trace his lineage. Even Turing didn't win a scholarship to Trinity. He'd had to settle for Kings.

It was not that big a thing, he said, waving that hand again. "In those days only five percent of people went to university. There was an automatic grant to all."

A math scholarship to Trinity? I said.

"I was good at maths," he said with a shrug. "I was good at it because I was interested in it."

However, by the end of his first year, "I was totally bored with math. In the third year I switched to moral science…all philosophy was called moral science in those days."

He did a math/philosophy degree (passing the Part I exam of two disciplines, the alternative to passing Part II of a single discipline, in three years) because "you can still do that at Cambridge." But he was already interested in cognition, though it wasn't called that then. Since the age of seven, he said, he'd been friends with another boy named Geoffrey Hinton and they talked about it all the time.

He gave me a toothy grin: he was checking for recognition, to see if I knew who he was talking about. I'm sure my face was blank because I had no idea who Geoffrey Hinton was or why I should care. No doubt that made him wonder how I'd found my way to him in the first place. I'd flunked test one. There would be others.

Geoffrey Hinton is a professor at University of Toronto, he explained, a leading figure in neural networks, "central to most current work in understanding mechanisms of intelligence." I made a mental note to look up Geoffrey Hinton. [161]

"We talked of these things," Harvey continued. "I made a robot at fourteen." He got that idea after reading a certain article in *Scientific American* by a scientist. He searched for the name.

"The mental neurons fail me," he said, but then he remembered the author as Ashby, W. Ross Ashby, a psychiatrist interested in how feedback works in the brain. He popularized the word cybernetics, first coined by Norbert Weiner. Ashby was a pioneer in general systems theory and well known to Alan Turing who, like Ashby, was a member of a group that met regularly to discuss it. When I looked, later, I found no robot article *in Scientific American* by Ashby: however, there were three by William Grey Walter who built those cybernetic turtles Turing went to see.

“We read it and decided to do it ourselves,” Harvey said. “We used a windscreen wiper motor, a small wheeled base.” The wheels could turn right or left and the robot had a photo sensor. “Bang, you have photo-taxis,” he said, meaning the light-following movement that plants display. “We used candles to make it follow, a mirror to confuse it.”

So why did you switch to philosophy? I asked, thinking maybe engineering would have been a better fit for him.

It was just about thinking clearly, he said. But then he decided he wanted out of that too in order to do “research in social anthropology.” With his grant, “my extravagant grant…I went on the hippie trail each summer to Afghanistan and Nepal.” The grant was worth 365 pounds a year, his scholarship was sixty pounds. (His curiosity was priceless.)

In 1966, Afghanistan was in a modernization and secularization phase. Nepal was caught by its geography among China, and Tibet, and India, political turmoil on all sides. Both Nepal and Afghanistan had once enjoyed British colonial attention, but by 1966, China, the Soviet Union, India, and the US had keen interests in these places, especially Tibet. They were hotspots in the Cold War.

“I wanted to research the moral relativity of tribes in Nepal,” Harvey said. So he did a postgraduate year at Cambridge to earn what was then called a certificate in social anthropology, now an MA. He hoped to go to Nepal for eighteen months. He was inspired by Professor Christoph von Furer-Haimendorf of the University of London's School of Oriental and African Studies who had been publishing on the tribes of the area (such as the Sherpa) since Nepal was opened to the world in 1953. Haimendorf, an Austrian ethnologist who'd sat out the war years in India, was also "the leading guy on Tibet." But during his certificate year Harvey also became "highly disillusioned" by social anthropology.

He found social anthropology to be "woolly." "The concepts weren't grounded. I was a verificationist then. It seemed to be hand waving."

"Hand waving," a favorite phrase of Harvey's, signifies the use of words—like emergent—to spread verbal white-out over difficult questions and make them disappear.

Harvey decided he didn't need to be working on a doctorate to go to the Himalayas. He could just go there. Though the normal visitor's visa was only for two weeks, he got one for eighteen months. How? "The Soviet Embassy used me as a translator. I taught the Economic Attache English." He went to Nepal and set up a mapping business and produced "the first trekking map of Nepal, it's still in print," he told me with satisfaction.

I wrote down these words, but thought they made little sense. Why would the Soviet Embassy hire a lad fresh from Cambridge to translate for them?

He explained that his girlfriend whom he had met at Cambridge and later married (not his current wife whose steps I could hear upstairs) had been hired to work at the British Council (an educational/cultural organization) in Kathmandu. She had a visa, but he didn't, so when the Soviet Embassy told the British Council in Nepal that they wanted an English teacher/translator all was suddenly possible. The drill, as he put it, was that one had to go to the Soviet Embassy and get a paper signed, and take it to the Ministry of Foreign Affairs where he could get a visa of any length.

"I leapt on the bandwagon," he said, brightly. "Then I set up an import-export business out of Afghanistan for eighteen years."

By this point in his story, I had almost forgotten why I came to see him. It didn't seem to make sense. Why would a man so very smart have immersed himself in the mundane details of import-export for so long?

He said he went to Afghanistan first in the summer of 1967, at the end of his first year at Trinity. He went to Afghanistan again in the summer of 1968 and stayed for three months. "I brought things back and sold them," he said. "It paid for the vacation."

He bought Afghan coats, Afghan socks. He looked at me and waited. Another test. I had friends who wore those Afghan coats—embroidered skin on the outside, raggedy sheep wool on the inside, as well as thick and colorful woolen socks shaped like mukluks. They could be bought in little boutiques along Toronto's Yonge Street in those days.

"I supplied the whole world," he said. "I employed 2000 people to knit the socks in Afghanistan. I had people who collected them and brought them to me."

After his eighteen months in Nepal, "I came back to the UK in late 1972, maybe at Christmas, and set up a business called Nomad Traders in the Cambridge market." He brought in things from Afghanistan and Nepal until he got tired of running a retail business and started selling instead to wholesalers.

"I brought big quantities from Afghanistan, from China."

You were in China too? When?

"In Red Guard times," he said. "During Mao times."

Wasn't that harder than getting in and out of Nepal?

"I gamed the system," he said, with a smirk. "It was dead easy."

Well maybe if you were a trusted Maoist, I muttered, but even then it was tough. Canadian journalist Jan Wong recounts in her memoir, *Red China Blues,* that though she was a determined Maoist of a well-known Chinese family in Montreal, and Richard Nixon had been to China just a few months earlier, it still took her a long time to get a visitor's visa to China (and once there, only relentless badgering won her the permission to study at Beijing University). An Asian Studies classmate at McGill, a white woman from Australia who applied at the same time, was denied entry.[162] According to Wong, so few people from the West had been permitted to go to China before 1972 that when she got there she felt it was her duty to record everything she saw.

"Incidentally," Harvey said, as if he was following my thoughts, "my friend Geoff Hinton's father (Howard Hinton) was a

Stalinist. His uncle (a cousin of Howard Hinton's) was in the UN....He was in China during the Revolution and stayed. William Hinton."

So William Hinton helped you?

"I didn't use that at all," he said.

So how *did* you do it? I asked.

Anybody could have done it, he said, with a grin. There was a Chinese trade fair in Canton every six months, he explained. "All trade with mainland China was through state trading corps." If one had papers printed up that declared one to be an importer of things these corporations exported, one could get invited to that fair. "You get an invitation to go, cross the border from Hong Kong for three days, and exit again."

It surely helped that in June, 1972, Britain let the pound float and imposed currency controls across the entire sterling area (composed of countries that used sterling as a currency or pegged their currencies to it). Those forced to submit to the new currency controls included colonies such as Hong Kong (but not Ireland, the Channel Islands, and the Isle of Man). The point was to prevent capital outflows as Britain jockeyed at the time for entry into the European Community.

After going to the trade fair twice, Harvey realized that Chinese visas were so tightly restricted that they even stated the ports of entry and exit. "If you thought of an exotic point of entry and exit, the Chinese Embassy in London put it on the visa."

Now why would the Chinese Embassy in London do that, I wondered? Because China was desperate to get its hands on hard currency, I answered myself, especially after June, 1972.

Harvey put Burma down as the exit on his next visa request. There was only one plane per week from China to Burma, leaving from the city of Kunming, in China's Yunnan Province. Yunnan province borders Tibet in the northwest, Burma (now Myanmar) in the west, Laos and Vietnam in the south. Burma and Laos border Thailand. These countries make up the Golden Triangle, the point of origin for the world's biggest crop of opium (until beaten out by Afghanistan in the twenty-first century). Those who commanded the opium trade belonged to two ethnic/religious groups of Chinese origin whose members lived on all sides of the various borders and traded across them.

Evolutionary robotics and dynamic systems fled from my head as this tale of eighteen years trading in Asia during the peak of the Cold War unfurled. It seemed to me he must have been either a drug smuggler, or a spook.

At the Canton fair, Harvey continued, he told the Chinese authorities he needed to buy a plane ticket for his flight to Kunming. They told him Kunming was a closed city. "I said my visa says I have to leave from there. That worked. The last plane ticket sold to a foreigner had been two years earlier. So I flew to Kunming, the forbidden city, no foreigners. They didn't know what to do with me."

He was taken to a hotel from the airport in a car with antimacassars on the upholstery driven by a chauffeur. No one spoke English at the hotel.

Harvey went out for a walk.

"I was followed by thousands, they'd never seen a foreigner. I was arrested by the first person who spoke English."

He was supposed to have stayed in that hotel for three days until the plane left for Burma, but since no one spoke English, no one told him.

"The next time I gamed it to Peking," he said.

Thanks to relationships with the state companies, Harvey went to Beijing half a dozen times. Later, he heard about a trade fair that used to be held on the northern border of China "where once a year, Tibetans came on yak trains." Americans used to come up there from Calcutta, he heard, to buy yak tails. Why? "The Americans shipped them to the US to use for Father Christmas beards till WWII, when it was disrupted. I heard of it. Felt it was my moral duty to resurrect it."

He failed.

But he sojourned on, to the Philippines, Sri Lanka, Syria, Nepal, always coming back to his home base in Cambridge. His timing was very interesting. He got to Laos just six weeks after the last American helicopter ingloriously lifted off the roof of the Embassy in Saigon as their war in Vietnam came to an end. He arrived in Vientiane the very day the Pathet Lao, backed by North Vietnam, took the reins of power and Laos turned into another falling domino.

He's got to be a spook, I said to myself.

"You're spook, right?" I said.

He just laughed. "Went to Mexico for hammocks, Colombia, but the mainstay was Afghanistan."

He was in Afghanistan when the King was ousted by his cousin, and when the two Communist parties battled it out in the Saur Revolution. When the Russians invaded in late December 1979, he was in Cambridge. He and a Canadian friend were due to go back in January, 1980. To go in the middle of an invasion, or not to go, they asked each other. They decided, "shit, we'll go." Their plane from London's Heathrow carried ten Afghans and "one other foreigner, an American guy. We were suspicious. We thought he was weird."

Weird how, I asked.

"CIA. At London airport we decided to stay clear of him. We landed at Kabul. The next night on Afghan TV we see he's arrested as an American spy...He was weird. A bit mad and weird. These places attract weird people."

And you never worried about getting arrested yourself, about being taken for a spook?

He could see how it might have looked to others, but no, he hadn't worried at all.

Why not?

"It sounds terrible but this is my rationalization of the event. If you go to Cambridge, you think you have a charmed life. You assume....you'll get away with it."

A flick of a hand.

What about your wife, she must have worried, I said.

Well, he said, she was *concerned.*

The mujahadeen began firing at the Russians with American made SAM (surface-to-air) missiles. The main Russian base was at the Kabul airport. At first "they just shot at military planes. The Soviets shot flares to confuse the missiles. Then they shot civilian planes." Still, he stuck with it, until 1987. That's when Harvey's Canadian friend called to say he was just back from Afghanistan. When his plane landed in Kabul, the rockets started coming in and he had to run for it. Just as he reached his friend's car, its windows blew out. Altogether, eleven people were killed. His friend said he wasn't ever going back. Harvey never went back either.

At the same time, he split up with his first wife.

"It was a midlife crisis," he explained.

I thought he was bit young for midlife crisis, only thirty-nine. On the other hand, twenty years of crossing fraught borders, juggling controlled currencies, and being taken for a drug smuggler would take a certain toll.

You were talent-spotted at Cambridge, weren't you, I said.

"The closest I got was being employed by the Soviet Embassy," he replied.

Come on, I said.

"MI-5 and MI-6 recruit at Cambridge," he said. "My supervisor, Dr. Denis Marrian,[163] was a talent spotter who recruited two people I knew. Not me."

Really, I said. But I wasn't sure I believed him. For one thing, how would he know who the talent spotter was and who was spotted if he wasn't among the anointed himself?

Were you disappointed?

"I would have liked to be *asked,*" he said. "Of the two, both were unsuitable. One committed suicide and the other was gay—in those days that was a barrier, a blackmail risk. So Marrian's judgment was not very good."

What did the authorities make of you, then, coming and going from all those iffy places? I asked.

"I was mostly suspected of drug smuggling, yeah, sure," he said. "There was a stage where I was stopped three times on leaving. I gained the impression I was flagged."

So you closed up your business in Afghanistan and left your wife. To do what?

To study neuroscience, he said. "I'd always been interested in neuroscience....I read a lot. *Scientific American* and *New Scientist*....I kept in touch with things. I'd been thinking about neuroscience when I was at school. I was interested in artificial intelligence type stuff."

He was also interested in how it might be improved. As his friend Geoffrey Hinton put it in an interview he gave for an oral history of neural networks, in 1986 he sent Harvey a new book by David Rumelhart and James (Jay) McClelland concerning the state-of-the-art of distributed parallel processing. Harvey was so unimpressed that he told Hinton that if that's as far as they've got he should go into the field himself. He decided to go back to school and take on the challenge.

Specifically, he was interested in evolutionary robotics though the field didn't exist at the time.

"In *Scientific American* there was a mathematical column by Martin Gardner," Harvey said. Gardner's column was always ahead of the curve: he wrote about important developments, often before they appeared in scientific journals. Harvey always read it. He was intrigued by a column about people evolving creatures in "what would be called artificial life." Also, home computers were then becoming available. He got one for fun. "My first...had one KB of memory. I got a sixteen KB pack to boost it and it cost ninety-nine pounds. I was writing genetic algorithms before I went back to school."

Until he split with his wife, he had been shuttling between his homes in Cambridge and Scotland. "I had a castle in Scotland at the time. A sixty-two room castle," he said, but then hastened to assure me that his castle had cost a mere 72,000 pounds so it was cheaper than a two-bedroom apartment in London. His former wife and kids remained there. He settled in at Cambridge where he had access to the library and could find academic journals to read.

Were you reading about genetics, about biology? I asked.

No.

Evolutionary theory?

No.

It just seemed "blindingly obvious", Harvey said, that artificial intelligence "crudely speaking, designs artificial means to do something similar to what humans and animal brains do....So evolution did it. It seemed to me a way to design artificial brains."

There were only two universities in the UK where he could study artificial intelligence, Edinburgh and Sussex. He liked Sussex better because it had an interdisciplinary angle. His friend Hinton had done a postdoc there too. He wasn't sure he could do a whole PhD, so he applied first to do a Masters in AI. As it happened, he really enjoyed it. Sussex was good, it was open, there were lots of other mature students like him.

But he "was shocked by the out-of-datedness of what I was taught."

No one at Sussex was teaching about artificial neural networks. It was 1988 or 1989 before the school brought in someone who could, though "I'd talked of them with Geoff Hinton twenty-five years before....I thought here I am in one of two UK universities, out of academia for eighteen years and in my shallow way in terms of direction to go I had more relevant background than anyone in the department."

All that changed when Phil Husbands came to Sussex as a Lecturer and became Harvey's tutor. "He had a poster about the genetic algorithm conference. He had just finished one of the first PhDs in the area. He supervised my doctorate."

Harvey spent the next four years working out the basics of evolutionary robotics. One aspect of his doctoral thesis dealt with the artificial evolution of anything—aircraft wings, timetables, objects you're optimizing, as he put it. And he got involved as well with *neutral* networks.

What's a neutral network? I asked.

Evolution can be understood as a search across a fitness landscape, he began, fusing Turing's terminology to that of neo-Darwinians. Mutations might happen constantly and yet not have any effect at all on fitness.

Ah, I said, a change that doesn't matter—until it does. A presenter in Montreal at the Summer Institute had given a paper about how consciousness might be a "spandrel," meaning an accidental byproduct of other developments that might eventually acquire value of their own.

But where does a network come into it, I asked.

He stood up, started moving the chair, the footstool, and the floor lamp around. He wanted me to imagine that fitness is a place on the high point of a three-dimensional landscape, on the top of a mountain or on top of the stairs of his house. We poor creatures dwelling on the ground floor had no way to get up there where things would be splendid. How could we get to the top of the stairs if we didn't have legs to climb with, for example? He stood beside his chair, pointing out how a series of imaginary planks—or random changes within a constrained system—could link the footstool to the chair to the lamp, then to the doorframe, and from there to the staircase. It was a matter of one random change creating a bridge to or a base for another, each having no effect on fitness yet finally, in a cumulative way, making a path to the peak of the fitness landscape. He'd done work along these lines showing how random mutations, one after another, in the messenger molecule, ribonucleic acid [RNA], might generate useful, meaningful properties. One paper demonstrated how this sort of neutral network might be used to evolve the word 'word' into the word 'gene.'

This amounted to yet another argument about how evolution really works. Darwin said slow, random, but useful changes in individuals add up to changes in populations, which eventually generate new species. Stephen Jay Gould said evolutionary change can also be sudden, often brought on by disasters (like the meteor that crashed into the Gulf of Mexico sixty-five million years ago, wiping out the dinosaurs and making room for mammals). Richard Dawkins argues that change happens through the mutation of genes which operate in their own interests: we are just hapless meat carrying genes through the world. As R.C. Lewontin described this theory:

> "It is really our genes that are propagating themselves through us. We are only their instruments, their temporary vehicles through which the self-replicating molecules that make us up either succeed or fail to spread through the world. In the words of Richard Dawkins, one of the leading proponents of this biological view, we are 'lumbering robots' whose genes 'created us body and mind.'"[164]

Population geneticists such as R.C. Lewontin, on the other hand, argue that evolution is a statistical phenomenon only observable in populations. Lynn Margulis showed that, especially for the first few billion years, evolution works also by merger and acquisition, cooperation and sharing.

All these theorists critiqued each other.

Well before he finished his doctorate, Harvey was hired as a postdoc at Sussex to help set up its Center for Computational Neuroscience and Robotics. "An evil name," he said. "I hate the whole notion of computational neuroscience, it's a buzz word...hence you can get funding since it's a nonsense phrase.... Wisely, I was overruled."

He continued as a postdoc there for seven years. His group was situated in the AI department. It was mainly composed of psychologists and computer people though they often talked to biologists especially those studying ribonucleic acid. He was interested in the search for fitness, in how mutations could take place within constraints without affecting the functionality of RNA, for example, until—suddenly—one last step produced a useful difference. He thought such a process could be mimicked to evolve control systems for intelligent machines "We were like cockroaches, in between the cracks of two schools," he said.

He worked mainly with Dave Cliff and Phil Husbands. "The three progenitors were us....The three musketeers. We put evolutionary robotics on the map. The only others working on this were Dario Floreano and Stefano Nolfi."

He said, "When I first met Dario it was in a castle in northern Italy." The people at that first meeting were the only ones in the world working in the field. There was no journal, no society, no one who could review their papers for existing journals or for grants. All the other folks in the game "had their antennae out for funding," he said. But not Harvey. He didn't need money. He mainly worked with pencil and paper. They only made a few rough machines at Sussex due to the fact that there weren't a lot of engineers around. In the beginning, his colleagues put his name on joint grant applications, but when he became a Lecturer he didn't employ a single postdoc. He did pull in one grant for a workshop on Daisyworld, but that was an exception. He won't go to big conferences on anything, "only small workshops in new subject areas, if it's big numbers I don't go."

"The EU is brilliant at funding the wacky, the flaky—the EU explicitly sets out to find that, and they pay people like me to do that.…The bureaucracy is a mess, but the fact of funding is adventurous. They ask: is it risky? If not, they don't want to fund it.…I am a reviewer for these projects," he said. He is on the reviewing panel responsible for making EU grants to the best Future and Emerging Technologies. In phase one, he does detailed reviews of some of the projects, and then he and the rest of the panel review all shortlisted projects and make the final selections.

His wife came downstairs to announce she was on her way out. Time for lunch, he said. So we shuffled back down the steep cobbled hill, commiserated with the policewoman still guarding the decapitated parking meter, and climbed into the van.

As we walked, I mulled over what he'd told me. It seemed to me that after he quit traveling, he'd replaced physically exotic frontiers with an intellectual one, substituting one form of intelligence gathering for another. And his timing in this was perfect. When he jumped into artificial intelligence, the old methods of trying to create intelligent machines had stalled. He and Dario Floreano and their colleagues adapted ideas from other disciplines that worked much better. As a leader in a new field who didn't want grants himself, he positioned himself as a Dr. Marrian of a better kind—not a talent spotter, but a talent enabler who helped to decide which bright boys and girls would bring forth smarter machines faster, and sent resources their way.

He said he wanted to eat at a particular pub outside the nearby village of Firle. We drove along a narrow valley road hedged by small woods, stone fences, and droops of ivy: the grassy hillsides showed big pockmarks of white. The white stuff is chalk, he said, this whole area was once covered by a great sea and the chalk is what remains of the shells of the sea creatures that lived and died in it.

Harvey turned his van into a small parking area beside the Ram Inn. It was warm inside, with a crackling fire and low beams. A crowd of well-dressed, cheerful people drank beer and wine and ate artisanal breads and cheeses off wooden platters. All that was missing was a hobbit.

The sun was getting low in the sky, which meant my time was running out. I had come here to find out where Gaia fits within evolutionary robotics' version of intelligence. I'd been so distracted by his fascinating back story that I'd still failed to ask him anything about it. It seemed to me that, in a sense,

Daisyworld amounted to a critique of evolutionary robotics. As Dario Floreano had acknowledged, their computer models hadn't fully replicated actual evolution, so they'd done a slimmed down version and employ a cheat—the god-like programmer who decides what fitness is, in any given situation, and selects for it. The Daisyworld models seemed to show that one can get intelligent, adaptive behavior out of a system without any evolution whatsoever, never mind a programmer God.

When did you start looking at Daisyworld? I asked.

Not sure when, he said, though he was pretty sure he had been skeptical. "It seemed teleological and woolly."

Now that's the pot calling the kettle black, I thought.

But, he said, he was intrigued because "I was an ex-mathematician and couldn't figure out how it worked. I assumed Lovelock wasn't lying about the results…."

You mean some people thought he was?

"Well you could see how he could fiddle the system. Why should it naturally arise without God feeding the parameters?" he asked. "That is the main objection."

Explain, I said.

Well, he suggested, certain biologists who studied it, including Richard Dawkins, believed there are only two ways a system like this could get to homeostasis, where the system maintains itself within certain parameters. One explanation would be "the modeler has cheated and put in the right signs when there is no reason to think the natural world has got that." The second of Dawkins' objections is "'okay, the only way I know things get designed to work adaptively in the natural world is through evolution.' So Dawkins assumed that yes, the parameters were fixed on appropriate values but evolution wouldn't have fixed it, you'd need a population of Gaias reproducing to get evolution working but that's clearly impossible so evolution is not available as an explanation...." This failure to explain how or why Gaia models could arrive at self-regulation without being part of a population was not helped by "Jim Lovelock's naïve biological thinking, his 'I'm a chemist but this is lifelike, I'll call it alive in the sense that biologists call it alive....'"

For Dawkins, the model couldn't work unless there is a cheat. In Harvey's view, Dawkins' reasoning was sound "if *only* evolution produces homeostasis."

Why wouldn't moving planets in a changing solar system count as a population? I offered.

He ignored me. He said he concluded there could be an explanation, though most who studied Daisyworld simply didn't understand it.

"As far as I'm concerned, the issue is resolved, but few read my paper and a small proportion understood it. My strategy was to use the second version of Daisyworld...."

I had just reread his paper, so I thought I knew what Harvey was going to tell me. Also, I knew my time with him was almost up. I checked my watch once more as he began to speak, just to see how much time I had left before

I had to get to my train. This annoyed him beyond measure. He thought I didn't want to hear what he had to say. He had already arranged a group of knives and forks and placemats on the table, and borrowed my pens to indicate graphs shaped like witches' hats. He had tied them to each other by a bridge of knives (the rein control part of the process his paper referred to.) He looked at me with a baleful eye.

"It took me five years to understand it and I do now fully understand it and spent a lot of time working on how people object...."

Okay, I said. Sorry. Please tell me.

Jim knows intuitively why it works, he said. He is clued up on cybernetics whereas Dawkins "is a single-minded crusader, he sees everything through that one prism."

Lovelock's interest in Daisyworld, Harvey explained, was tied to the operations of the real planet. Harvey's interest in Daisyworld was to determine if mathematics could describe it: after the whole thing has been simplified to its nub, he focused on the properties that remain, "not the bathwater I threw out."

He pointed at his two connected witch's hats made of my pens. "Say the viability of an organism, the black daisies or the white, is tied to temperature," he said. "Then the amount of daisies feeds back to what the temperature is. Black daisies, temperature goes up. White daisies, it goes down....The amount of daisies is the *cause*, not the effect. It's a linear function. If it's positive, and the daisies are black, it slopes one way."

One of the two witch's hats lying side by side described the effect of temperature on daisies, the other the effect of daisies on temperature. The two were also linked with two knives. Where the lines cross is the condition of equilibrium. They cross in three places on the graph and the dead point—no daisies— is at the bottom of the graph.

"A bit of math tinkering shows me the bottom and the top is stable, the midpoint is unstable equilibrium.... So, if white daisies slope one way and the black daisies slope the other, stable is the mirror image, regardless of the feedback effect you've got a stable equilibrium. So cheating is not needed. Both ways work. This insight shows either way you get stability."

I was beginning to get it. Once such relationships get started, they keep on going until something somehow kicks out the linkage, or otherwise does away with it.

I apologized for not paying closer attention. Hope I didn't piss you off, I said.

"Nobody understands it," he said holding his head in his hands.

Well, what about Lovelock?

I think Lovelock probably understands it, he said.

So is Gaia smart, then? You do define intelligence as adaptive behavior, don't you?

No, he didn't, he said, because that definition, like emergence, is meant to gloss over things. Intelligence is not a unitary property, he said, *intelligence is defined by the body*. For example, bacteria are good at making molecules and "if you put them in the Sargasso Sea they'd be fine, whereas if you dropped me in there and left me, I would die. But they're not so proficient at sending emails."

He illustrated this idea with reference to a proposal that had come to his funding advisory panel at the EU, the one that evaluates flaky but maybe extremely useful proposals. He and his colleagues had liked one from a group that proposed to make a communications interface that would allow two entirely different kinds of living things to send sex signals to each other. About ten million Euros had hung in the balance as they considered it, then funded it. You might read of it in *Nature* or some such four or five years from now, he said.

Well why did you consider it? I asked. Why would the EU care?

It had to do with being able to manipulate such communications, he said.

And all of a sudden, I grasped his point. What if a plant's molecular signals could be used to change the sexual behavior of things entirely different from plants? And vice versa? What if, eventually, there could be such a thing as a truly universal translator, a Turing machine for signals?

"This is *serious* flaky stuff," he said. It would be of no interest to splitters who focus on the differences between human intelligence and the way octopuses think, for example, instead of the points of commonality *which could then be used.*

"Personally, I am a lumper," he said. "I wonder what's common."

I was hastily packing up my notebooks when I remembered to ask him if he thought evolutionary robotics has run into a complexity barrier, as Floreano had said. He said he didn't believe it, and probably Dario didn't either.

But maybe there is a subtlety barrier, I said. Maybe you guys haven't built in enough of the means nature uses to make change go faster, like symbiotic merger and acquisition, for one.

There *is* subtlety in the gradations of where intelligence can be found, he said. There is a spectrum. It's not like there's a line somewhere, it's all shades of gray. Tornadoes have adaptive behavior.

There was a noisy group at a table nearby. I thought I'd misheard him. Surely he meant tomatoes. Say that again? I said. Did you say tornadoes?

He had.

Explain, I said.

"They fit the basic criteria...." he said. "They need a certain surface temperature—say twenty-seven degrees centigrade water temperature and once they set up, they perpetuate themselves."

He was talking about a certain kind of a wind over warm water exhibiting adaptive behavior, as if wind, a moving body of air, could also be a particular embodiment of intelligence. Having grown up on the prairies I can attest that when you lean into the wind you're never sure whether it will hold you up or strike you down, but you do know it has will and a strength —almost a body—far greater than your own.

"They perpetuate themselves, drawing energy from the temperature differential…and produce the tornado, the cyclonic hurricane…that keeps feeding itself on the temperature differential," he said.

Yes, but it's not alive, I said, reflexively. And then I thought, there you go again, drawing a boundary line. *Listen up.*

"They have a birth and a death," he said. "They don't *evolve.* They fit the Lovelock idea of life, not Dawkins'. They're fragile in some sense…they peter out when they hit land. They run out of food—the temperature differential. They have a *version* of life. Not the sophistication and subtlety of a human or bacterial life. A splitter would say they're different. A lumper would say, of course they're different but…."

But where's the adaptive behavior? I asked.

He spoke of a Japanese scholar, Hisashi Ozawa, who has studied the life cycles of hundreds of cyclonic winds as they move across the ocean.

"When the temperature varies, it has the option to go left or right downwind," he said. "If there is a temperature differential, it looks as if they tend to veer toward the food they feed off. I want to call it adaptive behavior. A form of intelligence. No high marks on an IQ test, but it is *tornado* intelligence.... It's foraging—at an incredibly unsophisticated level....And that's the place to start understanding—at the simple end of the spectrum."

13

The Microbe Man

I found James A. Shapiro, who studies the intelligent behavior of microbes, the way we find most interesting people nowadays—on that giant network called the Web. I was clicking my way from link to link in search of stories about Lynn Margulis, co-author of the Gaia theory. I found a six-page interview that *Discover* magazine did with her in the spring of 2011, shortly before she died. After explaining why symbiosis—cellular mergers and acquisitions— played such an important role in the evolution of life, and declaring cells to be "conscious," Margulis went after one of her critics, R.C. Lewontin of Harvard, a leading population geneticist. She claimed he had lectured to a class of hers and told the students, when asked, that although his population genetics' mathematics have not been confirmed in nature, or in the laboratory, he nevertheless continued to work with such models in order to get grants. Margulis characterized him as honest—though unethical. And then she denied that her own views were controversial, on the grounds that she was right.

To say this interview raised a substantial quantity of ire would be an understatement: the magazine's comment section roiled with the shrieks and cries of trolls, one of whom suggested her statements explained why Margulis' first husband, Carl Sagan, divorced her.

Then I came across a defense of Lewontin written by Jerry Coyne, a population geneticist from University of Chicago, on his blog. Coyne claimed Margulis had been corrupted by fame, that she had dishonestly reported what Lewontin said, that her symbiogenesis theory was not only mad, but fueled the crackpot creationists' attacks upon the theory of evolution.[165] Looking for more of Coyne, I found him dissing James Shapiro with similarly livid prose. Apparently, Shapiro had written things on his *Huffington Post* blog and in a new book which suggested microbial cells engineer their own mutations.

Better look him up, I thought. If cyclonic winds can decide to turn toward warmer water, why couldn't bacteria be that smart? Small, after all, is not the same as simple.

Shapiro's own postings made it clear he is part of what I was starting to think of as The Smarts Network—people who use the principles of dynamic systems and information theory to explain intelligence, and to mimic it. But Shapiro seemed to have a different take on the nature of evolution. He suggested it is not always slow, and most important, *not always random.* He appeared to believe that microbes do rearrange their genomes (the sum total of the DNA in their cells which order the creation of proteins) in order to meet challenges. He seemed to say that bacteria, like evolutionary robotocists, use evolutionary processes as a design tool.

Shapiro asserted that bacteria (which are so varied we have only identified one percent of their species) constantly alter themselves by rearranging their DNA, and by sharing useful bits of DNA from cell to cell, from species to species. This is how they solve problems, such as: beating the fast-changing immune systems of the plants and animals they live off of; warding off the viruses that invade them; making food out of things previously unknown to them.

He argued that bacteria should be studied as communities, not as single cells, because that is how they live naturally, and because their multi-cellular colonies display traits and capacities not shown by individual cells. This too made them sound like Floreano's surprising robots which invent how to do things together better than they can do them alone. Which raised these questions: do behaviors like learning, altruism, deceit, co-adaptation, and synchronization, also appear among the microbes? Or, to put it another way, are politics present even among the very smallest of living organisms?

> "Bacteria form complex communities, hunt prey in groups and secrete chemical trails for the directed movement of thousands of individuals," Shapiro wrote in an article in *Scientific American* in 1988.[166]

In that article, his descriptions of bacteria as multi-cellular systems that share molecular information sounded very much like Trewavas, Baluska, and Mancuso writing about plants. And sure enough, when I checked, I found his work cited by Baluska and Mancuso in a paper they wrote called "Deep Evolutionary Origins of Neurobiology." And when I looked for more of Shapiro's work, I found recent articles where he said flat out that microbes are intelligent.[167] In one, he also said that microbes use us the same way plants do—to advance their own interests. He put it in the abstract where no one could miss it.

> "Forty years' experience as a bacterial geneticist have taught me that bacteria possess many cognitive, computational and evolutionary capabilities unimaginable in the first six decades of the twentieth century....Contemporary research in many laboratories on cell-cell signaling, symbiosis and pathogenesis show that bacteria utilize sophisticated mechanisms for intercellular communication and even have the ability to commandeer the basic cell biology of 'higher' plants and animals to meet their own needs. This remarkable series of observations requires us to revise basic ideas about biological information processing and recognize that even the smallest cells are sentient beings."

Whoa, I thought, this guy is *brave.* Shapiro's eagerness to confront and reconsider dogmas, especially Darwin's, was definitely at odds with the behavior of most of his colleagues.

For example, about the same time as I began reading his blog entries, I also went looking for a lab where I could see the slime mold *Physarum polycephalum* in action. I called around to various biology departments in universities near me. But no one (except non-conventional computing people) seemed to be interested in slime mold anymore. Adamatzky was in the UK and not returning phone calls or emails, while Nakagami was in Japan, too far for me to go. I settled instead for peering through a microscope at an interesting fungus. The professor who allowed me into his lab studies fungal evolution by provoking mutations. He had once studied an individual fungus that cloned and cloned itself until it spread over fifteen hectares of forest, weighing in at about 10,000 kilograms. He established that this fungal individual/group is 1500 years old and still going strong. (It has since become a tourist attraction).[168] Another fungal species he studied has two nuclei that associate and cooperate. In one line of this species, if one nucleus generated a bad mutation, the other compensated by coughing up a good one. Some researchers I'd read had suggested that the way fungal hyphae spread in a network can be compared to a widely distributed neural network—an information network that learns. So I asked the professor: are fungi smart?

I got one of those looks. He said he doesn't pursue questions like *that.* "I just keep my head down and do my experiments, okay?"

I ordered Shapiro's latest book, *Evolution: A View From the 21st Century,*[169] published in 2011. The blurbs on its cover were from stellar people, including Carl Woese, the discoverer of the third domain of life called Archaea.

Shapiro devoted many pages and lots of charts to listing the known means by which various species reshape their capacities in response to challenge. He suggested that bacteria have needs and goals which they meet through the control of their own evolution. Bacterial cells respond to danger by hacking up and exchanging bits of their own DNA which may generate useful new proteins with helpful properties. Some of these responses are automatic, like reflexes. But when unexpected challenges are presented, Shapiro said, bacterial cells can act creatively to deal with them. He called the movement of elements around the bacterial genome *natural* genetic engineering. He said that bacterial cells are very complex dynamic systems embedded in communities of systems. He kept coming back to computer language to describe their every-which-way methods of molecular information exchange. He said the best metaphor for how this works is read/write memory, not read/only memory. In sum, Shapiro's book proposed a radically revised theory of evolution as nonrandom, not gradual, but *cognitive,* where the cell is the primary agent and the genome its mutating, transposing, acquiring, and rearranging slave.

In his bibliography, I found references to Trewavas, Baluska, and Mancuso (though not Inman Harvey or Dario Floreano). So I got on another plane and went to see him.

It was the kind of deceptive winter day that makes you believe in spring. The snow had vanished all at once. The exposed grass was yellow as old teeth. As I walked out of Chicago's Midway Airport, I peeled off my down jacket, grateful I'd left my winter boots at home.

The taxi driver was anxious about where we were going. I could see why. We rolled through mile after mile of awfulness on Chicago's South Side, past small bungalows with boarded up windows, past low-rise apartments and storefronts slathered in graffiti, past huddled men on street corners waiting for something to happen, anything at all. The spires of downtown Chicago glimmered like Oz, a world away.

Finally, I saw a sign for University of Chicago. And then the taxi stopped on a tree-lined street in a neighborhood undecided about whether it was on the way up or down. Grand old houses filled the gaps between apartment buildings. Shapiro's Victorian had a vast side veranda and a front door sporting an oculus so visitors can be inspected before it's unlocked.

I followed him through the hall past a living room with a grand piano, a dining room with formal table and chairs, to the arrestingly modern kitchen where he had his laptop open on a big tulip table. He is a tall, lean man, cruising close to seventy, with gray/blond hair, aviator glasses, and a mustache. He kept rolling up the sleeves of his shirt as if about to do manual labor, and sat ramrod straight in his chair. His wife, a tiny woman who worked in community banking until the funding was cut, made wry remarks about his word choice when he admired her sweater set. He called it cerise. She rolled her eyes.

"It's purple," she said. "Men."

By the time I walked through his door I had learned a lot more about Shapiro. He is one of the leading geneticists of his generation. He was the first, along with Jonathan Beckwith of Harvard, to actually isolate a gene. He was among the first to show that *Escherichia coli* [*E. coli*] bacteria can mutate by inserting pieces of their genomes into new locations, which changes what they can do. His work further confirmed by molecular means the astonishing insights of

Barbara McClintock (winner of the Nobel Prize in Physiology or Medicine for 1983) who long before had identified what she called controlling elements in the corn genome.

McClintock started her career in the horse and buggy days of genetics, peering at corn chromosomes through microscopes to correlate chromosome change with changes in the outward appearance of mutated plants. When she was young, geneticists had just started to use radiation and X-rays to provoke mutations so as to answer questions about how information is transmitted from generation to generation.

When Shapiro became a geneticist in the mid-1960s, the field had moved into its pre-modern stage. No one knew how DNA instructions are read, why there are mistakes, or why or how mistakes are almost always corrected. The code that DNA uses to guide the process of protein assembly was just being worked out. No one knew that the same protein could be arranged or decorated in different configurations which generated different properties. No one knew exactly what a gene is (and they still don't) or how units of heredity are read, or repressed, or if they are the sole blueprints for cellular and organism development. No one knew what directs the winding and unwinding of chromosomes to allow genes to speak with RNA, or how chromosomes arrange themselves in the right order as cells copy themselves, or how chromosomes are repaired by cells when they break. No one knew how or why the organelles in cells reconstitute after every cell division. But this overwhelming lack of knowledge didn't stop the eminent geneticists of the day from creating dogmas about most of it.[170]

When Shapiro began his work it was known that some of the DNA of *E.coli* include both the instructions for the making of a protein and also for repressing that operation. Later, it would be shown that these molecular bits operate together, though they may be located far apart on a long string of DNA, and that some of these units—mobile genetic elements—copy and recopy themselves when cells divide, multiplying and re-inserting themselves into new areas of the genome. These transpositions can make new and inheritable changes in how a cell works. And they can be shared: transposable elements can be dragged in and out of host cells by viruses.

Shapiro found that such elements naturally insert into genomes when studying what is called the *gal* operon of *E. coli.* The *gal* operon orchestrates the metabolization of a sugar called galactose. The *lac* operon does the same for a sugar called lactose. *E. coli* cells can ingest either lactose or galactose to power their operations, which is why provoking mutations that affect these functions allowed some to demonstrate how these operons work.[171] In short, Shapiro was in at the very beginning of a great revolution in biology which soon transformed itself from being a strictly observational science into one which studies organisms by messing with them at the molecular level and using computation to make models of their behavior.

Our conversation began where I often like to start, at the beginning. He said he was born and brought up on the South Shore of Chicago: his father was in business, his mother was a housewife, and he was the younger of two sons. He was definitely the kind of child who always wanted to know how things work, and something in the family dynamic, as he put it, led him to work hard to understand things for himself. Yet in spite of his interest in the how of things, when he entered Harvard he didn't study engineering, he studied English. He wanted to be a writer.

His English courses were a disappointment. They were all about "the New Criticism," which by then wasn't new at all. He described it as studying literature by chasing metaphors to their points of origin, though its perpetrators described it as an "objective" method for engaging with a text, divorced from the intent of the writer or the experience of readers. His plan was to graduate and go to medical school. He thought that as a doctor he could pick up lots of stories to write about. But that didn't work out either.

"It was a good thing I never studied medicine," he said. "I'm too curious and too idiosyncratic to be a good doctor, they have to follow best practices, they can't be too exploratory, they have to be conservative and predictable."

He grinned, as if to say I'm radical, and proud of it.

And yes, he was a brilliant student, though he characterized himself as merely getting "good" SAT marks and "good" marks at Harvard. He graduated as one of the Senior 16 *Phi Beta Kappa* (the top sixteen students of his graduating year) at number ten on the list.

It was 1964. The Viet Nam War draft was dragging in any healthy young American male who couldn't get a deferral. Somebody, he can't remember who, suggested he should apply for a Marshall Scholarship (created by the British government as a thank you for the US's postwar Marshall Plan). He applied, and he won, and was sent to Cambridge. He had just married. He got a small stipend but it was so tiny he couldn't have made ends meet if his wife hadn't worked as a teacher and his family hadn't chipped in. (After he became a successful professor, he served on the Marshall board and helped reorganize its financial structure; for these services the Queen awarded him an honorary O.B.E.) The only person he'd known who went to Cambridge before he did was James D. Watson who famously had worked out the structure of the DNA molecule with Francis Crick. Watson had been teaching at Harvard when Shapiro was an undergraduate. "In fact we competed over a woman," he said. "Jim comes from the same neighborhood as I do."

Shapiro thought he should study biochemistry at Cambridge as a preparation for medical school. But the biochemistry classes were full so he had to pick something else. He'd done a genetics course at Harvard, so he went to Cambridge's department of genetics and asked if he could do research in bacterial genetics. They said sure, and sent him off to the Medical Research Council to ask Sydney Brenner for some advice and for some strains of bacteria to work with.

Brenner was just beginning the work that would win him a Nobel: he was tracing the fate of every cell of the worm, *C. elegans*, as it developed. "I remember seeing his office strewn with photos showing the cell lineages," Shapiro said. "He also had these *E. coli* strains and he suggested a strategy for dealing with the *gal* operon which turned out to be very useful."

There were several chemicals then known to cause mutations in *E. coli.* Some were believed to generate what is called a frame-shift mutation which makes RNA misread the DNA strand by starting its translation one base pair too far to the left or to the right. Shapiro studied "spontaneous" mutations in *E. coli* that appeared on their own and sometimes reversed themselves. At the time, the only known reverting mutations were base substitutions or frame-shifts. So he tested for both, using mutagens that induce both and ought to have stimulated reversions. But the mutagens had no effect. At that point, no one had imagined that there might be naturally occurring insertions of significant pieces of DNA that could change behavior, or that these changes could be excised in order to revert to the original sequence and restore function.

"For some reason I thought of insertions," he said.

It seemed to him that there might be some sort of terminator sequence at work that said 'stop transcription' because that's what seemed to be happening. "I thought it put extra DNA in and transcription came to a stop," he said. That guess turned out to be both right and wrong. He thought that this stop signal was dependent on the size of the section of DNA being inserted because earlier work on the behavior of the *lac* operon had pointed in that direction. In fact, the inserting DNA segments contained specific stop signals. So he was wrong, but his hunch started him along the right road.

He worked away at this problem in Cambridge for two years, then spent a third year doing his PhD thesis under Bill Hayes at University of London. There were no doctoral course work requirements in the UK, he just had to do his experiments and write up his thesis and defend it. By 1968, he had a freshly minted Cambridge PhD. He'd actually defended his thesis in 1967, but put off getting the degree for another year "because of the war. I didn't want to be drafted." Hayes sent him to Paris and the Institut Pasteur to do a postdoc with the Nobel laureate Francois Jacob.[172] By then, Shapiro had become almost certain that the mutations he studied in the *gal* operon were the result of insertions. A colleague at the Institut Pasteur, Alex Fritsch, taught him a technique called equilibrium density gradient centrifugation. It allowed him to spin cells at very high speed in a centrifuge and then measure the different densities of the different particles that were separated out. He took bacterial virus carrying the mutated *E.coli* DNA, put it in the centrifuge, spun it and then measured for any change in the size of its DNA (as compared to the DNA of the same virus not carrying mutated genetic material). He found the spun DNA had greater length than it should have had, so he had proof of insertions.

"The bacteria showed me they could do things they were not supposed to do," he said. "One of the benefits of having done English literature instead of a biology degree was I was never taught what I shouldn't think....the bacteria gave me different answers than the conventional wisdom. I wasn't the first to find these mutations. But others had contrived elaborate explanations as to why *they must be* frame- shift mutations."

He went back to the US and did a postdoc with Jonathan Beckwith at Harvard Medical School. Beckwith had developed techniques to move the *lac* operon into different positions on the *E. coli* chromosome. This involved using viruses which invade bacterial cells' genomes. As the bacterial genomes copy and divide into new cells, they also copy the virus's genetic material as well. Shapiro wanted to see if he could control lactose metabolism with the lambda virus. First he had to somehow get a deletion, connecting the *lac* operon of the *E. coli* cell to a chromosomal copy of the lambda genome. It didn't work. "The garbage project, we called it," he said. But he and his colleague Karen Ippen did manage to clone the *lac* operon in the lambda virus. And eventually, he and his colleagues were able to isolate pure *lac* operon DNA by using the lambda-*lac* and Phi80-*lac* viruses.

"We published that on the front page of all the papers as the first purification of a gene—though I don't believe in genes—but I was not then so radical in my thinking. What we did is of no practical consequence now, but this was pre-restriction enzymes and pre-sequencing so the cells themselves had to provide the biochemical tricks needed to clone *lac* in the viruses."

Naturally occurring restriction enzymes cut chunks of DNA out of the DNA circles floating freely inside bacterial cells, which have no nucleus. These circles—plasmids—then reseal themselves. The restriction enzymes are tools the bacteria use to get rid of invading viral DNA. After their discovery, restriction enzymes became the major means used by geneticists to study how genes work by moving genetic material from one kind of organism to another. This recombinant DNA technology is now ubiquitous in labs, pharmaceutical, and agricultural companies.

"A lot of the technology involved letting the bacteria do the rearrangements for us. To me, this has always been part of the understanding of genetics. That the *cell* is in control of the structure of DNA, not the other way around," he said.

When he and Beckwith called a press conference to make their announcement, they used the opportunity to also denounce "the misapplication of science in the Viet Nam War."

But then Beckwith also expounded on the dangers of genetic engineering. "I didn't agree with that," said Shapiro. The Asilomar Conference, which led to guidelines restricting how genetic engineering research is done, was held not long after. "But it was based on false assumptions," said Shapiro. "Cells do recombination of DNA all the time."

And after the postdoc at Harvard?

"I taught in Cuba for a year and a half," he said, blandly.

He might as well have said he went off with Jane Fonda to visit Hanoi. The United States had slapped an embargo on Cuba after Castro led the revolution there in 1959, and it had tightened these restrictions further in 1962. No American could go to Cuba in 1970 without breaking the rules against trading with the enemy.

Why did you go? I asked.

He said he had been involved in "anti-war activities, where we were all talking of revolution." Cuba's revolution was supposed to be a kinder, gentler form. He wanted to see it for himself.

"It was one of those experiences that had wonderful and horrible moments," he said. "I learned Spanish and a lot of the complexities of revolutionary situations....If Fidel wanted it, good or bad it got done. The hierarchical nature of the society became clear. The revolution was good for the people—they got health care and education—but it was run in a top-down manner. It was not the democratic, participatory utopia it was made out to be by the American Left. People said, why go? We said to find out what it's like. They said 'we already know.' It's like science; it's amazing how many questions in evolution have *not* been addressed experimentally but people claim to know how they are solved."

By 1972, he was back in the US looking for a job. The only university that would hire him, tainted as he was by a year and a half among the Cuban revolutionaries, was Brandeis. "The only place with courage enough to give me a place," he said.

A few years later, he landed a job at the University of Chicago as an assistant professor in microbiology. At Chicago, he aimed first at what he thought was practical work. However, when he went to meetings and heard others talking about using transposable elements as tools in genetics, he said to himself, that's my turf, I've got to get involved. There was a meeting at Squaw Valley where he realized, as did others, that the subject of movable DNA elements was vitally important.

So he went back to working on *E. coli.* He began to ask fundamental questions about how DNA movements occur. In 1976, he helped organize a leading edge conference at Cold Spring Harbor Laboratory on Long Island on all the transposable elements then known. Cold Spring Harbor Laboratory, formerly home to the American eugenics movement, was where the newest ideas and techniques in genetics were discussed and shared. Barbara McClintock had lived and worked there on corn genetics since the late 1930s. That's where Shapiro met her and realized how important her ideas were. They shot an arrow straight through the heart of simple, neo-Darwinian notions of how change happens. There was a lot more to it than slow, random mutation and sexual recombination.

"I always felt that the assertions of random mutation were not well founded. I didn't know enough to challenge it, but I knew it didn't include the things I knew to be important," said Shapiro.

Yet Darwin's vision was very hard to critique. By 1980 a new theory that most DNA is junk—the silent relic of old mutations that no longer do anything—had begun to circulate. Shapiro and his colleagues working on microbes pointed out that bits of DNA often duplicate as they transpose, and then integrate into different areas in the genome; this had to have played an important part in evolution. The population geneticists who argue that selection acts on genes, which determine everything, didn't like that one bit. He went to a big meeting held at Cambridge in 1982 to mark the centenary of Darwin's death. "Dawkins spoke there. Stephen J. Gould, Lewontin, Francisco Ayala. I asked Ayala what about transposable elements? And he said 'transposable elements don't have anything do with evolution.'"

At that 1982 meeting, Shapiro went public with his new ideas when he gave a talk called "Variation As A Genetic Engineering Process." He had by then begun to spend as much time as he could with Barbara McClintock. By then too, evolutionary theorists, faced with the results molecular biologists like him were getting in the lab, had become "resentful" about such intrusions into their evolutionary studies. Certain ideas were totally excluded, according to Shapiro, such as "the whole idea of learning, control, teleology." He thought these theorists felt threatened by molecular biologists' observations because they had defined themselves as materialists and equated materialism "with random, stochastic change without a notion of other processes. They explicitly ruled out the role of life history in affecting the evolutionary processes."

The debate, in his view, harkened back to the vitalism versus materialism arguments at the end of the nineteenth century. Two students of the German biologist Ernest Haekel took opposite sides. Hans Driesch noted that a hydra can regenerate when pieces of it are cut off, and that he could get a whole sea urchin out of each of the first two cells of a developing embryo, and even out of the first four. He said there must be some inward knowledge of an overall pattern. He described this *entelechy*—Aristotle's word for it—as a mind-like life force. (Later in his career Driesch also became interested in parapsychology, specifically in telepathy, clairvoyance, etc.) Wilhelm Roux said all developmental mechanics must be explainable by the known laws of physics.

Roux's position became "the predominant idea behind molecular biology at the mid-century," said Shapiro. "The vitalists had no way to treat their ideas in a scientific way...It's the difference between necessary and sufficient," he continued. "Is something necessary for a process to take place? Or is it sufficient? Any complex process has many things that are necessary but not sufficient. Are physico-chemical processes necessary for life to succeed? Obviously, yes. But does that mean they explain all of life? I don't think so."

Are you talking about information or patterning as something different from molecular processes? I asked.

"I'm not sure I know what information is," he said. "Data and information are two different things."

He mentioned as an example the way light is transmitted from the eye to the brain. The brain reorganizes it and assigns it meaning. This whole other level of organization is not implicit in the chemistry and physics of perception. "Where are the properties of water in what we know about oxygen and hydrogen?" he asked.

You're hunting for the word emergent, I said. I was thinking I should introduce him to Inman Harvey who would be happy to critique its limitations.

"I dislike the word emergent," he said. "It's like the word gene, it's like natural selection, it puts a name on a black box."

The idea that cells are smart, that they represent a level of organization and behavior not predictable from their genes alone, went back to Barbara McClintock's Nobel lecture of 1983. She had distinguished between programmed responses with which the genome corrects errors as it copies itself, versus an organism's unpredictable responses to unexpected environmental or internal challenges, "leading a genome to reorganize itself." She said such rapid reorganizations may "underlie some species formations" though she also pointed out that even massive chromosome fusions did not necessarily result in new species. She went through a long list of the means by which cells repair damage, including the ability to sense the ruptured ends of chromosomes in the nucleus and to unite them so they are correctly oriented. Then she said:

> "The ability of a cell to sense these broken ends, to direct them toward each other, and then to unite them so that the union of the two DNA strands is correctly oriented, is a particularly revealing example of the sensitivity of cells to all that is going on within them. *They make wise decisions and act upon them.*"[173] [italics mine]

She cited all sorts of ways in which cells respond to stress by adaptation.

> "The sensing devices and the signals that initiate these adjustments are beyond our present ability to fathom. A goal for the future would be to determine the extent of knowledge the cell has of itself, and how it utilizes this knowledge in a "thoughtful" manner when challenged."

McClintock also pointed to other smart phenomena such as the signal interactions between different kinds of living things so they can cooperate with each other. She mentioned the different kinds of galls that form on the leaves of a certain plant. The galls will house three different species of insects' eggs until they hatch, each gall meeting that insect's specific needs. The plant makes these galls because the insects have somehow taken control of its genome at each location. She also referred to the galls that form on legumes: nitrogen-fixing bacteria put out signals that reprogram the plants' genomes, which then produce the galls these bacteria will live in. Both life forms benefit from this kind of smart interaction.

You use the word "cognitive" to describe cell behavior, I said to Shapiro. Why that word?

"I mean taking action based on sensory information," he said. "I understand 'cognition,' like all these words, has manifold meanings...."

Other people call it intelligence, I said.

"I want to stay away from the word intelligence because there's so much human self-love in it," he said. He was talking about bacteria, but he reminded me of Suzanne MacDonald talking about elephants and orangutans.

It was way past lunch time. We went to a nearby Mediterranean restaurant he favors. It was warm by the time we left the house. I was sweating in my long sleeved shirt, he was sweating in his.

Over the hummus, I asked him why he studied the behavior of bacterial colonies only later in his career.

Because microbiology had its origin in the study of disease, he said. The methods he'd learned first focused on isolating single cells in order to follow Koch's postulates.

Which are?

"You must isolate the microbe from the disease; you must grow it in pure cultures and then inoculate it into an organism and get the disease; and then re-isolate it and show that this microbe is identical to the original. That became the holy grail of the work."

However, by purifying a sample down to only one kind of microbe, he said, "you only get one percent of what's in the sample." All of us swim in a microbial stew called a microbiome, with microbes on us, in us, and all around us. Microbial colonies also have associations of many kinds. Purifying the sample breaks up the interrelationships of much more complex systems.

Still, his breakaway from these old methods was the unintended consequence of something else entirely. In the early 1980s he started experimenting with transposon-mediated genome fusions. He had told part of the story in his 1988 *Scientific American* article:

> "I was experimenting with a genetic-engineering tool, designed by Malcolm J. Casadaban of the University of Chicago and his students, in order to study enzyme expressions in *Psuedomonas.* The technique enabled me to join the genes for certain *Psuedomonas putida* enzymes with the DNA sequences from *Escherichia coli* that encode the enzyme beta-galactosidase.…It causes certain chemicals to change colour when they are exposed to it."

He grew the modified *Pseudomonas* on agar containing these chemicals. (Agar is a gelatinous substance derived from red algae and used as a substrate to grow microbes in Petri dishes.) The modified *Pseudomonas* ate the agar, and then its imported *E. coli* DNA sequences caused the *Pseudomonas* colony to change color, signaling that the insertion had worked. The "amount and distribution of the color is a measure of the beta-galactosidase expression."

For another study, he had to take daily photographs of the bacteria he was growing in order to document the results. He'd heard of a new high resolution Kodak film. "I tested the new film on the *Pseudomonas* fusion colonies. When I printed—I remember being in the darkroom and seeing the patterns come up, thinking 'my god, it's like a flower, like the patterns in McClintock's maize kernels.'"

He repeated the experiments using other naturally pigmented bacterial species. Each colony

> "...produced its own unique flowerlike pattern. The fact that different strains and species produced distinctive colonies...gave me reason to believe colony growth in bacteria is a highly regulated process and is under some form of temporal control....It is now clear to me that colony organization follows certain general rules, which help to explain the existence of general patterns."[174]

In other words, individual bacterial cells work together when grown in colonies and these colonies are much more complex than the sum of the individual cells. The colonies are dynamic systems with abilities and behaviors of their own. In fact, the individual cells within them develop new shapes and behaviors in these new contexts. These colonies of reproducing cells formed circles divided into pie-shaped sectors, separated by rings. The cells in the pie-shaped sectors had originated from one cell, some growing more than others, some expressing different genes than others. However, the cells in the rings separating these sectors shared the same expression but not the same ancestry.

> "Some system must exist that bestows common properties on bacteria within a ring and distinguishes them from bacteria in other rings," he wrote.

He thought there was a biological clock at work, too, because the rings formed at a specific time in the course of colony development. He soon made a link between what seemed to be simple "swarm" rules operating among these small living things, and those at work at our own end of the complexity scale. "Both biological clocks and the temporal control of development, previously unknown in bacteria, are important features of higher organisms."

By then he had begun to read the work of Sergei Winogradsky, an early twentieth century Russian soil microbiologist. Winogradsky had discovered how certain microbes get energy from inorganic compounds (work that made Lovelock's insights possible many years later) and how certain other microbes fix nitrogen in legumes. Winogradsky also realized that if soil is washed, one would fail to see how bacteria grow naturally—as biofilms.

As with Mancuso, Baluska and Trewavas, reading works from the past led Shapiro to the future. He decided to have a look at how other bacterial colonies behave. He watched how *Proteus mirabilis* colonies swarm over surfaces and make interesting patterns. He read the work of Louis Dienes (once chief of bacteriology at Massachussetts General Hospital). Dienes had learned that if two groups of *Proteus mirabilis* are put on the same Petri dish and grown, they form what is called a Dienes line and stay separate (suggesting self-versus-other recognition by bacteria). Shapiro kept thinking about the meaning of these patterns, about why they form and what regulates their behavior. He began to see these colonies as three-dimensional organisms with specializations and divisions of labor reminiscent of the tissues of 'higher' animals.

Show me, I said.

So we drove back to his house. In the kitchen, he swung his laptop around on the table to display his images. They were weirdly beautiful. They looked like pressed flowers or like shiny mosaics shaped like flowers.

"They spread from the inoculation point," he said, meaning the place where the bacteria are first dropped into the Petri dish. If two spots of *Proteus* are put down at the same time, the colonies meet each other in synchrony. If they are set down separately, each keeps its distinct rhythm. The colonies generate patterns, with terraces, or edges. The patterns they make depend on the strain grown. But he found he could alter these patterns by modifying the growth medium, by adding a chemical insult, by food withdrawal, or even by gouging troughs into the agar.

Each *Proteus mirabilis* strain also has its own particular swarming behavior. The colonies all start from one spot of liquid culture with short cells, but as they develop, differently shaped cells with long flagella appear, and move to the edge of the colony. As their flagella move, the whole colony spreads outward from the edge. These swarming cells work together, their flagella flailing synchronously, which creates oscillating waves as they go. As he described it in *Scientific American:*

> "Swarming is strictly a multi-cellular activity…an individual cell that gets separated from the rest is

> unable to advance over the agar unless it is engulfed by another swarmer group, at which time it starts to move again."

"Each responds to its own rhythm of population dynamics," he said. And a Russian physicist had modeled these dynamics. If one changed the agar or the glucose fed to them, it affected the development, but not the timing of the terraces they made.

Can you show me more in your lab? I asked.

Don't have a lab, he said. Not since 2000.

I was shocked. Why not, I wondered.

"I couldn't get funding for this, or for control of DNA rearrangements," he said. He sat back in his chair and grinned at me—still radical and proud of it.

But why not, I asked again.

He said it had something to do with issues of orthodoxy in the granting agencies. But then he grinned again. "I'm known as someone a little off the reservation. There are people doing *Proteus* research but they are into well connected networks so their work gets funded. My ideas on genome change, regulation, and evolution are not accepted."

Because you use the word "cognitive" to describe bacterial behavior?

But he wasn't even the first to do that, he said. Alexander Fleming, who discovered penicillin, had wondered "in print if *Proteus* could possibly have a nervous system."

Well that must have made people nuts, I said. When was the first time *you* put the cognitive idea in print?

He thought about that. It was before his book, it was certainly in a paper he published in 2006, maybe also in one in 2005.

It suddenly occurred to me that the oscillating behavior he had recorded in bacterial colonies was similar to the streaming motion of slime molds solving a maze with their bodies. So I wondered: are bacterial colony movements a kind of computation?

Hey, I said, do you know about slime mold computation?

Of course he did. He knew all about Adamatzky's work. He was in contact with Adamatzky's colleague Larry Bull. In fact, he was way ahead of them. He'd started thinking about cellular computation as far back as 1984 when he was studying the effect of starvation on the genome of *E. coli.* He'd been using Malcolm Casadaban's technique for fusing genetic material to the *lac* operon with the transposable virus known as Mu—which inserts itself just about anywhere in the bacterial genome. When he starved the cells, he found he had to wait what seemed to be a long time for the colonies to start growing. As he had been told by a former student who used Casadaban's method, he needed to use thicker agar plates. It could take up to two or three weeks for the colonies to appear. "Then they started to accumulate more and more rapidly, more and more every day until the Petri dish was saturated with colonies."

"Starvation triggered a process that led to the formation of these fusions," he concluded. The fusions didn't form without starvation, but once they occurred, colonies could form in two days.

He studied them for several years, publishing a paper on what came to be called an "adaptive mutation." Selection due to stress was at work.

"When you put bacteria on a selective medium (one exposed to oxygen) the stress they encounter—oxidative starvation—triggers the DNA change systems, and some of the changes allow bacteria to form on that medium. I was using a system that I selected for specific rearrangements. Others looked for other kinds of DNA changes stimulated by selection, and found them. It is *not* a random process. It can be triggered by a given set of conditions. Susan Rosenberg's group [at Baylor's College of Medicine] tried to figure out how many different proteins are involved in regulating this response. It's at least ninety-four. Some are completely unknown in function."

In other words, the bacteria use a complex cell system to "sense danger," as McClintock put it, and rearrange their genomes in order to deal with a challenge.

In 1991 Shapiro published a paper called "Genomes as Smart Systems." It was a really radical notion, yet it got no comment at all. He clicked on to PubMed to see if the paper had been cited by anyone since. "No, never cited," he said.

In 1995 he published a review paper, articulating what he and others had learned from the study of the adaptive behavior of bacterial colonies and setting out where future studies ought to go. He described how computer scientists were starting to model the behavior of colonies grown in subpar conditions, to correlate shape changes with certain transition points, some of which were reminiscent of transition points seen in inorganic systems. Challenge by starvation produced all kinds of interesting group behaviors.

He pointed out that multi-cellular behavior starts as soon as a single cell divides to make two. These two cells then lengthen and stay side by side as communication between them begins. Just as two colonies will fuse together and operate as one, these two cells transfer information cell to cell and, as they divide, take advantage of specialization to perform the colony's labor. As he put it:

> "...Science is undergoing its greatest intellectual revolution since the time of Descartes. The new emphasis is on connections, complexity and the behavior of systems, rather than the inherent properties of the smallest possible units. Living systems are the key to this new approach because well-established biological concepts like homeostatic feedback regulation, sensory information processing, behavioral responsiveness and hierarchical integration are applicable to all complex systems."[175]

I got him to explain what he meant by that—in everyday English.

"The point is the cell has this elaborate set of circuits to detect permutations and stresses," he said. "It regulates biochemical activities that can restructure the genome. It can make big or small changes. We know cells can turn on this machinery to change the genome in response to stimuli like aerobic starvation. We don't yet know whether the changes that occur are related to the stress applied."

Well if the change is related to the stress, would it mean they are in effect computing the right response, or even remembering it, I asked.

"That was claimed by John Cairns of Harvard," he said. Cairns got a lot of attention for the claim of "directed mutation.... People in the field have suffered for his deeds."

Cairns and colleagues had developed a strain of *E.coli* that could not metabolize lactose. Then they starved the bacteria of everything *but* lactose. They claimed the bacteria soon found a way to eat lactose. Cairns and his colleagues published their article in *Nature.*[176] "He claimed the cell could sense if it needed to grow on lactose and change itself to suit. There were all kinds of problems with the experiments that scared people off serious studies that do need to be done."

Because, Shapiro said, when an *E.coli* cell is starved of glucose, for example, it produces an internal chemical signal in the form of a molecule called cyclic adenosine monophosphate (AMP). These molecules are known to carry messages but also to greatly amplify signals. When *E.coli* is starved of amino acids, it makes a different alarm signal, a molecule called guanosine pentaphosphate (ppGppp), which inhibits RNA and protein synthesis to conserve amino acids that are already present. "It would be interesting to compare mutations of one to the other," said Shapiro. "The question is, do they both stimulate mutation and in the same way? I'm not aware of anyone asking."

Someone at Baylor had done preliminary work on this question sixteen years earlier. Someone else spent twenty years developing an *E.coli* strain that can utilize citrate and found there had been a slight genome duplication that activated a transporter.

"Big deal," Shapiro said, with a dismissive little shrug. "This is considered a *significant* evolutionary experiment. It's kind of ridiculous."

So what's gone wrong here, I asked. Why aren't these questions being pursued?

People are afraid, Shapiro said. They have given lip service for 100 years to "a mistaken view of how evolution occurs. A lot of the experiments are taboo. Present them and people say 'that's outrageous. How can *E.coli* discriminate between stresses?' Cairns was disposed of in 1988 [actually in 2002 and others have since disputed this disposition] so we don't have to pay attention to it. But the evidence is so overwhelming that certain conditions lead to major changes in genomes, major changes in evolution....When you present facts and references they claim it's nothing new."

And these things tell you...what? I asked.

"They tell you that cells have tools to modify their genomes," he said.

And you think it's *intentional*, I said, getting back to where we'd started.

"Intentionality is anathema," he said "Teleology is considered verboten. They don't stop to think we know lots of examples of intelligence and teleology in nature all the time. The complex cell cycle, finding food, reproducing, are all goal-driven activities. But somehow, genomic processes of change have to be kept separate....There's a deep confusion that Darwin shared....Does selection occur before or after variation? He describes it as after. But we've discovered that selection, or stress, can also *trigger* change."

So far as he was concerned, neo-Darwinian evolutionary theory is riddled with these sorts of chicken/egg confusions. Another example: it is known that significant chunks of genomes are very resistant to change, which is interpreted to mean these sequences are so vital "they are not free to change by mutation...."

"People doing genome comparisons use terms like 'constrained' to avoid attributing it to natural selection," he continued. "These are hyper-conserved regions that don't change at all. But you can test it, knock it out, and the animal survives....We need answers and we don't have them."

He found it sad and ironic that on the one hand, people want random-change-followed-by-natural- selection to explain everything, but on the other hand, the metaphor Darwin chose himself to explain natural selection is the action of the careful farmer trying to get a better breed. “Darwin’s metaphor is the *soul* of intentionality,” he said.

His next book would be about how the cell rewrites the genome. “My idea is to point out all we know of these processes,” he said. “It takes the cell from being the passive recipient of random events to an active agent in control of its genome.”

But how *does* a cell sense, with what tools, and how does it decide, I asked.

“We know how electronic circuits do it,” he said, “but don’t understand how molecular circuits work.” And yet, he said, single-celled bacteria, or bacterial cells working together as a colony....

Or, *C. elegans* larvae deciding whether to hibernate, I interjected.

“...can tell what is in their environment and decide to give up the ghost or keep on trying.”

It was time to go. But before I left, I wanted him to say out loud what he’d said in print. Why? Because he hadn’t done so, which made me wonder if he’d lost his nerve.

So microbes are intelligent, I said. Right?

“My point is that they are taking action based on sensory information,” he said. “How do these tiny organisms do that?”

Had he waffled, or hadn't he? While I waited in the airport for a low-lying fog to clear, I reread a paper he'd published in 2007. He'd closed it by proposing computer metaphors to help us understand what bacteria do. "Our digital electronic computing systems are far simpler than the distributed analog processors in living cells," Shapiro wrote.

> "The take-home lesson of more than half a century of molecular microbiology is to recognize that bacterial information processing is far more powerful than human technology...bacteria display astonishing versatility in managing the biosphere's geochemical and thermodynamic transformations, processes more complex than the largest human-engineered systems. This mastery over the biosphere indicates that we have a great deal to learn about chemistry, physics and evolution from our small, but very intelligent, prokaryotic relatives."[177]

In his book on evolution, Shapiro had referred to a Canadian scientist named Sorin Sonea. Back in the mid-1980s, Sonea published a theory that all prokaryotes (species without a nucleus such as bacteria) also form an association much bigger than any colony. He argued that they trade information in the form of molecules through an incredibly complex meta-genome stretching around the world. As with plants and animals, there's a whole lot of cross talk going on up, down and across Gaia's complexity ladder.

When I got home from Chicago, the newspapers carried a report that made Sonea and Shapiro seem prescient. First, viable bacteria had been found in the upper atmosphere. But also, groups at the universities of Western Ontario and Arkansas had just published a study showing that the behavior of some bacteria in the guts of human children may play a role in autism (by impairing mitochondrial function and fat metabolism). Autistic behavior has also been induced in mice by messing with their gut bacteria. The mice change their behavior when certain bacteria are killed off, and change it again when the right bacterial colonies are re-introduced. This work demonstrates that brainless bacteria—like brainless plants— affect animals like us even in our most sophisticated processes. Just as some bacteria can force a plant's genome to make a useful gall, others can influence our behavior, including the way our minds work.

14

Spaun

The traffic had finally thinned out, which was good, because, as I drove the main highway across southern Ontario, with one eye on the clock, I was also shouting at the radio. I was listening to a political affairs program while on my way to an appointment at the University of Waterloo. The usual pundits were going on about the Canadian Senate scandal, a petty, venal affair concerning public money improperly sucked up by some senators as repayment for housing and travel expenses. Three stood accused of claiming housing expenses while in Ottawa as if they didn't live there, though they did. Senator Mike Duffy had been told he owed about $91,000. Duffy, who'd famously called a Canadian senator's job the "taskless thank," had piously claimed it was all a misunderstanding and he'd borrowed from a bank (he named the bank) and paid it all back. But then the public belatedly learned that Prime Minister Stephen Harper's very wealthy Chief of Staff had paid Duffy from his own account so that Duffy could repay what he owed—without losing a nickel—and continue to be useful to the prime minister and his party.

The whole thing was really about our prime minister, our 'Nikkie' of a prime minister. But the radio commentators kept missing the larger pattern. Such as the fact that Prime Minister Harper had ignored the constitutional requirement that senators must reside in the provinces they represent, though when he appointed two of them they did not. Such as the fact that the energies of the political class were now focused on a few hundred thousand dollars plucked from the public purse by the undeserving while the prime minister ripped *forty-five billion dollars a year* out of government spending by some calculations, nevertheless claiming that services remained as before though refusing to show a full set of books to the Parliamentary Budget Officer, Kevin Page. (Page sued the government in Federal Court, but still failed to get the information.)[178] As a country, we were failing the Democracy IQ test. We were kowtowing to our prince, but not insisting on getting a proper return. Pay attention to what *matters*, I was shouting. *So stupid!*

I took a deep yoga breath, in, then out, and plunged back into the here and now. A mist, like a shroud, appeared out of nothing; you could say it *emerged*, dancing across the face of the Niagara Escarpment, a giant limestone ridge sprouting from southern Ontario's great plain like the backbone of a titan. The soft green blush of the shrubs and trees thrusting out of the Escarpment's crevices colored the mist acid green. It swayed in the breeze. As if alive. Or, as Inman Harvey would say, as if it had a *version* of life.

Hey, *you* pay attention to what matters, it said to me. Pay attention to the fact that you are not just you, that you are enmeshed in all kinds of other living, sort-of-living, and non-organic systems going about their business with their own forms of intelligence and purpose. Things you can't see crawl over you, fending off assaults from other things you don't notice either that could change you or impede your continuance.... Smartness is everywhere, intelligence is at work in *everything*.

The mist vanished. Poof, gone.

I was on my way to the University of Waterloo to meet Chris Eliasmith and his smart machine. From what I could make of it, Eliasmith had found a way to mimic the characteristic behaviors of neurons in order to generate a whole new form of artificial intelligence. He called his *in silico* Frankenstein Spaun.

Eliasmith is a man of many disciplines: philosophy; systems design engineering; computer science. He is also Director of Waterloo's Centre for Theoretical Neuroscience and holds a Canada Research Chair in same. He runs a lab within the Centre called the Computational Neuroscience Research Group. Eliasmith and his colleagues had published a recent article in *Science* describing Spaun's astonishing capabilities.[179] Their display of a Spaun-directed "bot" had been mobbed at the 2013 meeting of the AAAS in Boston. In 2014 he would be elected to the Royal Society College of New Scholars, and in 2015 he would be awarded the Natural Sciences and Engineering Research Council of Canada's (NSERC) prestigious Polanyi Prize for the year's most outstanding scientific advance. Even the relentlessly anti-science government of Stephen Harper had claimed credit for funding Spaun in a recent federal Budget.

I had read and re-read the *Science* piece trying to understand what all the hoo-hah was about. Spaun was said to model an eye, a brain, and an arm in order to be able to copy numbers, letters, and words and to answer questions. It could even do a human IQ quiz. But Spaun is not a robot with a body. It's a simulation running in a supercomputer.

So how could it be as intelligent as claimed, I wondered? If there was one thing I thought I'd learned during many months chasing after the what and the who and the how of intelligence, it is that smarts arise—so mysteriously—from particular bodies enmeshed in the struggle to continue. It followed that for any artificial intelligence to be intelligent in a human way, it would have to have a human-like body. Intelligence, Harvey had said, is embodied action.

Nevertheless, Good Old Fashioned Artificial Intelligence kept leading the news. Google had made a big fuss about how its cluster of 1000 parallel computers taught itself to recognize faces from 20,000 unlabeled images downloaded from the Internet (cat faces got a big mention).[180] It was only 15.8 percent accurate, yet this had been touted as a major advance. I thought it was pretty pathetic after fifty years of effort. I was even arrogant enough to think that I understood why they were failing. No body!

Then Ray Kurzweil, the inventor and futurist, announced that Google's founders, Larry Page and Sergey Brin, had hired him as director of Google engineering. Kurzweil had just published a book called *How to Create a Mind: The Secret of Human Thought Revealed.* He said he was going to lead a Google project to generate a new artificial intelligence modeled on a reverse-engineered human brain. Google signaled that this brain project had been part of the company's master plan from the earliest days.

So, I thought, they will have to get a robot to embody this brain. Yet they didn't mention that in their press releases. I got in touch with a PR person at Google and asked to interview Kurzweil (though Inman Harvey had advised that the real deal at Google was Peter Norvig, Google's Director of Research).

The Google PR person promptly offered me an interview with someone else instead. When I said, no, I need to talk to Mr. Kurzweil, she had asked what *exactly* I was interested in. I told her I wondered what sort of robot body Google would use to house this new 'brain' they were developing so that it could learn by interacting with the world. I was told I would get an answer, but it might take a while to get things set up. I waited patiently. But no Kurzweil interview ensued. However, a few months later Google announced its purchase of a leading US robotics company, and then another, and another. By January 2014, the number of robotics companies bought by Google had risen to eight, including Boston Dynamics, makers of Big Dog featured on DARPA's site. This caused one person commenting on the tech website Gizmodo to ask if only he found it "scary" that one company was becoming an oligopoly of robo power and robo patents.[181]It seemed to confirm that Google also believes autonomous bodies struggling in the world are the *sine qua non* for generating an artificial intelligence as smart as we are.

And yet Eliasmith had made a very smart machine with no body involved at all—or at least not much of one. He claimed that Spaun was not only able to learn and remember, but also able to predict what might come next using pattern recognition and inferential logic. How could a bodyless Spaun work better than anything else out there?

The more I tried to wrap my head around Eliasmith's *Science* article, the more it felt like I was trying to fry a ball of water and the less confident I became that the phrase 'embodied action' defines intelligence any more than words and phrases like 'emergent,'or 'adaptive behavior' do the trick.

My questions had begun to multiply again like tornados in hot weather. Is intelligence the same no matter how it's expressed, or is it particular to each body? Is there a fundamental difference between human mathematics and *C. elegans*' computations? Is the slime mold's ability to anticipate the regular return of a zap of current the same as the orangutan's ability to forecast when a fruit will be in season? Could humans really be getting smarter, as suggested by rising IQ test results worldwide, or are we just getting better at doing IQ tests, as our modern analytic culture spreads at a startling rate (as suggested by New Zealand IQ scholar James R. Flynn)? Is the amazing stability and complexity of Gaia—so much greater than the sum of the interactions of all bodies—a towering intellectual achievement?

As a simulation of eyes, neurons, and muscles, Spaun seemed to crouch on a gray bridge linking computation and behavior, the actual with the virtual. Using their own software, called Nengo, Eliasmith and his colleagues "instantiated" [meaning they created examples of] the individual characteristics of 2.5 million neural cells, arranged in brain-like regions and linked by networks to each other, to a sensory-like visual input, and to a physical arm output. They describe this as a complex, distributed, *analog* computation system.

The *Science* paper pointed out that real neurons in real brains are not identical. There are many types, and within types each one displays particular characteristics. Their individual behavior depends on their speed, on their connections, and on their placement in the brain's many layers of tissues, and its various regions. Real neurons don't use a simple on/off code like digital computers, they react to *gradients* in electrical and molecular signals, generating characteristic electrical spikes to move information along. Some spiking reactions take twenty milliseconds, some take up to 200 milliseconds.

This natural variation in individual neurons' behaviors has been incorporated into Spaun. The parameters of individual 'cells' can be randomly modified by the Nengo software. As a result, Eliasmith and colleagues can generate many individualized versions of Spaun, each slightly different from any other. Each individual version of the model can then be compared to all the others, just as psychologists test the intelligence of individuals against populations of people like them.

And so: using fewer than .01 percent of the actual number of neurons of a real human brain, Spaun is able to do eight human-like thinking operations, from copying a shape it has never 'seen' before (a test Google and others use as a security measure to distinguish humans from net bots), to answering questions from a Raven's Progressive Matrices IQ test for "fluid" intelligence. Spaun is not pre-programmed to deal with variations in these challenges but is flexible enough to handle them all.

The Raven's test requires significant working memory and the ability to recognize and compare visual patterns. The Raven's test-taker has to infer from a sequence of related patterns the visual logic that determines what a missing pattern in that sequence should look like. My mother would not score well now on the Raven's (although she can still beat me at bridge.)

I remember doing what must have been a Raven's as part of the selection process for the gifted class. I did it very quickly. When I was young, as soon as I saw a pattern I seemed to inhale its salient features without consciously looking for them, and I could easily manipulate them in my head through change after change. I loved doing physical puzzles and I loved doing mazes. When I became a journalist, this became more than fun. Being able to sort relevant, meaningful features quickly from a mass of information, and being able to see the same patterns repeating themselves in various situations, are among a journalist's most important skills. But that capacity diminishes with time, at least it has with me. These days, I *think* I see relationships and then I second guess myself to be sure. I carefully tell myself the story of the pattern to be certain it's really there. Does that caution make me more or less intelligent than I was thirty years ago? It depends who you ask but God knows what I'd score on a Raven's now.

The *Science* article claimed Spaun had performed at just below the average adult human level on a portion of a Raven's test. Eliasmith and colleagues also claimed that the aim of their research was to explain "the rapid flexibility of biological systems." So Spaun was either an incredible leap toward truly intelligent machines, or....

I had questions.

The University of Waterloo is on the edge of the Ontario city of the same name, its dark brick buildings utterly graceless amid pretty wooded hills. I parked the car and found Eliasmith in his office in the humanities building. I was grateful he'd found the time to see me because Eliasmith had been extremely busy for months. He had been hired as one of the presenters for a television series called *Brave New World With Stephen Hawking*, so he had been flying around the US doing stand-ups on camera. He had also been reading grant applications in a field called neuromorphics (building computer hardware that works like a brain) for the EU. He had also just done a TEDx talk.

There is a photo of him on his website: I had thought it must be an old shot because he looks only about thirty in it. Yet this man in front of me looked pretty much the same: he had the same shoulder length sort-of- blond- sort-of-curly hair and the same expression—an odd blend of puzzlement and confidence. He was dressed like a musician in a black t-shirt embellished with two white guitars. He didn't look like a professor of anything.

How old *are* you, I wanted to ask.

Copies of his newly published book had been set out in a stack on the table in front of me. It was titled called *How to Build a Brain: A Neural Architecture for Biological Cognition.* The cover illustration showed a human brain with little workmen crawling all over it. In this, his second book, he lays out the basics of Spaun's structure—which he calls Semantic Pointer Architecture—using his Neural Engineering Framework.

As I settled in, he was talking about the film shoot he'd just done. It was in a warehouse where robots fetch and carry things. These robots are smart enough that they will make most warehouse workers of the world obsolete in the near future, which is why Amazon just bought the company for "three-quarters of a billion dollars." There are five presenters on the show. "I'm the only neuroscientist, there's a biologist, a physicist... the other hosts are hair and teeth people, they have agents," he said, dismissively.

At first, I had no idea what he meant by hair and teeth people.

He gestured at his hair and his teeth. Then I could see exactly why the producers had grabbed him. How many neuroscientists look like a front man for a band? He was perfect for TV.

"They found me on a Ted talk," he said.

So tell me about yourself, I said.

Born in Ottawa, he said, Dad had been a pilot in the Canadian airforce, and Mom a housewife and teacher. He had an older brother, an engineer, who now lived only a few blocks from him in Waterloo. The whole family moved every few years from military posting to military posting. The last one was to Germany. He'd lived in Moose Jaw, Toronto, Winnipeg, Ottawa again, then back to Winnipeg. By the time his parents moved to Germany, he was in university, which meant he had a place to crash when he traveled around Europe.

Were you in a gifted class at school, I asked.

He shuffled his feet and shifted in his chair. He said his mother had read to him when he was a kid, maybe too much, because he had been slow to read. Nevertheless, when he was in grade three, in Winnipeg, he was selected for a gifted class. Oddly, when he moved to Toronto in grade six, the tests said he wasn't smart enough to make the cut. And yet when he moved to Ottawa, he was again put in an enriched program. When he moved back to Winnipeg to finish high school, he was also placed in a gifted program. Though he did science fairs, like his brother, he hadn't thought about being a scientist. Actually, he said, as a kid, he'd wanted to be an accountant.

Why, I asked.

Because I liked numbers, he told me.

His brother was already at Waterloo doing mechanical engineering when Eliasmith graduated from high school in 1989 and was admitted to University of Waterloo, as well. His brother told him to study systems design engineering because "he knew I didn't know what I wanted to do," and "systems is the arts of engineering," perfect for engineering students "who don't want to commit."

Engineering was his *only* choice, at least so far as his parents were concerned. It guaranteed a good job. But he had other interests. He was musical. He had played classical piano for ten years. He was in an alternative band, had been in a band all the way through high school, and still has one with his grad students. He had done a tape, he'd got it on the local radio. He wasn't the kind of student who liked doing work for work's sake, but when he arrived fresh from high school in Winnipeg (which ends at grade twelve) and had one week to catch up on Ontario's grade thirteen math curriculum, he'd managed it well enough. He also signed on for a lot of extra courses that had nothing to do with engineering, courses in psychology and philosophy. By his third year, he found psychology so fascinating he wanted to switch. But his parents and brother said "finish that engineering degree, you can get work-study jobs to pay for school, and in the end it guarantees a job in *something.*"

"And it was probably the best decision I ever made to stay in engineering," he said. He did get jobs, one coding switches for Bell Northern, another working with standards and safety for a management consulting firm. But they were both boring, while philosophy and psychology were not boring at all.

"In your twenties you tend to introspect a lot," he explained. And he thought maybe he was like his grandfather who used to sit in the mall watching people go by, just to see what makes them tick. Psychology had tools for exploring the self and the behavior of others. And there even seemed to be an overlap between computer science technology and psychology. The technology that interested him the most was artificial neural networks.

"They were just coming into engineering. But I decided they weren't how the brain works."

By then, Eliasmith knew that is what he wanted to explore—how the brain works, specifically the human brain.

He graduated in 1994. "I decided I had done the practical degree and could do what I wanted, which was cognitive science." He decided to do his MA in philosophy with Paul Thagard.

Thagard (a winner of a 2013 Killam Prize) had been working for years on the philosophy and science of mind at University of Waterloo. Under Thagard's direction Eliasmith immersed himself in the philosophical debates and questions regarding mind versus brain versus consciousness. Is the mind an emergent product of any embodied system? Or does the human biological brain do something very special and irreducible because its neurons somehow make representations of the world? Eliasmith was interested in the eighteenth century arguments between the rationalists (Descartes, Spinoza, and Leibniz) and the empiricists (Hobbes, Locke, Berkeley and especially Hume). He wrote a paper for another professor on the relationship between matter in the head, and the soul, exploring what happens when a brain is damaged and a change in character—"soul"— results.

The 1990s were a time of ferment for anybody interested in either artificial intelligence or what actually goes on in the human brain. The dynamic systems/evolutionary robotocists like Floreano and Inman Harvey may have thought they'd dispensed with philosophical ideas about representations and made the hard problem of consciousness go poof, but Eliasmith didn't agree. He thought representations were of central importance. How to explain language, or visual or auditory symbols, without them? Those who worked with dynamic systems focused on behavior, not on how symbols attach to things, how the word rock comes to stand for rockness, but surely that process is what makes the human brain so very interesting.

"In my thesis I argued that dynamic systems were not new as tools and the way they were used was bad. The notions of representation and computation that they got rid of lost them the ability to explain things like language…."

Eliasmith wanted to know about meaning, about how the brain creates concepts which can be manipulated and then shared with others. He could see that those who thought they could build an artificial intelligence from layers of neural networks were having trouble with basic things like language translation. Human languages are loose and fuzzy. Words take on subtly different meanings depending on context, tone, and timing. How to capture that fuzziness in an algorithm?

Eliasmith's first publication, in 1996, a critique of the ideas behind the dynamic systems approach to intelligence, was based on his understanding of the mathematics sustaining the theory. "I knew what the claims were and whether they made sense." And he didn't think they did.

But what about evolutionary robotics, I asked. It makes use of that point of view and it works. I was thinking: what would Inman Harvey make of this guy?

Eliasmith saw evolutionary robotics as "off on a side channel" from the main debate. "To work, it has to have a god," is how he put it, meaning a clever programmer managing the selection and/or critical design decisions.

Yes, he knew Floreano's work, and he knew Brooks' Roomba and his new robot Baxter (which can learn to do factory jobs). But way too much human decision-making went on in the design of their machines. And more to the point, the method only produces an insect-like intelligence. He believed it would take forever to get to autonomous, human-like machines if people kept marching down the dynamic systems path.

But the human form of intelligence is not always the target, I said, thinking of Adamatzky and the slime mold.

"Brooks saw the human as the target," he said. "A huge number see it as the target."

To get a scholarship to do a PhD one must apply for it when starting the MA, and Eliasmith hadn't done that. He had a year to kill after his MA before he could hope to get into a doctoral program. Paul Thagard challenged him to find a project that he could do while he was waiting. "It had to be a cool project," Eliasmith said. He decided he should try to build "a model of analogy-making in humans using distributed representations."

I tried to parse that sentence in my head. I thought he meant he wanted to build a model mimicking how the brain forms the idea of dog, which can then be applied to all doglike creatures. How do we distill dogness from creatures as varied in size and shape as a teensy chihuahua and a great dane?

We carry models of things in our heads that tell us if an "input comes in and it acts in a certain way, yes, it's a dog," he said. So Eliasmith asked: where in the brain does this analogy-making occur and how does it happen? How can a bunch of neurons in different regions activate to make analogies? He focused on what kinds of representations and computations could be done to mimic 'brain-like' analogies.

This was a novel idea. No other cognitive scientists were working on analogy using brain-like representations when he started. Some neuroscientists were recording single cells firing in animal brains as they performed a specific behavior. Others were looking at the variability in neurons' spiking behaviors and trying to develop computer models that reflected "biological reality." And "there were theorists trying to understand information processing in a single or a small population of cells," he said. But nobody was bringing these disciplines together. "They were worried about seeing a single visual edge, not the whole concept—like dog," he said. And some are still stuck on that, I thought, remembering several presentations at the Comparative Cognition meeting in Florida.

As he worked on the project, he also looked for a doctoral program where he could combine his interest in mind with cognitive science. He signed up for a program in philosophy run by Bill Bechtel, then at Washington University in St. Louis, Missouri (now at University of California, San Diego).

Soon after his arrival in St. Louis, Eliasmith gave a talk about the project he'd been working on with Paul Thagard. A man in the audience asked Eliasmith if he knew about the course he offered at Washington University in computational neuroscience. He suggested Eliasmith should come to his lectures.

"He said 'you can find out how the brain works'.... Charlie Anderson became the most influential person in my research career."

Charles H. Anderson had had two big careers before he got to Washington University. As an experimental physicist, he had worked for twenty-five years at RCA developing flat screen TVs and had become fascinated with how vision works. Then he'd joined the Jet Propulsion Laboratory in California and worked on a sensing and perception system for its Mars Rover. That's where Anderson met David Van Essen who had appointments in anatomical neurobiology and physics at Washington University. Van Essen had been mapping brain activity for years. Anderson began to work with Van Essen on the visual system of macaques. The result was a vision-neuroscience lab at Washington University.

"Most of Charlie's course was on vision, how neurons are arrayed, their visual sensitivities," said Eliasmith, all of which Eliasmith needed to know. He learned in that course that there were were equations available to describe how neurons work, and a statistical framework for understanding neural computation.

Later, Eliasmith put these skills and methods to use in his doctoral thesis called "How Neurons Mean: A Neurocomputational Theory of Representational Content." It was a philosophy doctorate: it was not about the underlying mathematics. In fact, he hid the math but used the ideas it represented to talk about the how and the where of representations in the brain. He coined a phrase for the process: neurosemantics.

Eliasmith worked every summer with Anderson in Van Essen's lab. That led to his first book, *Neural Engineering: Computation, Representation, and Dynamics* which he co-wrote there with Anderson while a postdoc, though he had already been given a position at Waterloo even before his doctorate was complete. The book was published in 2003.[182]

Writing a book as a postdoc is unusual for philosophers, let alone neuroscientists. People he met for the first time at conferences kept expecting him to be older than he was. But getting a much more senior scientist to let his name come second on the title page is rarer still. "Charlie wasn't a big fan of writing and didn't worry about publishing," Eliasmith explained, though Anderson thought it was important to get the ideas out there for others to use. "We were generalizing these simple models we built using three principles, the same principles used to build Spaun. We call them the neural engineering framework."

So just to be clear about how Spaun works, I said. It's a digital representation of an analogue system.[183] Right?

He didn't say yes and he didn't say no. "We never think of our project as digital, but instead we describe analog systems as they are in the brain. What we want is a mathematical description of how neural systems function—many layers do interesting information processing changing energy into useful forms. When you write mathematical descriptions down, you can simulate them in a computer—digital or analog."

This answer was no clearer to me than the *Science* article. I had been confused. I was still confused.

In the 1980s, at around the same time as parallel processing was being invented by Daniel Hillis, Carver Mead, a visionary American engineer and computer scientist, created something called a neuromorphic silicon chip—an analog chip which translates the way neurons signal each other. Neuronal signals are not on/off. Their gradients convey information. As Mead put it on his own website, from the 1970s until 2000 he worked "to bring about a general awarness of computation as a physical process rather than a purely mathematical one." His first neuromorphic chip was a "physically based computation inspired by the neural paradigm...visual and auditory computations of enormous complexity can be carried out in a minimal area and with minute energy dissipation compared with digital implementations."

Yet Kurzweil, the man hired to reverse engineer the human brain for Google, had long argued that it would be better to emulate neurons in a digital, massively distributed system because digital systems work at the speed of light, not the much slower speed of life.[184] Kurzweil wants to create an artificial intelligence that feels, emotes and thinks *better* than humans do and he's in a hurry. At the end of his last book he defined intelligence as "the ability to solve problems with limited resources, in which a key such resource is time..."

Eliasmith could see that I was confused so he tried to explain.

"Say the brain's function is a trajectory through a high dimensional sphere but you can understand it by looking at the path projected onto slices, reducing the dimensions. So, you can give a description in lower dimensions. Giving these descriptions captures the majority of the interesting behavior...you can then map this back to the full dimensional description.... Finding these mappings gives you a deeper understanding."

I thought he was talking about a method that reduces complexity and then restores it.

Try again, I said to Eliasmith.

An animal like a goldfish has forty neurons to control the horizontal movements of its eyes, Eliasmith said, looking at me as if he couldn't quite believe I didn't get it. "You can look at the behavior and record the neurons. So there are forty neurons—a forty-dimensional space—related to eyes moving left and right—a one dimensional space. We can describe the behavior in one dimension and see how forty neurons can do that. We can test our ideas by damaging the neurons and see what happens in the real fish and in the model." And eventually, he continued, you end up with many examples, understanding constrained bits of behavior.

"So we developed a method to apply to all bits of behavior. And that's where the first book stops. It let us build more complex models than before...small, respectable models, but models from all different parts of the brain."

The *Science* article had described how many dimensions are pushed down to two, as had an article he published in 2006 about working memory in monkeys. I had read the words but simply failed to grasp their meaning. As the 2006 paper made clear, earlier models of the brain's behavior didn't properly reflect the all-over-the-map way in which individual neurons fire because they assumed neurons were all one-dimensional. Yet their variability is how they get their jobs done. It is as important to capture their specificity, and the time each takes to fire, as it is to know how they are physically connected to each other. As Eliasmith put it in the 2006 paper, the heterogeneous way working memory neurons behave *over time* can be modeled best in two dimensions (think of a graph in which a vertical line meets a horizontal line), not one. This gives a model in which "all the categories of responses…can be captured."[185]

Spaun ties individual simulated neurons with their individual behaviors into simulated groups of neurons organized in networks. Signals move through these networks which are organized into three compression hierarchies, an action-selection mechanism, and five other subsystems. The model employs [square brackets mine, parentheses Eliasmith's]

> "…spiking neurons that implement neural representations that we call semantic pointers using various firing patterns. Semantic pointers can be understood as being elements of a compressed neural vector space. Compression is a natural way to understand much of neural processing. For instance, the number of cells in the visual hierarchy gradually decreases from the primary visual cortex [where visual signals are received from the eye] to the inferior temporal cortex [where they are finally processed] meaning that the information has been compressed from a higher-dimensional (image based) space into a lower-dimensional (feature) space. This same kind of operation maps well to the

> motor hierarchy where lower-dimensional firing patterns are successively decompressed (for example, when a lower-dimensional motor represention in Euclidean [flat, two-dimensional] space moves down the motor hierarchy to higher-dimensional muscle space.)"
>
> Compression is functionally important because low-dimensional representations can be more efficiently manipulated for a variety of neural computations.... The model's functional states are not hard-wired...."[186]

As you can imagine, it took me a *long* time to digest that. Eventually, I came to see that Spaun, which has only a minimal body, embodies several thousand years of philosophy and mathematics, plus a whole lot of contemporary work in biology, neurology, and computer science. Plato famously argued that there is a true world of unchanging Ideas which underlie our everchanging versions of reality. We humans can only apprehend these immutable Ideas by their dim reflections—like shadows on a cave wall. Eliasmith's successful model suggested Plato got something almost right: our sensory systems only take in reduced versions of the real world, and successively squash even that information down to the neural equivalent of gesture drawings which stand in for the real thing. These compressions must then be decompressed—reversed—to provoke and guide action in much more complex three-dimensional reality.

But surely this also meant that Spaun is a dynamic system. While Eliasmith may have dissed dynamic systems theory in his MA thesis, it seemed to me that he had become a convert and put it to work in a new way. And he agreed: "I love dynamics, I just don't like getting rid of computation and representation and I care about implementation."

Eliasmith's assistant professorship in philosophy at Waterloo became a tenure track position after he completed his PhD. "I got a good job after all," he laughed.

But he found himself in a bit of a crisis. He wanted to go beyond what he'd done with vision in St. Louis and model the behaviors of the rest of the brain. So how was he going to attract students? He wasn't sure if he should tell prospective students what *he* wanted to do, as the big European professors do, or ask what *they* wanted to do. He decided to ask. "I get students by asking: what part of the brain would you like to model?"

This turned out to be a great strategy. His students just naturally chose to work on different brain areas which could then be integrated into a larger whole.

His first Masters student got his degree in 2005. His first PhD. student, Bryan Tripp, defended his thesis in 2008. "Tripp was interested in how basal ganglia work," he said. The basal ganglia region of the brain distributes signals throughout the motor system as a whole, which makes it critical to movement. "He wanted to understand how simple actions combine to complex action," Eliasmith said.

Tripp's background was in kinesiology (the science of movement). He'd also studied biomedical engineering at University of Toronto, and then worked for five more years on computer programming and engineering systems. Yet Tripp's thesis turned out to be more theoretical than Eliasmith had expected.

"We didn't get a basal ganglia model by the end of the Ph.D. But we got other fantastic insights on how information could be encoded and processed." As Tripp went off to do a postdoc at McGill, Eliasmith's first postdoc came on board. Terrence Stewart "developed the basal ganglia model we use in Spaun," he said.

Stewart had done a systems design engineering degree at Waterloo. Then he'd studied cognitive science at Carleton in Ottawa. "So he came here to work with me and Paul Thagard, maybe in 2007…."

Student by student, structure by structure, the lab as a whole got closer to a larger integrated simulation. Stewart's basal ganglia model was critical to making the other models work as an integrated whole. The basal ganglia model controls the order of Spaun's operations, changing the information flow between subsystems.

"I had a student working on working memory, and another who did vision who went to Geoffrey Hinton's lab, one who was doing motor control, another learning basal ganglia, and a student—Dan Rasmussen—who was working on intelligence tests. Cognition, human-like thinking, is his goal and he figured the best place to get cognitive data in humans is intelligence testing."

I found myself thinking: why on earth would you care about human intelligence tests? You're the guy who was declared very smart in grade three, not so smart in grade six, pretty smart in junior high, and then very smart in high school—all on the basis of IQ tests.

I cocked an impatient eyebrow and said I didn't think much of IQ testing.

"*He* was interested in that," he said, a bit defensively. "I had built a model of inductive reasoning...."

Eliasmith's model infers the unstated rules connecting a series of related instances, and then applies those rules. This is known as inductive reasoning. The *Science* paper had mentioned that Spaun could perform part of a Raven's IQ test almost as well as the average adult human, something no other brain model can do. And most of the other brain models available can not do the other seven tasks Spaun has mastered either. Spaun can: copy a drawing; recognize an image, meaning it can write a number in Spaun's own default writing style that is presented to it in another style altogether; choose which of three possible choices will produce the greatest reward; reproduce a list of any length; count or add; answer questions about what is in a specific position in a list, or, if given a number, declare the position that number should have in a list; complete a complex pattern after being presented with a few examples. Most of these things require the capacity to infer something beyond what is given.

"Spaun...performs tasks in any order without modeler intervention," they wrote. Meaning, there is no selecting god in the Spaun machine, no human programmer telling it what to do, when.

What they didn't say in *Science* is that they had also carried out more complex tests of Spaun's reasoning power. They had also measured the performance of a version of the model on the Raven's Progressive Matrices and compared it to the performance of an average group of undergraduate university students. This model's performance was similar to that of the students; so were its errors.

They also used this model to connect behavior to neurophysiology. When they manipulated Spaun's parameters to test two different theories of the causes of intellectual decline through aging—neuron loss, and "representational de-differentiation"—Spaun's results matched those of aging humans.

This paper had been sent to a journal, he said.

Long pause.

And, I asked.

Like many other papers he and his students have written, they had been getting flack from reviewers, he said.

Why?

"One reviewer thinks the work on intelligence is hard to discern because of the long explanation of how the model is built. Another thinks there isn't enough detail about the model."

But that was only part of the problem. The bigger issue is there's nothing out there to compare it to. "Which makes it hard to publish," he said. "It's a lot of work for reviewers to understand. And after they read through it there are still basic misunderstandings."[187]

That puzzled look had come back again, but this time he also looked slightly anxious, as if he worried that maybe these problems were somehow his fault. I thought, all things considered, he'd done pretty well at getting his paradigm-cracking ideas into the world. It's not every day a lab gets to publish an article in *Science.* But underneath the puzzlement and the anxiety there was also some other emotion he was trying to supress. Anger? Yes, he was mad.

He began to tell stories. He and his students had decided early on that their software should be made freely available to anyone who wants it—anyone who at least isn't a company. They could have patented it, they could have tried to get rich from it, but they had taken a vote as a group and decided to take the high road: it would have so many useful applications they thought it should belong to everybody. Yet others seemed to assume the worst about their ethics and motives. For example, he'd been asked to participate in a panel discussion at a cognitive science meeting in Seattle. He'd presented early results on Spaun's Raven's test performance.

"One panelist says, 'just a minute, how do you tell it what the rules are.' Eventually the panelist was more explicit. He said, 'Just a minute, where's the cheat?'"

The whole point of Spaun is that it infers rules for itself. It had taken Eliasmith many hard years to arrive at a working model of inductive reasoning. Yet the questioner assumed that all models are cheats, everybody knows it, so this one had to be a cheat too.

“I didn’t know what to say,” Eliasmith said.

And, though his group had been able to publish on the small bits of the model without difficulty, when they began to put the small models together into a single system, reviewers just balked. On any given paper one reviewer would ask, ‘why all this neuroscience?’ while another would say, ‘I’m a neuroscientist, and all the bits of what they’re doing, they’ve been done before, so why is it new, just because it’s all together?’

“Yes,” Eliasmith would answer. “There’s nothing on this scale that works.”

Just as Stefano Mancuso had grown frustrated over the glacial pace of peer-review in big journals, Eliasmith decided he had to put it all together in book form. That’s when he came up with the name for the whole system, semantic pointer architecture unified network, or Spaun. He thought if he explained the model in detail in a book, when he sent out papers in the future people would take the how of things on faith after reading the book.

Good luck on that, I muttered.

When he sent the paper to *Science* in 2012, eight months passed before it was accepted. One reviewer asked why they hadn’t published earlier in more high profile journals, as if that would allay fears that there was something wrong with this work that the reviewer just hadn’t figured out. Eliasmith’s answer was it wasn’t for lack of trying. So when it was finally published, "It was a huge breakthrough for us," he said.

Okay, I said, but does it really explain how neurons make meaning? Do you know how your simulated neurons infer a pattern?

"We do not have a perfect understanding of the function of the model," he said, carefully. "There were surprises."

So some behavior just *emerges?* I pressed, thinking, dammit, here we go again, a buzzword instead of analysis.

"Emergence is another way of saying surprise," he said. "The accumulation of parts becomes something new that's not the parts," he said. "Structure in the relationships gives rise to behavior that's novel, but we can go back and understand why....Call it emergence, it's fine by me."

Spaun's structure has flexibility built in. Each simulation configures itself for the right neural connections. Because the simulations capture particularities of behavior and structure, they also emulate the messiness of real, physically embodied action.

"The vast majority of ways of connecting neurons give you nothing, but some ways gives you a person. It's not enough to have the elements," said Eliasmith. This reminded me of Dean Falk who had said about the hobbit brain that it's not the size, it's the organization that matters. Or put another way, while anyone can learn to read musical notes quickly without touching a piano, playing the *Moonlight Sonata* well requires about eight years of serious practice.

Okay, I said, please show me.

We walked over to his lab. I knew it was ridiculous, but I was secretly hoping to see a long robotic arm going scratchy-scratchy as it drew an answer to a question. I had grown so attached to the idea of intelligence as embodied action, and I was reluctant to give it up.

We walked down a hall. A group of young men (why is it always young men?) were hard at work on their laptops in a room on the left. On the right, the door to a big seminar room stood open: a group of tables had been formed into a U. Another young man sat there alone, working on a laptop. He was slim as a willow, with fair hair hanging down his back, and granny glasses perched on his nose, like a refugee from the back-to-the-land movement circa 1970. He was introduced as Terrence Stewart, who modeled the basal ganglia.

And it turned out there was a body to watch after all. Stewart was putting what he called a Spazbot,[188] a small wheeled robot, through its paces. Using his laptop to send it wireless signals from a small portion of the Spaun computer model, Stewart was asking it to turn right or left. The Spazbot had to take in the signal and make its wheels and motor do the proper turn using only the model's spiking simulated neurons to tell it how.

The Spazbot rolled around and around, vroom, vroom. It was made from off-the-shelf Lego bricks, three pairs of wheels, an interface, and a motor. It also had a scoop-like attachment with a sensor, making it look like a cross between a dinky-toy car and crane. They use these robots to study the way signals flow back and forth between the simulated brain and muscle.

Eliasmith and his group want to model the way a human decides to lift her head and then does so, the way an animal does the same. In other words, they want to know how the ephemeral controls the meat. "We gotta do weird computations so we can use our methods on these problems," Stewart said.

The screen on his laptop showed the twitching rise and fall of little boxes along the bottom of a graph. These twitches reflect the individual spiking neurons of the Spaun simulation, signaling and receiving as the Spazbot moves. The Spazbot was making use of only about 1000 of Spaun's millions of 'neurons.'

"In a real brain," Stewart said, "you'd have ten to twenty thousand neurons connecting to serve a muscle. Each has a little current—a kind of summation happens. If lots of current comes in, you get a big muscle contraction."

He pointed at the graph. The bottom line represents what we want the robot to do, he explained (in this case, turn left). "This is reasonably similar to the representation in the motor cortex when you are moving your arm left or right."

"We need three million neurons to do cognition stuff in Spaun," he added.

The current computers running the Spaun model are way too slow. It takes two and a half hours for them to simulate one second of the full model's 'neuron' activity. Getting a higher number of simulated neurons to work together faster is the next goal: they want to get 25 million neurons modeled over the next five years and eventually a billion. They have been working with a group at Stanford University headed by Kwabena Boahen—one of Carver Mead's former students—to develop a neuromorphic chip much more powerful than anything available, but also much smaller. Scaling up to a model that has a thousand times more neurons than Spaun is a problem of size and power.

"A human brain has 100 billion neurons," said Stewart. (According to a Brazilian researcher, it may be only 85 billion. To come up with the new number, she turned four dead human males' brains into soup, took samples, counted the neuron nuclei in the samples, and then scaled up.[189]) It would take about half a gigawatt of power to run all their interactions in a standard digital computer at the same time, yet the human brain uses only twenty-three watts to do the same work. People in Zurich, the US and in Germany are trying to figure out how to reduce the power draw by several orders of magnitude, Stewart said.

Twenty-three watts? I said. That's what it takes to turn on a small lightbulb. It's a ridiculously small amount of energy, considering what brains do with it.[190]

That's it, said Stewart.

And, said Eliasmith, who had taken up a watchful post beside me, "it opens up being mobile too." His colleagues at Stanford wanted to make chips so small they could be inserted into a human brain, but the power draw would have to be low enough, and the chip would have to be shielded well enough, so as not to disrupt the brain's operations.

Until that moment I'd been thinking of this work as a mock brain reduced to a chip: it hadn't occurred to me that one could put such a mock brain *into* a *real* brain. But why couldn't one augment the thinking capacity of a real brain? Add to its memory? Change its mood? Make it much better at drawing inferences? Kurzweil had been writing about that sort of possibility for years, so I shouldn't have been surprised, yet the audacity of this idea left me gasping. And why couldn't it also run any other kind of movable machine?

The possibilities were practically limitless. I could almost see them. At this point, my heart rate kicked up. My hands were sweaty and my mouth had turned dry. A psychologist would have taken one look at me and said I was experiencing some kind of terror, or some kind of thrill.

They began to talk about the way they'd shifted Spaun's parameters to model things going wrong in an aging human brain. I was taking notes, but my own brain had shifted into some other mode well beyond mere listening.

I felt as if I was trying to drink water from a fire hose—what they had done, what they were doing, was overwhelming. The journalist in me knew immediately that their dementia work alone could be vitally important. Dementia will afflict a large proportion of the aging population in developed countries over the next thirty years. What if all that misery could be slowed down or stopped altogether with a silicon chip? Another part of me was thinking about my mother who feels so crushed by the indignities an aging brain has unleashed upon her. *I'm so stupid*! she says when she can't remember that she just asked the same question a moment before, a question she wouldn't have asked when she was her old self and exactly the opposite of stupid, before her brain/body began to fail her. And what about my brilliant dad who, before he died, made me fly home, ostensibly to make sure he had enough money to meet his bills, but really so that I would take note of the changes going on with his wife, my mother, and understand the rate of her decline without him saying out loud that his beautiful wife of so many years was no longer someone he knew. And then I remembered the class action lawsuits that were moving through Ontario's courts to get justice for those with IQs that had been deemed insufficient by government authorities. They had been locked away as children in special residential schools where they had endured beatings, filth, disease, and rape, all under the Government of Ontario's tender care. What if a small chip had been available to help them do better on IQ tests? What if, in the future, getting smarter became just as easy as getting a pair of glasses?

What if a tiny machine could give me back what is already slipping from me, that instant pattern recognition of my youth?

I fell abruptly in love with this work. Oh, I feared it too. A little voice muttered warnings about Hidden Helpers taking over.

Shut up, I said to myself. And just like that, my reportorial objectivity vanished, like the mist on the Niagara Escarpment. Poof, gone.

How much money do you need to get this going as fast as you can, I asked.

I was thinking: who can I call?

Then I remembered: my husband Stephen had contacts in Silicon Valley, people who back things like this.

Then I thought: why should Silicon Valley get this? These are Canadians, this is a Canadian innovation, the prime minister already claimed it, Canadians should make it fly.

The two of them looked at each other. They were reticent to talk about money, though money is a *sine qua non* of science and everybody knows it.

Giving away the software is going to cost you, I said. Money likes patents.

But, they answered, they had applied for patents on the architecture, to keep others from stealing it or hoarding it, and they were still going to make businesses pay to use it. One member of their group is a serial entrepreneur who came back to school to get an MA. He was helping them to think through the problems of finding investors.

How much do you need, I asked again.

Stewart said about $10 million over five years would be good.

And yes, they had already checked out obvious places, like Google.

"I talked to Google before I submitted it to *Science,*" admitted Eliasmith. He had sent his proposal to Peter Norvig, who sent it on to the Google brain/mind people. "I was asked to come out...nothing came of it."

"A lot of their focus is pattern recognition," said Terry.

"I thought they would be interested in this—it does stuff they don't have yet...." said Eliasmith.

But Google's work would go in the opposite direction if they followed Kurzweil's belief in the virtues of digital models, I thought. And I figured that even if Google had been interested, the company wouldn't like it that the software was already freely available to academics and hackers, as Elisamith put it. Google barricades its research projects in an out-of-sight skunkworks. It makes its money by selling everything known about each of us and what we do on its many networks to advertisers, while keeping its own secrets as long as possible.

How much did the government give you? I asked.

"We must be politic," said Stewart, looking alarmed.

"Our research was in our most recent federal budget. One of their examples was our team from Waterloo," said Eliasmith.

But there were no details given, just the mention, I said. So how much did they give you?

And now another look, not a happy one, passed over Eliasmith's face. Spaun has never been funded directly. Eliasmith is the recipient of $30,000 a year for five years Discovery grant which he won in a competition administered by NSERC. He doesn't use that money for Spaun. He uses it to get his graduate students around to conferences. Spaun gets money indirectly from a grant from the US Office of Naval Research which supports the Stanford group's development of the neuromorphic chip and software.

Did you try DARPA? I asked. These words left my mouth without volition on my part. I wanted to shout at myself: what are you saying, why would you want DARPA in on this?

"We did apply once to DARPA—we got this ONR grant…it helps the guys at Stanford build the chip and us build the software—building the chip is the expensive part. It's about $1 million over five years. I have two postdocs and ten grad students…two of them are unfunded. The way most money comes in is with students who win their own NSERC grad student funding. Except for two I don't have funding for," said Eliasmith.

Besides, he said, there are problems with DARPA. It's not just that it's military, but it's the paperwork, they constantly review progress towards specific goals and what makes his lab work is bottom-up free inquiry.

What about the European Union then, I asked, thinking if he had got money via Stanford from the US Office of Naval Research, why not?

No, he said, they don't fund non-Europeans. "They have tons of money for cognitive systems, I review for them."

So what would it take to go fast, I asked.

We'd need about twenty people, it depends on whether they are also grad students, said Eliasmith. It takes about two years to train each student to work on the methods.

If you were in the EU, what would you get?

About $3 million a year, he said.

By the time we'd walked back to Eliasmith's office, I was in a rage, a humming rage. The Stephen Harper Government had used these people as poster boys, but given them next to nothing. (Even the $250,000 Polanyi Prize would be a drop in the bucket compared to what is needed.) I've got to help, I was thinking. Goddammit, someone has to make this happen.

So did you try the Blackberry people? How about Mike Lazaridis? Lazaridis, a co-founder of Research in Motion which created the Blackberry, funded the Perimeter Institute to pursue important physics questions.

No, said Eliasmith. The deal at Waterloo is that only people high up in the fundraising department can pitch him and they don't think he'd be interested. His immediate problem concerned NSERC which funds MA students for only one year of a two-year program. Yet he has to guarantee full funding for any student he takes on. So how was he supposed to do that? The Feds had cut back on science funding....

And that's when I crossed the line completely.

"I know people who might be able to help," I said.

Okay, he said.

In the car, driving home, the sensible part of me commenced screaming: what are you doing? Are you senile? *Journalists don't get involved in the story, that's hugging the third rail.*

But the part me who once read pulp science fiction all night under the blanket with a flashlight to keep my mother at bay was far too busy drooling over future possibilities to take heed of such rules. There were so many ways in which this technology could change things, so many ways in which it could make the world so much better. A whole new narrative about what intelligence is, and how we might augment it— so that we might be repaired, even redeemed!—began to write itself in my head.

I was also thinking that these people in Waterloo had captured the essence of whatever-it-is that we call intelligence. It is not a little homunculus tucked away in each one of us humans, making some of us smarter than others. It is a process triggered by many interacting systems involving cells arranged in a particular way, smart cells interacting with each other and the world. Spaun transduces an organized array of signals, which turns sensation to thinking, thinking to acting. That some of their results had surprised them convinced me more than anything that they'd got it right, that they'd actually harnessed the-more-than-the-sum-of-its-parts-will-o-the-wisp that arises from living things chattering, remembering, forgetting, learning, and solving problems presented by wave after wave of experience.

If Eliasmith and his colleagues can get the help they need, I was thinking, Spaun will find its way into an unimaginable array of complex machines, and even into a complicated organic machine like me.

Turing would have loved it.

For a little while, I even forgot to fear it.

15

Of Living Machines

A sun-washed July afternoon, 2013.

I stood on the doorstep of my brother's house in Richmond, British Columbia, my luggage in hand, my rental car at the curb, waiting for him to answer the bell. He was going to put me up while I attended a meeting of the Society of Plant Signaling & Behavior at the University of British Columbia in Vancouver. I was writing about it for *Maclean's Magazine*. Behind me, the mountains hovered in the distance spreading right across the whole horizon, yet utterly ephemeral, like an illustration for a fairy tale or an objective correlative for the nature of intelligence.

One year had passed since I was first dazzled by the use of evolution as a design tool for smart machines, and only months since I fell in love with Spaun. Yet I was entranced no more. My return to objectivity (and prudence) could be blamed on a fellow named Bill Joy, or the wages of failure, or both. Failure has always been good to me: it usually drives me back to safer ground after doing something stupid. If I'd actually succeeded in finding investors for Spaun, ethics would have required me to leave it out of this narrative and you wouldn't be reading here about machines that make representations of the world and so can reason, the be-all and end-all for those who want to outdo human intelligence.

But first, I'd pitched an old friend over lunch. I didn't do it well. I sprang Spaun on him without warning, with the coffees. He'd looked at me with surprise and even pain on his face as if to say, how can you even *think* I would care about this, and what the *hell* do you think you're doing? "Try a hospital foundation or something," is what he said out loud.

Then I'd tried a very wealthy frenemy who cleverly evaded my pitch by failing to reply to emails and phone messages.

That's where Bill Joy came in.

Joy is a brilliant programmer who helped found a brilliant company, Sun Microsystems, and then became a venture capitalist trawling for disruptive inventions. I thought Spaun qualified. Joy had inquired once about a project of Stephen's, so I pressed Stephen to connect Chris Eliasmith with Bill Joy. Of course it came to nothing, which I should have expected.

Joy had warned against the advent of intelligent, autonomous machines in an article he wrote for *Wired* in April, 2000. Joy had pointed to the convergence of three technologies—genetic engineering, robotics, and nanotechnology—which could render humans utterly superfluous and/or extinct. He wrote this jeremiad after meeting Ray Kurzweil, the man now in charge of building the Google 'brain,' at a conference. He'd been sitting in the bar talking with John Searle, the philosopher of consciousness, when Kurzweil joined them.

Joy was intrigued enough by what Kurzweil had to say about humans fusing their minds with robots, and robots attaining consciousness, that he read a galley of Kurzweil's about-to-be published book, *The Age of Spiritual Machines.* It quoted Unabomber Ted Kaczynski's manifesto which argued that if such technologies were permitted to merge, either the machines would rule, or a tiny fraction of humankind would gain control, reducing the rest of humanity to the status of domestic animals. But Kurzweil had looked at the Unabomber's facts and seen Valhalla not dystopia. The convergence of these technologies could allow him to upload his mind/soul/consciousness into a robot and live forever.

At first, Joy thought both the Unabomber and Kurzweil dwelled in Sci-Fi La-La Land. But then he asked his physicist friends if this stuff might be for real. They told him that nano machines were pretty close to physical embodiment. He began to think about unintended consequences—because modern science and technologies have generated so many of them. He learned that nanotechnologists talked among themselves about the problem of self-replicating nano machines escaping into the world. Tough, bacteria-like nano machines could out-compete real bacteria for resources, for example, "and reduce the biosphere to dust in a matter of days," he wrote. "Among the cognoscenti this threat has become known as the gray goo problem."[191]

"We are on the cusp of the further perfection of extreme evil…" he wrote, "a surprising and terrible empowerment of extreme individuals."

When I re-read Joy's article, it was like a sharp, clarifying whiff of ammonia straight to the brain. Suddenly I could see that everyone I'd interviewed, almost everyone I'd read, had focused on the virtues of building intelligent machines by copying nature. No one had mentioned unintended consequences.

Which raised the question: what perfect evil could an extreme individual or group do if armed with Spaun?

My brother finally opened the door. Good to see you, mind the cat, sorry, gotta watch this, he said, and ran back to the family room where the Lac-Megantic derailment disaster, a tragic mash-up of unintended consequences, was burning up his television screen. Oil fracked from North Dakota's Bakken formation, much more volatile than other crude, had been accepted for shipping to a refinery in New Brunswick by a hard-pressed railroad company. The crude was carried in old-fashioned, weak-walled tankers not yet regulated out of existence. The train was manned by one person serving as engineer, conductor, and crew because Canadian railroad regulations had been relaxed to allow that. He stopped for the night at Nantes, Quebec, parked on a descending grade of the main track, set the train's air brakes but too few mechanical ones, and went to sleep in a hotel, though the train's main locomotive was smoking. The local fire department was called, the firemen put the fire out and shut down the main locomotive, which killed power to the air brakes. After the firemen went home, the train began to inch, then roll, then gather enough speed to derail on the ill-kept railbed in the center of Lac-Megantic, right by a bar/restaurant/patio where people were enjoying drinks under the summer stars.

The tankers broke open. The oil exploded. The center of the town was vaporized by raging fire. Officials could not say how many had died. It would take weeks to get that number —forty-seven.

My brother couldn't tear his eyes from the screen. He has been an aficianado of everything to do with transportation since he was a boy. Railroading is an old and well understood way to transport dangerous goods, yet out of nowhere, it had produced this holocaust. He wanted to understand who was responsible. The answer would be that many people had made poor decisions that accumulated into this horror.

Then the regular newscast came on. It featured the month-old Snowden affair, an unintended consequence of Turing's legacy. Now *I* couldn't tear my eyes from the screen. Turing and his colleagues helped save our democracies from tyranny by creating the first digital computer to break machine-encrypted signals. But Edward Snowden said our democracies now use these tools like tyrants: they surveil the entire world, starting with us. I have been fascinated by the way governments gather our private and personal information ever since I wrote a profile of a Solicitor General of Canada who was then reorganizing the Canadian intelligence system. I had finished my interviews with him and was checking in for a vacation flight when I was paged at Pearson International. The official at the counter said: "The Solicitor General wants to talk to you." But I hadn't told the Solicitor General or his staff that I was going anywhere, let alone when, or by what means.

The newscaster reported that Edward Snowden was still hunkered down in the international transit zone of the Moscow airport. He'd fled there from a hotel in Kowloon, Hong Kong, where he'd handed over a trove of stolen, top secret documents to selected journalists who'd then published a series of front page stories. Since Snowden outed himself on a British newspaper's website, a horde of journalists had been haunting the Moscow airport, trying to get a quote from him or even a glimpse. Still no luck.

It seemed to me that our Smart era could now be divided into sub-eras—before Snowden and after. Before Snowden, our smart machines and networks— our phones, iPads, tablets, computers, the Internet, search engines, Wi-Fi, social media—were indispensable means to share ideas and opinions (especially when organizing revolutions). After Snowden, they became traitors snitching us out to governments and corporate powers.

The Snowden affair had erupted in early June, 2013. *The Guardian* and the *Washington Post* had each carried articles showing how the National Security Agency (NSA) and the Anglo spook alliance called Five Eyes, hoover up information on everybody, everywhere, mostly without warrants, usually from the smart phones and other devices and software we all use every day.[192]

It was not news that some governments snoop. But this was supposed to be the specialty of the bad guys. The US and Canadian governments had been complaining for years that Chinese military signals intelligence groups and/or their criminal hacker friends were burrowing into important corporate, media, and governments' computers to steal intelligence.[193] The Citizen Lab at the Munk School for International Studies at the University of Toronto had caught them in the act. They had set up what they called a 'honey pot' computer (a variation on John le Carré's honey trap). They infected it with a malicious type of Trojan horse code that their target hackers had used on Tibetan government-in-exile computers: it had enabled the hackers to rifle through documents via a network the Lab dubbed the Ghostnet.[194] The Lab noted that this code permitted hackers to read every keystroke of an infected computer, and also to listen to and watch its user even when the computer was turned off. The Lab warned that this software 'tool' can be downloaded from the Internet by anyone who wants to be a do-it-yourself cyberspy. The Citizen Lab then traced the honey pot hackers to a Chinese government-owned computer on Hainan, an island that is home to the Lingshui signals intelligence facility and the Third Technical Department of China's Peoples Liberation Army. These revelations made the front page of the *New York Times* in 2009.

But these new articles based on Snowden's purloined documents made it clear that the West is as intrusive as the East. One story in *The Guardian*, for example, was based on a "secret Presidential directive, signed by President Obama...ordering the Pentagon and related agencies to prepare for a series of aggressive cyber operations around the world." Another, on the NSA's 'Boundless Informant' program, showed that US officials had lied to Congress when they claimed they had no means to quantify how many communications they were collecting.

These stories ran just before President Obama met the new president of China, for the first time, in California. Each story was more damaging than the last.

The first appeared in *The Guardian* under Glenn Greenwald's byline. Greenwald, an American, is a former constitutional and civil rights lawyer who had become a very well known blogger: he often wrote about the US government's lawful snooping on Americans which he regarded as unconstitutional. A few years earlier, he had moved to Brazil to be with his lover, David Miranda. Greenwald had recently been hired as a columnist/blogger by the British based-*Guardian's* New York outpost, an online operation.

Greenwald's story was based on a classified document. It was a secret order, issued by the FISA court, requiring Verizon, a major US telephone company, to hand over to the US National Security Agency all their subscriber information about all foreign and local calls on an ongoing basis for ninety days, starting April 25, 2013.

The secret FISA court was set up to protect ordinary Americans from warrantless intrusions into their private communications. These had been carried out without legal authority by the NSA, the FBI, and the CIA for many years—until exposed by the Watergate scandal and follow-on inquiries. The Senate's Church Committee inquiry led to the passage of the Foreign Intelligence Surveillance Act of 1978, because, as Senator Frank Church himself had put it, without safeguards, no American would have any privacy left "such is the capability to monitor everything....That is the abyss from which there is no return."

The Act created the FISA court to hear evidence about, and to deny or issue warrants for, the secret interception of Americans' communications with dangerous foreigners. Innocent American persons were supposed to be free of surveillance. But after 9/11/2001, during the presidency of George W. Bush, the FISA Act evolved. Amendments were passed which made it legal to snatch, without a warrant, any Americans' communications that began or ended in a foreign country. In addition, a change in the government's interpretation of the law, expressed by a secret presidential directive in 2002, meant that gathering the billing and contact information of all phone calls and emails— metadata—requires no individual warrants either. Metadata is not a record of the communications content, but it shows who you talk to, for what duration, through what means, at what time, and where you were when you made the call or sent the email.

Soon, secret warrantless demands by government agencies for metadata became so routine that the US government began to pay partner telecoms and Internet companies to provide them. But the telecoms and Internet providers became afraid that if customers found out about this handover of their information, they'd be sued. And they did find out in 2005 in a front page story published by James Risen and Eric Lichtblau in the *New York Times*. By 2008, the telecom and Internet service providers (ISPs) had lobbied for, and got, an amendment to the FISA which made it impossible for them to be sued for handing over metadata *under a court order.*

The FISA court, according to Greenwald, has denied only a handful of government requests for warrants. Its rejection rate is .03%.

In 2014, the Stephen Harper Government would tack similar provisions onto a Canadian cyberbullying bill. By then it would be known that the Canadian government's demands for such metadata had also become voracious. Canadian telecom/ISPs had been handing over the same sort of information in reponse to millions of government requests since 2006.[195] Was it legal? In June, 2014, the Supreme Court of Canada would rule *unanimously* that the handover of metadata by internet service providers without a warrant is an improper search forbidden by the Canadian Constitution.

Still, it was shocking that the court had ordered Verizon to turn over *all* records. Obviously, all of Verizon's customers are not plotting terror with foreigners.

Barton Gellman, a Pulitzer Prize-winning journalist and author, published the next story in *The Washington Post.*[196] Gellman shared byline credit with a woman named Laura Poitras, an American filmmaker and winner of a 2012 MacArthur Foundation Genius award, who was working on the third of a trilogy of films about government surveillance from her base in Berlin. Gellman and Poitras's *Post* story outlined an NSA intercept program codenamed Prism.

The Guardian's story on the same subject was posted on *The Guardian's* website only twenty minutes later.

Both stories were based on the same classified NSA PowerPoint provided by the same secret source (Snowden), described in the *Post* only as a "career intelligence officer."

The Gellman-Poitras story in the *Post* claimed the NSA had direct access to the server systems of nine globe-spanning US 'partner' corporations: the story named Google, Facebook, AOL, Microsoft, Yahoo, Skype, YouTube, PalTalk, and Apple. The PowerPoint said the NSA is able to grab their customers' actual communications, "including the contents of emails, search histories, live chats and file transfers."[197]

The Guardian's story on Prism, bylined Glenn Greenwald and Ewen MacAskill (a *Guardian* senior reporter), focused on the fact that Britain's GCHQ had access to these companies' networks and their customers'content too, though there are strict rules in Britain about privacy. *The Guardian* story included furious denials from the named companies that they even knew about Prism, let alone had given access to the NSA or GCHQ.

As both papers uploaded these stories, Poitras, who had inveigled both Greenwald and Gellman to look at Snowden's documents, was in Snowden's five star hotel room in Kowloon with Greenwald and *The Guardian's* Ewen MacAksill. They were interviewing Snowden for the video she was shooting in which Snowden identified himself as the source of *The Guardian* and the *Post's* documents.

This was a very odd sort of competitive journalism. But both stories made it clear that most online privacy had completely disappeared. And more stories with more revelations, dappled with code words like Muscular and Upstream, made it obvious that the Five Eyes governments together intercept almost all of the world's Internet and telephone traffic. In March, 2013, according to one document, NSA had collected 97 billion data points world wide.

How was this achieved? Almost all Internet traffic starting anywhere in the world at some point runs through servers and fiber optic cables located in the US, and/or through undersea cables which come ashore in the UK, all of which can be intercepted, and are. In addition, by 2010, the NSA had broken most commercial encryption systems, or had bullied communications and software companies into handing over encryption keys, or building 'backdoors' into new products. By that point, all intercepts could also be stored for later use. If need be, a warrant could then be obtained for any individual's communications, which could be retrieved by government agencies (or the contractors working for them). This system, which had spread everywhere, would never forget anyone.

Snowden also insisted that access to one form of communication opened the door to all the others. He said he could have listened in to the cell phone calls of the President of the United States if he had his personal email address.

All impersonal, just metadata, no real privacy invasion going on, the NSA and the Communications Security Establishment of Canada (CSE) insisted as these stories rolled out. But each national government treats the citizens of other countries as fair game, and who could say how material is shared after it has been grabbed? Could I be targeted by the NSA? Why not? The Five Eyes partner governments tried to deflect public ire by blaming the public. For example, while insisting that CSE never does anything outside the law, CSE's former director John Adams later implied that any destruction of privacy was the unintended consequence of people posting too much personal information on line. As he told a Canadian Senate committee considering a new oversight bill for CSE, half of all Canadians are "stupid" about how they manage information, "and the other half is stupid." Apparently this is because Canadians are more active on Facebook than any other nation.[198]

Greenwald and Poitras also doled out Snowden's documents to other partner news organizations in other countries. They too would make headlines: they would show that no one, low or high, escapes Five Eyes, no matter how unCanadian and therefore smart they are. Angela Merkel, chancellor of Germany, and Dilma Rousseff, the president of Brazil, had been listened to for years as they talked on their private cell phones.

In short, NSA and Five Eyes made the East German STASI look like unimaginative children.

The Poitras video on Snowden was posted on *The Guardian*'s website on the Monday following the Chinese President's visit to the US. Edward Snowden said into the camera that he is an American, age twenty-nine, and that he took the documents from his employer, Booz Allen Hamilton, a contractor to the NSA. He said he'd worked for the company as an "infrastructure analyst." He said he revealed these documents so that Americans would know what was going on. He thought the people should decide what to do about it.

The first consequence of revealing himself was that the American government—which pulled his passport—set up a global hue and cry to arrest him. The second was that every reporter with an expense account began to track his story which soon raised many more questions. How had Snowden, a high school dropout, gotten a CIA job with diplomatic cover in Geneva? Why couldn't the NSA even say how many files he'd stolen? Booz Allen Hamilton director Mike McConnell said it could be as many as 1.7 million, that's how many documents Snowden had accessed in his brief time with the company. But no one knew for sure.

The third consequence, perhaps intentional, was fame. A story introducing Poitras and Greenwald as heroes (interviews done in Greenwald's home in Rio), would appear in the *New York Times Magazine* with record speed. Another, about the life and times of Edward Snowden (with side bits about his pole-dancing girlfriend) would be published in *Vanity Fair.*

And then there would be glory. The Pulitzer Prize would be awarded to the *Post* and *The Guardian.*

But it wasn't what Snowden said so much as how he looked that drove home the hard truth beneath his narrative. In his on-camera debut (there have been several more appearances since) he looked impossibly young. With his barely-there goatee, shortish fair hair, and pasty, mole-flecked, computer-geek complexion, he seemed to have just stepped out of an undergraduate computer science class. To look at him was to know that Five Eyes' secrets—all our secrets—are not protected by grizzled, prudent, wise-in-the-ways-of-the-world spy folk working for the public good. To look at Snowden was to know that NSA and its contractor partners had been utterly careless with everybody's secrets, especially their own.

If the enemies of the Five Eyes alliance are so dangerous that all our privacy had to disappear, how could these folks have been so remiss? Snowden was the living, breathing proof that the grand intelligence system protecting us from disaster *is* the disaster. He was its rebuke. He was the unintended consequence of all our smart machines, all their parallel and dynamic systems. He was the unintended consequence from Hell.

The University of British Columbia's remarkably pretty campus knows its place in the natural order. A dark swath of provincial forest curves like a vast scimitar along its western boundary. There are long, grassy, shaded allees with gardens and water features at important junctions. In the distance, the mountains fade in and out like the Cheshire cat's grin. It is a beautiful, calm place, but as I walked along the Main Mall in search of the Plant Signaling & Behavior conference, I was feeling anything but calm. I could not rid my mind of Bill Joy's warnings. I kept saying to myself: Elaine, settle down. Granted, Spaun could be a big problem, but what sort of unintended consequences could possibly erupt from the study of plant signals and behavior?

Yet something inside whispered: watch out, no one ever sees it coming. No one who helped build the first Turing machine foresaw that its descendants would enable governments to eavesdrop on everyone. Even Turing, who saw so far into the future, failed to consider that machines with good memories might be hooked together via globe-straddling networks. The World Wide Web didn't exist even in imaginations until the 1980s. No one suggested then that cellphones could also become computers, Internet access devices, entertainment platforms, cameras, and personal assistants. These properties just accumulated, one after another, until they added up to something so much smarter and so much more dangerous than the sum of the parts.

The members of the Society of Plant Signaling & Behavior had already begun their morning session by the time I found the right hall in the basement of the Irving J. Barber Learning Centre. About 100 scholars had come to hear each other talk. The young and not- so-young sat in swivel chairs behind long curved desks, their laptops and smart phones at the ready.

This was the first meeting of the Society since 2011, the first to be held in North America ever, and it was critical that it go well or the Society would fade away, according to Susan Murch. A professor of chemistry at UBC-Okanagan, Murch was the Canadian organizer of the event. She had been a member of the Society for Plant Neurobiology, aka Plant Signaling & Behavior, from the beginning.

As I looked around the lecture hall, it seemed to me that the Society was in no danger. For one thing, like our favourite food crops it had achieved global distribution. There were people from Japan, India, Germany, Spain, France, the UK, Israel, Australia, the United States, and a big gang from Italy. There was no sign of Anthony Trewavas, the Society's intellectual godfather, but sitting alone on the far side of the room I spied Mancuso's partner, Frantisek Baluska. Mancuso himself had taken a chair in the center of the action. It was clear from the abstracts of the talks to be given that he was still locked in battle with the forces of orthodoxy. Few Society colleagues were prepared to explore plant intelligence beyond a stimulus/response kind of model. The wider world of science had barely budged.

The paper Mancuso had described to me in Florence about teaching the hyper sensitive *Mimosa pudica* plant that being dropped a short distance does no harm, remained unpublished. Its first author, Monica Gagliano, a marine ecologist at the University of Western Australia, told me it had been turned down ten times. She'd tried *Nature, Science, PNAS, PloS 1* and *PloS 2,* etc., and none had even sent it out for review. The editors had balked at the phrase "animal-like learning" in the paper's title which she refused to change because she believed it is a correct comparison.

"Good luck to you," I told her. I meant: your work is terrific, don't give up.

"Is that a curse?" she asked.[199]

When Murch announced that there were six reporters in attendance, I was curious enough about my competitors to ask the young woman at the registration table who they were. She said there was someone from Australia on an arts council grant; someone from a nature magazine I'd never heard of; a freelancer with no specific assignment. Nothing to worry about there, I thought. And then she said: And there's someone from *New York* magazine.

Who's that? I asked, thinking, whoa! If *New York Magazine* is interested, I've got to tell *Maclean's* to move this story fast.

A guy named Pollan, she said, after flipping through some papers.

Mancuso had mentioned that Michael Pollan, author of *The Botany of Desire: A Plant's Eye View of the World*, etc., had been to see him in Florence. But Pollan writes for the *New Yorker*, not *New York Magazine*.

Back in the main room, I spotted Pollan right away. He had chosen a seat near Mancuso's chair. He looked as if he followed his own diet advice. Eat food. Not too much. Mostly plants. His jacket was way too big for his fleshless frame. He had his laptop open and he pecked at it from time to time—until he stepped out with people he wanted to interview. According to Murch, Pollan had been invited to this conference by Mancuso. Not so, Mancuso said when I checked, he had not *invited* Pollan: Pollan had come on his own.

The first person I saw Pollan disappear with was Jack Schultz, Professor of Plant Sciences at the University of Missouri, after Schultz's amusing talk about the hullabaloo his work had generated thirty years before. In 1983, Schultz co-published a paper about plant communications. He found that when the trees he was studying came under attack, they released volatile molecules which rapidly caused trees in the neighborhood to also change their outputs. To Schultz's surprise, the *National Enquirer*, usually more interested in UFOs and Area 51, did a good job of reporting on his work. But his colleagues were not amused by his then revolutionary notion that plants signal each other about dangers and respond to warnings by changing the molecules they produce.

While Pollan grilled Schultz, I cornered Susan Murch. She had been invited to the Florence symposium in 2005 because her work in the late 1990s strongly supported the notion of plant neurobiology. Murch was the first to show that the same neurohormones which have a profound impact on human minds and feelings are also made by plants. Now she holds a Canada Research Chair in natural products chemistry. She also has another project on the propagation and distribution of breadfruit trees as a famine food (for tropical places were food is scarce and disasters frequent). Mainly, she detects and quantifies chemical compounds produced by plants, especially medicinal plants, and studies those that affect human behavior.

As she points out on her website, the average leaf sample reveals about 30,000 different chemical compounds, the average cup of coffee contains about 8,200, the average glass of Merlot has 6500 (while human brain tissues generate only about 800 small molecules). Plants make chemical compounds that act in humans as neurotransmitters, neuroregulators, and neurotoxins, such as melatonin, hyperforin, and BMAA (short for beta-methylamino-L-alanine) a non-protein amino acid.

I wanted to hear how she came to make her discovery and what she thought about it. We sat on a bench outside in the sunshine and she told me her story.

She was almost fifty, with fair hair, thick glasses, and an earnest, open demeanour. She grew up in southwestern Ontario, the youngest of her siblings. She had wanted to be a writer, not a scientist, so she entered the journalism program at Carleton University only to find that she hated everything about it. She was about to quit school altogether when someone reminded her that she'd been very good at chemistry. So she joined the chemistry program at University of Guelph and found her intellectual home. She put herself through school doing what she called "odd" jobs, such as working on the Hostess potato chip factory line and washing dishes in a chemistry lab. She married young, and with her husband, a sculptor, had a son. Then her husband's parents required care. All of which made her career journey much slower than usual.

After her first degree, she took a job in a Guelph lab helping a researcher who was working on amino acid metabolism in humans, and on the amino acid requirements for low birthweight babies. Just before her thirtieth birthday (last chance!) she signed up for graduate school in plant biochemistry. She'd been working in the lab of her professor, Praveen Saxena, who studied growth regulators in plants. She became fascinated with the whole subject.

Murch did her MSc, but then took another job because she didn't think she was smart enough to do a PhD. Yet once she took the plunge into a doctoral program, she went at it with a will. In 1997, just one year after she started, Murch published in *Lancet* her groundbreaking discovery of the human neurotransmitters melatonin and serotonin in such plants as feverfew, St. John's wort, and Huang-qin. These three have long been used as folk treatments for depression, and for sleep/wake disruptions. It had just become known that in humans, melatonin affects the sleep/wake cycle, and in humans, serotonin affects mood.[200] Which begged these questions: why would plants, which have no neurons, produce neurotransmitters? Did they make them to influence animal behavior? Or for themselves? Or both?

As with Schultz's work, the press came running. *Penthouse* wrote about her *Lancet* article in its Men's Health column, which annoyed her professors who worried that media interest would undercut the seriousness of the work. To study the role these compounds play in plant behavior, she devised a fascinating method to alter serotonin and melatonin levels in developing plant embryos. She exposed them to SSRIs (Selective Serotonin Re-uptake Inhibitors), pharmaceuticals that affect human moods—specifically Prozac.

SSRIs in humans inhibit the normal accumulation of serotonin in pre-synaptic brain cells.

In plants, they affected development. Increased serotonin levels in embryos corresponded with increased shoot formation. Decreased melatonin levels reduced root regeneration. She concluded that melatonin and serotonin play general developmental roles in plants. In addition, plants, like mammals, have sleep/wake cycles that are regulated by melatonin.

So what does it mean? I asked.

She thinks plants may do more sophisticated and complicated things with these compounds than the human brain, she said.

"How does a plant keep track of time, an eight to twelve hour day...." she asked. They have to rely on the production of chemical compounds, whereas humans have more options for survival and more protective mechanisms to deal with challenges. A plant must convince animals not to eat it by changing its flavor, or by making a poison, or by "manipulating people into taking care of it," said Murch.

These interactions between people and plants drove her to investigate plants used in traditional human religious ceremonies—plants that cause hallucinations, or changes in mood, or disrupt sleep patterns. She went to Hawaii to do a two-year postdoc with Paul Alan Cox, a famed ethnobotanist, then the Director of the National Tropical Botanical Garden in Hawaii. He had been named as one of eleven "Heroes of Medicine" by *Time* in 1997, and would go on to win a Rachel Carson award among many other honors. In Hawaii, she helped Cox solve a medical mystery about what causes the Chamorro people of Guam to die from an affliction similar to Parkinson's, Alzheimer's and ALS (Amyotrophic lateral sclerosis).

They followed an earlier researcher's path and looked for a toxic agent in the Chamorro diet. They soon found the neurotoxin BMAA in cyanobacteria that live as symbionts in the roots of cycad trees. The BMAA concentrates in the cycads' seeds, which are then eaten by various animals, including flying foxes, a prized food of the Chamorro people who boil them in coconut cream. Murch and colleagues found BMAA in the tissues of preserved and trapped flying foxes, and determined that BMAA is not destroyed by cooking. Later, they also found BMAA in the brain tissue of Canadian patients with Alzheimer's who did *not* eat flying foxes, suggesting it bioaccumulates through more than one pathway.[201]

Little wonder, then, that it saddened her that the Society had backed away from using the word neurobiology in its title. She believes it is useful.

"If you apply extraordinary stress to plants you can see very clever and creative responses," she said. "And that leads to the idea of intentionality....The word neurobiology stimulates us to design experiments differently to see plant survival not as an accident, but as a deliberate process."

The next day, I popped outside again to talk with Frantisek Baluska. He was dressed all in somber black and exuded frustration with the way the talks given so far had divided between plants-as-preprogrammed robots versus plants as intelligent systems. He had given a paper on the first morning with the word 'neurobiological' in the abstract. It was about how plants sense, and how they transform information into behavior. He had pointed to the plant hormone auxin as a plant version of a neurotransmitter. These ideas had not gone down well with a contingent of tanned and assertive Germans who made their contrary views clear throughout the meeting.

I asked him too about the Society's name change: did he regret it? He told me that if he and Mancuso had insisted on the word neurobiology, the Society "would now only have five members."

Inman Harvey would have considered that a plus, I almost said, but held back because Baluska could not be expected to have Harvey's Cambridge-shaped confidence. Baluska was born and educated in Czechoslovakia before it broke away from the Soviet bloc and split into two states. He had been associated with the Institute of Botany in Bratislava, Slovakia since 1981. He had been at the University of Bonn, from 1994 without tenure until 2006.

Yet he had bravely got in touch with Anthony Trewavas after Trewavas published his *Nature* article on plant intelligence in 2002. Baluska had just been intrigued: he thought Trewavas' thesis explained data that did not fit within 'classical concepts,' as he called them.

Baluska and Trewavas had actually gone further out on a limb than Mancuso by delving into the realm of plant consciousness, an idea that found no favor whatsoever in Germany. "They don't support this kind of science," he said. "They're convinced it is nonsense....There is a really deep cleft. The mainstream sees plants as automata, stimulus/response, almost no intentions, motivations, almost robots, like a non-feeling system...."

The Society's members were mainly interested in the signaling aspect and maybe behavior, he said, "only slowly they accept that plants are living systems. Most of them are not convinced regarding feelings, or that they are conscious."

But Baluska certainly is convinced.

"For life, plants must be online," he said. "They must screen hundreds of parameters. The mainstream accepts *that,* but not that there must be *something* that processes the data in order to respond to sudden change...."

In other words, in his view, every plant has some sort of central processor, some kind of computing system. In his view, this ephemeral, distributed *something* has the capacity to react and decide.

But before this idea would be accepted, Baluska said, with a bitter grin, "a generation must be *dead*."

Right across the mall from the bench we sat on, a lovely stand of trees swayed in the breeze at the edge of a park, with an ugly sculpture plunked in the middle. Students sprawled there on the grass, half in the shade, half in the sun, eating lunch. The students were certainly aware of the trees and of themselves. But did that work in reverse? No matter how long I stared at them, Baluska's conviction that those trees were as aware and responsive to the environment as the students, just refused to take root in my head.

Like Mancuso, Baluska believes this reluctance to accept what is perfectly obvious is due to a human psychological foible.

"We are dependent on plants. If they can be conscious and manipulate humans, this makes for bad feeling. We think still that we are the top of evolution, not plants...."

Was he right? Does our delusion of centrality in nature blind us to the beingness of plants? Or do plants relate to the world *without* what we call a self and do we somehow sense that? The night before, I had watched my sister-in-law's cat rub against a catnip shrub in her garden. The cat had seemed utterly helpless to pull itself away from the catnip's emissions, which I could not smell at all. It kept pulling itself across the plant's leaves, over and over, almost abasing itself before the plant's power to give it pleasure. According to Wikipedia, a chemical compound that catnip makes attracts butterflies as well as cats. Yet it also repels termites, cockroaches, and mosquitoes (more effectively than DEET) and has medicinal uses. Little surprise, then, that humans tend catnip. But is the plant's behavior deliberate and intentional? Does a catnip plant calculate the effect of its chemical outputs on animals, and decide to produce them when a cat wanders by? How would you begin to test that?

Some people argue that bacteria are intelligent, too, I said to Baluska. Isn't that more difficult to swallow than that plants are smart and conscious?

"Not so controversial," he said. "They are too small—so they don't threaten our dominance somehow. Plants live longer and are independent of us. If humans disappear, most plants would survive. Not corn but…if plants were gone, we could stay alive maybe a few years…."

I brought up Floreano's robots. His tiny machines solve problems without a complex central processor. Maybe plants don't need one either, I said. Maybe small networks of identical cells are more than enough, I said.

Baluska did not agree: the difference between the semi-intelligent robots in Floreano's lab and plants came down to the ability of plants to act *autonomously*. Baluska didn't think there would ever be truly autonomous, intelligent machines because real intelligence requires consciousness, and consciousness could never be instantiated in machines. For this you need millions and millions of years of evolution, he said. "A machine will not be able to make decisions suddenly if they are not programmed in. For this you need biological evolution…."

No, I said, you just need better chips. I told him a little about Spaun. But he didn't think that a neuromorphic chip would make a big difference either. He thought hybrids between living things and machines might work, but insisted machines alone would never make decisions on their own.

He's just inventing another barrier, I thought, one that sanctifies living things above all else. It's just vitalism by another name. Get him to define what he means by consciousness. Bet he can't do it.

So what do you mean by consciousness exactly, I asked.

It is not defined, he said. But it is nevertheless real and necessary. "We need it. If we lose it, we are not surviving."

Meaning?

He looked at me as if I was an idiot.

"If no consciousness, you can't respond to the environment. Same with animals. And with plants."

What do you mean, same with plants? Do you mean plants can be made unconscious? How?

Well, he said, plants make the same substances we use to affect our own consciousness. They make ethylene which used to be used as a surgical anaesthetic. "It is flammable so they stopped using it in humans in the 1970s though it is one of the best. No bad headaches when you return to consciousness." Ethylene is produced and used by plants when they are wounded or under stress, he said, and plants are almost always under stress. "They produce ether as well….These substances are produced locally….They synthesize a lot of painkillers."

Susan Murch had said that, too. She'd explained that the compounds produced when a plant loses a branch are akin to a human scream.

So the question is why do they make these compounds, Baluska continued. "Either they synthesize them *for* humans, or, *they* feel pain and they use it to survive all the time in stress."

But you said they can be unconscious. How do we know that?

When they are not conscious, he said, just like animals or humans, *plants do not behave.*

He said the pathway and products involved are known. I asked for any papers he had on the subject. He said he would send them, which he later did. "Stefano bought an apparatus to measure volatiles given off when roots are exposed to light," he said. "He measured a lot of ether released from the root."

He wanted to point out that these studies of what happens to *Arabidopsis thaliana* roots exposed to light mean something entirely different to a plant neurobiologist than to a plant physiologist. When exposed to light in the lab, the roots of *Arabidopsis thaliana* seedlings grow very fast. Physiologists interpret that growth as a plant's response to optimum conditions. From the plant neurobiologists' perspective, exposing the roots to light causes the seedlings stress. The plant increases root growth in order to escape the light and find its way back to the optimum condition—darkness.

I went back into the meeting room. I was still thinking about what Baluska had said about plants and consciousness, which is why I forgot what he'd said about hybrids between living things and machines. I picked at what I took to be Baluska's poor logic. Just because a plant makes ether or ethylene in places where it's been wounded doesn't mean the plant as a whole is conscious, as we are, or that it loses consciousness as we do. And as Harvey had pointed out in Montreal, we humans experience different degrees of consciousness and unconciousness. Do plants? Trewavas had proposed that plant intelligence arises from information processed via diffused networks of similar cells. So why would plants experience pain the same way we do? Animal nervous systems are made from *different* types of cells, firing at different speeds, arranged in different regions and tissues. That is the dynamic systems lesson Spaun teaches. Surely these differences between plant cells and animal brain cells mean something?

I was also beginning to think that my concerns about unintended consequences from this science must be hogwash. Nothing discussed so far seemed dangerous in any way. Plants will signal, I thought, and behave. So what if people make machines that work as plants do? What possible danger could there be in trying to figure out how they do what they do?

And yet something prickled at the back of my skull.

A man from Japan, Tomonori Kawano, led off a new session. One of his colleagues had been unable to get a visa to enter Canada, he explained, so he had agreed to give both their talks. The title of the first was supposed to have been "Micro-biorobotic controls of green paramecia under artificial lights." Since paramecia are not plants, but single-celled critters with a big nucleus and lots of little ones, I almost skipped out to get a coffee. Paramecia are boundary buster species that were brought into existence by symbiotic mergers, the kind that interested Lynn Margulis. They are protists—neither plant nor animal. They are eukaryotes with a nucleus, not prokaryotes, without nuclei, like bacteria. Their outer skin-like pellicles are covered by little oarlike structures which they use to beat their way through the fresh water they live in. I was introduced to them way back in the gifted class.

But for some reason, I stayed to listen anyway.

It's a good thing I did. The work Kawano described would have sent Bill Joy screaming from the room.

Kawano is an associate professor in the department of Environmental Engineering at the University of Kitakyushu. He was an early member of the Society, had organized two of its meetings in Japan, and had spent a year working out of Mancuso's lab in Florence. A few months before this meeting, Mancuso's LINV lab had opened an outpost in Japan at Kitakyushu's Science and Research Park. In Switzerland and Italy Mancuso and his colleagues were building robots that mimic plant behaviors. Mancuso would describe these efforts later in the week. (One of his colleagues would show a video of a robot that moves like a root, swelling on one side and then on the other to mimic the way roots grow on one side, then the other, to bend around barriers.) In Japan, at Kitakyushu, Mancuso's colleagues, led by Kawano, were doing something else. They were turning living things into robots.

The Japanese LINV lab, Kawano said, works with "a huge number of companies," whose logos flashed across the screen, trying to develop what Kawano called a plant mediation system—a way to signal plants in the languages they understand. Since plant leaves use light as a source of energy, and roots are repelled by light, they had first set up an LED-equipped lab to control plant behavior and development with light signals.

Kawano explained that scientists realized early on that plants bend toward light by growing more on one side than the other. But in order to use that knowledge to direct plant behavior, there had to be better light technology than was available in the nineteenth century. First came better candles, then gas lighting, then electric light bulbs, then, in the late 1950s, light emitting diodes. The first LEDs emitted light only in the red range of the spectrum, then orange, then yellow, and finally, in 1993, along came the first High Brightness LED, now used to illuminate tablets and phone screens. These LEDs can produce light ranging from ultraviolet to infra-red.

"Enter Claude Shannon…," Kawano said.

Shannon, a contemporary of Turing's, understood very early that the world is made of matter and energy, and that both are information. Shannon did his pivotal work at Bell Labs. His ideas on how to communicate information efficiently in noisy channels, and how to organize it in switchable packets, were vital to the development of high speed telephone networks and digital computer networks.

The plant, continued Kawano, echoing Baluska and Trewavas, is essentially a processor of information. There is a similarity between what plants do and photovoltaics, between artificial photosynthesis and what photosynthetic organisms do, and finally between the communications going on in living cells and a computer language. Plants process information like computers, Kawano said. Plants compute with light and color, using a Boolean algebra all their own.

My notebook records that I had trouble capturing these words, both because Kawano had a strong accent when speaking English, but also because his ideas at first seemed startlingly new, until I realized they are a variation on Nakagami's theme. Nakagami and Adamatzky argue that slime molds compute with their bodies the answers to topological problems, such as how to find the minimal path to reach the oat flakes in the maze. Kawano's point was that plants also compute with their bodies the answers to the problems they face. If the sun is on the left, they grow to the left. If a predator approaches, they try to put out a specific chemical compound that predator won't like, which also alerts their kin to change their own chemical outputs, not to mention neighbors that happen to be eavesdropping. Plants have an array of chemical photo receptors attuned to the different colours of light, from red to dark blue, and they keep a history of the last exposure. The net result: one could make plants compute—grow—as desired by means of controlling light signals.

"We propose plants as computing matter....They could be considered as automata and their outputs as a function, so plants can be considered as *machines*...and the behavior of living cells can be described with Boolean models," he said.

This was the opposite of Mancuso's doctrine that plants are intelligent, that the tips of their roots and their root networks amount to a kind of brainless brain. It also seemed to fly in the face of Susan Murch's assertion that plants' solutions to problems are creative and flexible. Kawano was saying that plants are as reliable as a Rude Goldberg machine and will always produce the same behavior if we give them the same input.

Photosynthesis, Kawano continued, is the "most important chemical reaction on earth," one which transforms carbon dioxide into starch when energized by light. Differently coloured LEDs, when pulsed quickly, will force plants to do this conversion more quickly and efficiently than under normal conditions. But there are also small, single-celled organisms which get energy through photosynthesis. As an example, he cited green paramecia. Increasing LED light pulse frequency can drive up the efficiency of all their photosynthetic reactions by as much as 50 percent. Pulsed light with computerized control can therefore be applied to food crops to make them grow faster.

Smart, I thought. *Really* smart. Imagine what greenhouse growers could do with a technology like that.

But using LEDs to increase the efficiency of photosynthesis was only one part of Kawano's message. He moved briskly to the next, the micro-biorobotic control of green paramecia.

As he explained it, he and his colleagues had become interested in six different species of green paramecia. These species live in symbiotic relationships with green algae, so they straddle the boundaries between plant, animal, and protist. They self-propel through water and they live off bacteria like other paramecia. But they are also energized by the photosynthetic reactions going on inside the small green algal symbionts they carry within them. Because of their algal companions, green paramecia can survive on photosynthesis alone.

Green paramecia, often referred to as swimming neurons because of the constant communication going on between the paramecia and the algae they carry, synchronize their separate functions, including their cell divisions. And yet, these symbionts can also be separated from each other and both will survive.

"We can make the paramecium free of the algae," Kawano said, "and freeing space, can put other things in the cell."

Put in other things? I muttered as I struggled to keep up with my notetaking. Like what? To what end?

"A machine that moves," he said, as if he heard me.

He should have said a machine that senses and moves on its own without a motor or a program, a smart, autonomous, living, self-replicating machine.

In 2009, he and his colleagues had shown that cells of *Paramecium bursaria*—which each host about 100 green algae—can be made to transport things. First, they used a well known method to get rid of the algae. Then they ran a series of experiments. They offered the emptied paramecia a few nanospheres made of different materials. The paramecia ingested these spheres as if they were bacteria.[202] (In a later experiment, they offered them cyanobacteria which they ingested also).

Then, they got the paramecia to deliver these beads to a specific destination.

Then, the paramecia were made to carry both nano and micromillimeter sized beads through narrow capillaries.

And then, they were induced to swim along smartly, one by one, through the paths and gates of microchips.[203]

How? Because Kawano and his colleagues had found a way to make the de-greened paramecia respond to "traffic signals."

It has been well known for many years that paramecia are attracted to electrical currents, said Kawano. This is known as galvanotaxis: paramecia are attracted to the anode—a positive electrode. He and his colleagues found they could make them switch direction by changing the polarity of electrodes. But this required care: too strong a signal and the cells died. So they followed up on work done with their colleagues in Florence—Mancuso and Baluska—studying the avoidance behavior of Arabidopsis roots exposed to light.[204]

Green paramecia also respond to light signals. First they tried fluorescent lights, but they were too strong in the UV range. Then they tried blue, green, and red LEDs. They found blue LEDs are good drivers of cellular movement. Eventually, they got the paramecia to go get the ball (a microsphere) and to bring it back, by guiding it with the blue light.

"The next target was to use the organisms in a maze game, like a Pacman game, to eat the beads," he said.

Eventually they ended up with a living, swimming, micro-robot, able to thread its way through a maze while carrying a payload and obeying a light signal, a robot that can replicate itself indefinitely. A robot that is much smarter than the William Grey Walter cybernetic turtle.

Images of swimming paramecia floated up on the screen behind him. I caught a glimpse of their cilia beating away like oars on a Roman galley.

Then, he said, they induced the paramecia to swim one by one through the complex pathways of a microchip, as if they were electronic signals carrying information.

"They must choose a path. They go left," he said. "They sense the presence of others. They avoid each other. That's it."

I was stunned. This was an innovation as important as the first intentional recombination of bacterial genes back in the early 1970s. It would be much easier to use intelligent living organisms to transport nanoparticles or microspheres around in living bodies than to build nano machines and program them to behave well when introduced into someone's veins. It would take many, many evolutionary program iterations, as Baluska had said, to teach any inorganic machine how to do what paramecia can do from the start: sense the presence of others and avoid them, respond to light, move toward an electrical anode, and ingest and carry objects along the best path through a very small maze.

Because: paramecia have already been designed by evolution to be intelligent—meaning adaptive and flexible in their behavior. They don't need a battery to power their movements. They just need bacteria, which are everywhere. They also don't need to be taught how to hook or unhook at the right moment, nor do they have mechanical latch issues. They are utterly self-sufficient. Yet they can be put to work by humans. Possibly *in* humans.

In other words, intelligent, living, autonomous animal/plant hybrid 'machines' exist already, which is what Baluska had been trying to tell me.

Soon these LINV researchers would also learn to control the direction and speed of plant growth by introducing other kinds of signals besides light. Such as gravity. Such as particular molecular signals. And it wouldn't stop there. If researchers can figure out the languages of plants, why specific plants make a particular molecule in response to a particular signal, plants could become living pharmaceutical factories.

This was not the Second Copernican Revolution that removed humans from the pinnacle of evolution: this signaled the start of a Great Restoration putting humans back on top. In the distant future, a few humans might be able to control or influence the behavior of ninety-eight percent of the living things on this planet.

My hand kept writing his words in my notebook, but at some point I realized I had forgotten to breathe. I had been so focused on what might happen if machines become intelligent and autonomous, I had not even considered that smart living things might be made into robots—enslaved by us. But wouldn't the plants fight back? Wouldn't they try to evolve new means and methods to control their own destinies, as resistant bacteria have done when challenged by new antibiotics?

And there are no rules to restrain this work, I thought as I scratched down his words, there is nothing to hinder it.[205] There had been no meetings of scientists to work out safety practices as happened at the dawn of molecular biology at Asilomar, California in 1975. Back then, because leading scientists had witnessed the horrible consequences of the atomic bombs in WWII, there had been a ding-dong public battle about the dangers of swapping chunks of DNA across species. Some scientists had argued this work was extremely dangerous and had to be stopped. Others had been outraged at the very notion of constraints on their ideas. The freedom faction had been on the verge of walking out of Asilomar until the lawyers explained about the massive lawsuits that would be brought if something went wrong in their labs. In the end, no one regulatory agency was put in charge of the men and women in lab coats: in the end, there were only guidelines. As James Shapiro had said, DNA swaps happen everywhere in nature, so no big deal. But at least it was discussed.

But this machining of living things had not drawn even that level of concern. There are no restrictions on nanotechnology, and none on autonomous robots either. While individual scientists and corporations and governments have been working hard on artificial intelligence since Turing's day, no lawmakers have laid out basic rules about what kind of intelligence should not be made. Who was asking: is it safe, never mind ethical, to turn smart green paramecia into micro versions of Big Dog? Is it wise?

Some might argue that plants use animals to serve themselves, so why shouldn't we use them? And others would say we have always enslaved others, ever since we planted the first seed, corralled the first horse, fed the first wolf, carried off the first slaves after the first war. So why not do this?

But surely, just because we've always done something doesn't mean we should continue, especially if we do not know what might result.

As I caught up on my notes, I kept one ear cocked for the question someone would surely ask Kawano about the consequences of his work, intended or otherwise.

Silence.

Well, Michael Pollan will say something then, I thought, leave it to him, he's the senior journalist here.

More silence.

I finally looked up, looked around, and realized Pollan wasn't in the room. You do it, I ordered myself. But by then, Kawano had made way for the next speaker.

So there were no questions about the coming elevation of humans as overlords over ninety-eight percent of the living things on this planet.

Not a single one.

16

Here Come the Lawyers

Most writers of nonfiction works have a narrative arc in their heads as they begin a project. My plan was to open this book with my class reunion and to end it with a reunion of a different kind—a trial. Why? Segregation of high IQ children in gifted classes like mine was only part of the eugenics movement's 'scientific' program, the part that would advance the so-called brightest and best (high IQ *white* children) and win the Cold War for the good guys. Another part of the program entailed preventing "defectives" from giving birth to more of their kind, and genetically swamping their betters. The eugenicists were nothing if not thorough. They didn't just lecture on the so-called dangers of interracial sex (miscegenation was the scientists' preferred term for that), and press for the legislated power to destroy the fecundity of morally unhygienic "idiots." They also attended to the "special" children whose IQs were one or two standard deviations below the norm. The power of the eugenicists' intelligence narrative continued to be exercised long after the science of eugenics had ceased to be. What happened to these other children was about to be displayed in a public courtroom thanks to the Huronia class action lawsuit.

A sprawling story about the nature of intelligence should be held together by these human/political bookends, I thought. The Huronia class action trial would show—again—how bad ideas about the nature of intelligence retain an almost magical power to destroy human lives. I thought the fates of the other kind of special children would make a good close. (Are you laughing out loud, Frans B.M. de Waal?)

Early on a chilly Tuesday morning in September, 2013, the local radio newscast woke me up. A reporter was talking about the class action lawsuit brought against the Province of Ontario on behalf of all those interred at Huronia Regional Centre between 1945 and 2009. Children with low IQs had been imprisoned there (and in its predecessor institutions), generation after generation for more than a century. Their case had been scheduled to go to trial the day before. But it hadn't. On the evening newscast, one of the two lead plaintiffs, and the woman acting as legal guardian for the other one, had complained that they had been kept in the dark.

I'd been keeping an eye on the case since it was launched. I knew the lawyers acting for the plaintiffs, and how strongly they felt about what had been done at Huronia.[206] And I hereby confess: I had slid into the ridiculous habit of thinking about my years in the gifted class and the years of the plaintiffs in Huronia as morally equivalent. I told myself that during this trial the Huronia class members would grapple with disturbing memories, and the hope of redemption, just as I had at my reunion. They would mourn losses, as I had.

The members of their class were dying, just like the members of mine. Two more of my classmates had passed away since the reunion: genius boy Alexander, who defied all rules and drew Roman soldiers marching over autumnal hills, was one of them.

But after I finally got around to reading the legal pleadings, just before the trial was to begin, I became completely ashamed of myself, and the ludicrous equivalence I had imagined. Equating my experience to that of the members of the Huronia class was like asserting that the Nazi concentration camp commandants shared the experiences of their inmates.

I was surprised to hear that the plaintiffs didn't know that their own trial had been delayed. I had emailed the law firm myself on Sunday to ask when the trial would start and in which courtroom: I had been told it had been postoned for twenty-four hours. Then I was told not to bother coming down to the court until so advised. That advice had still not been forthcoming, but listening to that newscast, I had a hunch *something* important was about to happen in that courtroom.

So I threw on my clothes and drove downtown.

It turned out that Kirk Baert, Celeste Poltak, and their colleagues at the Toronto law firm Koskie Minsky LLP had wrung a settlement out of the Ontario government at the eleventh hour, the first of three they would win on behalf of the former inmates of Ontario's facilities for the intellectually challenged, the Huronia, Rideau, and Southwestern Regional Centres.

For more than 140 years, children and adults deemed too stupid to struggle with life had been stripped of their civil rights and held in these isolated prison-like places under the complete control of directors and staffs. The Huronia Regional Centre had begun as the Orillia Asylum for Idiots in 1876. After that it had been called the Idiot Branch,[207] the Ontario Hospital for Idiots, the Hospital for the Feeble-minded, and penultimately, the Ontario Hospital School. It was neither a hospital nor a school. As the plaintiffs' statement of claim asserted,"evidence will show there was no educational component to Huronia." [208]

According to scholars Kate Rossiter and Analise Clarkson who published an article on Huronia (and Rideau and Southwestern) in the *Canadian Journal of Disability Studies*, these places were modeled on the Victorian era's "medico-pedagogic" theory. Children consigned to these places were supposed to be protected from society while they learned practical skills under social (read sexual) control. Hiding them away from the rest of the community was considered to be a progressive solution.[209] At first, no distinctions were made between intellectual deficits and psychiatric issues. But after Galton's eugenics theories caught on, and the pre World War I invention by Binet of an IQ test purporting to identify mental 'defectives,' tests led to diagnoses, and diagnoses led to life-long institutionalizations. As IQ tests claiming to measure the intelligence potential of children became available, the number of kids stashed in these institutions just kept growing. They became dumping grounds for all those who scored poorly on tests, but also for those who were desperately poor, and/or orphaned, and/or difficult.

As Rossiter and Clarkson put it: "In Canada, the impact of testing led to the growth of specialized institutions such as the Orillia Asylum...urban centers such as Toronto established education systems based on new 'scientific' methods of intelligence testing which helped to rank and separate 'defective' children...perpetuating the popular notion that 'feeble-mindedness' was an increasing urban threat."

Huronia and Southwestern and Rideau looked from a distance like imposing estates surrounded by agricultural lands but they were decaying pestholes. Staff was never sufficient. Those inmates who had parents or families rarely saw them because their relatives were discouraged or prevented from visiting. They might as well have hung signs over the entryways reading *Arbeit Macht Frei.* The smartest took care of the slowest, the strong worked in the fields, or made clothes as part of a cottage industry (earnings clawed back in to the government's pocket), or cleaned the overcrowded and rundown sleeping quarters, playrooms, and washrooms. Those who worked were mostly paid nothing. The inmates had no privacy, no autonomy, and no safety. One inquiry into Huronia noted: "...those capable of being absorbed into society constituted a...labour force for the institution and were too valuable to be released." So they were kept as slaves. In these places, justice was as Thrasymachus defines it in Plato's *Republic*—the interest of the stronger.

The inmates were assaulted by all the diseases associated with overcrowding and poor sanitation. Inmates were also subjected to rape, molestation, and beatings by other inmates and by staff who were rarely disciplined, let alone charged. They were victims of crimes without recourse, entombed until the government let them go, or forgot about them, or until they succumbed to early, sometimes violent deaths.

In 1960, the great journalist/author Pierre Berton visited Huronia with a friend who had a child there. Outraged by what he saw, Berton published his findings in the *Toronto Star.* He described the filth, the decay, the stench, the cruelty, the appalling overcrowding and understaffing, the enslavement of the defenceless. He wrote that after WWII, many Germans said that they had been unaware of what the Nazis were doing in the concentration camps. As Berton put it, Ontarians no longer had such an excuse about Huronia, because "you have been told."

But no one did anything about it for *fifty years.* The government finally fixed Huronia, Rideau and Southwestern in 2009 by shutting them down in. That's when the lawyers took up their paper cudgels.

Baert, Poltak and their colleagues had asked for damages for negligence and breach of fiduciary duty by Huronia in the sum of $1 billion, and for $1 billion in punitive damages. (They asked the same amounts in the other two cases as well.)

The real surprise was that the government held out for so long. It had admitted so many abuses, including: the failure to meet recognized standards of care; the use of inmates for unremunerated labour; taking money from inmates' accounts; failing to teach inmates; failing to protect them; failing even to keep the buildings clean never mind treating the people living in them with dignity. The government's argument boiled down to this: none of that mattered since the inmates were damaged before they got there.

In Huronia, no one had rights to: sleep in an uncaged bed (if staff so decided); keep their own teeth in their heads (if staff so decided); bathe in a tub instead of being splayed out on a table and sprayed (if staff so decided); freedom from parasites and infectious diseases. A Huronia document declared that seventy to eighty percent of inmates contracted Hepatitis B and that almost every inmate was likely to get it if interred there for more than two years.

One expert explained the damage done to every inmate just by being in Huronia for more than a year, regardless of whether they had actually been tortured or beaten or raped or made to work for nothing. If ever there was a clear proof that intelligence is an emergent by-product of interaction with a particular environment it lay here, in the list of what being in Huronia did to these children:

> "impaired language development;
> difficulty in social relationships;
> episodes of aggression;
> impaired ability to form attachments or bonds;
> lowered self-esteem;
> impaired gross and fine motor skills;
> impaired mental performance particularly verbal intelligence;
> symptoms of learned helplessness."[210]

The lead Huronia plaintiffs had each been selected for this special education at the ages of six and seven, too young for any IQ test to predict with any accuracy their school potential. One of the plaintiffs had been removed from her home by the Children's Aid Society and later sent to Huronia. Both plaintiffs had been dumped into a world of pain: for the crime of refusing to eat, one of them was beaten with a radiator brush and held upside down with her head in ice water. She was also dosed with dangerous drugs as punishments for speaking out, and was generally made to live in quaking fear for the duration of her time at Huronia. This abuse continued when she was moved to a group home. The other lead plaintiff was sexually abused when she was sent to live in a group home supervised by Huronia. There was no one she trusted enough to complain to, and anyway, she was more afraid of being sent back to Huronia. She attended public high school and then went to College. These plaintiffs were lucky. They got out of Huronia alive. Many did not. Huronia had a field of unmarked graves. The lead plaintiffs in the other two cases had endured similar cruelties.

Their intelligence had been no more fixed at birth than mine: mine had been buoyed by my favourable circumstance as theirs had been hammered by years of abuse. When released, they had struggled, they had learned, they had built lives for themselves, and now they wanted real justice.

Three television trucks had parked in front of Toronto's iconic Canada Life Asssurance Company building where the Superior Court of Ontario has courtrooms. By the time I found courtroom five, the chairs in the visitors' gallery had been filled, save one, which I grabbed, while the court officials scrambled to set up more in the aisles and in the hall.

There were a lot of gray heads in the gallery, people with the thickened bodies of late middle age, like mine. It wasn't easy to spot who had been a former inmate, who was a family member, or a friend. Only a few appeared to have any disabilities. Several women talking among themselves may have been born with Down's syndrome, and there was one woman with extremely short legs. A young man strapped into the wheelchair beside me was asleep. A tall man spoke in a very high, querulous voice.

A middle aged woman in the gallery clutched a small black and white photograph of a small child. She was Marilyn Dolmage, the legal guardian for the plaintiff Marie Slark. Dolmage's brother Robert became an inmate at Huronia at age three. He had died there at age eight from untreated pneumonia. She took a job at Huronia at the age of nineteen and eventually became Slark's social worker there. They became friends.

As the clerk opened the door for the judge, the people in the gallery became very quiet, their faces alive with expectation.

Justice better not be the interest of the stronger today, I thought.

Madam Justice Carolyn Horkins made gracious remarks about how welcome we were in this court which belongs to all the people of Ontario, and it was too bad that some had come the day before expecting a trial and been disappointed. There would be an explanation. She went out of her way to make everyone feel included in the dignified, measured ways of the court, and above all, to feel respected.

Kirk Baert got to his feet. He had a settlement to announce. He said he was sorry about those who had come the day before, expecting a trial, but the lawyers had been negotiating and so no announcement could be made until they reached agreement or impasse. They had signed an agreement an hour earlier.

The judge asked him to lay out the details.

First, said Baert, there would be an apology issued by the government to all former residents of Huronia. Next, there would be a cash settlement of $35 million to be shared by members of the class.

That was a very small number compared to the original demands of $2 billion. People squirmed in their seats and began to whisper.

Baert described the claims process. The lawyers estimated that there were 3,000 members of the class still alive. Those who claimed the general harms of just being locked up in Huronia would each get $2,000. Those who claimed specific harms would be put on what he called a bell curve: maximum payout for maximum harms, anything from severe beatings to rape, could possibly mount to $42,000. The province would care for a cemetery on the property in perpetuity, unmarked burials would be found with a GPS survey and listed in a registry, and there would be a placque erected to remember those who suffered at Huronia. The 65,000 documents produced in the action would be made available to scholars and to the members of the class under the rules of the Freedom of Information and Privacy Act. If money was left over, then some would go to help families coping with disabled children, or to help those who had suffered tell their stories.[211] However, this settlement would also extinguish all future claims against the government over Huronia.

There were more grumbling sounds as people tried to figure out how much they would get and how they felt about the cancelled trial. They had wanted to use it as a truth and reconciliation commission, to show their wounds so as to heal them: telling their stories in public was supposed to have been a very significant component of justice. The judge reminded the room that the lawyers had worked long and hard for a settlement but the members of the class would have the opportunity to object to any parts they were not pleased with at a later date. The gallery emitted a communal sigh.

"I want to thank you all for coming," the judge continued, "...and I hope it will provide closure and satisfaction for every member." She rose from her chair and left the room.

The man with the high voice stood up. He began to cry. "We won," he squealed.

Patricia Seth, a lead plaintiff, was also crying.

Are you happy or sad? I asked her.

Both, she said.

On one hand, they had wanted to tell their stories. On the other, there would be a little money, an apology, a plaque. For once, the members of this special class believed they were going to be treated as full-fledged members of that marvelously intelligent species perched atop the smarts pinnacle.

Outside, on the court steps the cameras were waiting.

"We won," the high voiced man kept saying, "we won." His picture appeared in the papers the next morning. While the money was much less than hoped for, his arm was held aloft, like a winning boxer's, by his daughter. This was the real victory. If the eugenicists had had their way, he would never have been allowed to have children.

But even as that moment played out in front of me, I knew I had not reached the end of this narrative. What about the Great Ape Trust/IPLS soap opera? What about those bonobos– especially Kanzi, the Bonobo genius—owned by a charity run by a woman described as not well, who had been keeping it going with her own funds?

Throughout 2013 I had checked into various websites to try and keep tabs on what was going on. The Iowa Primate Learning Sanctuary (as it is properly called, but also known as IPLS, or the Great Ape Trust) seemed to be where Sue Savage-Rumbaugh hung her hat since her reinstatement as resident scientist in 2012. But she also seemed to work for the Art for Bonobo Hope Sanctuary (legal name: the Bonobo Hope Initiative, Inc.) At a certain point I sent another hello? hello? email to the IPLS media page.

At last, I got a response—peppered with enthusiastic exclamation points– from a person named Tami. She forwarded my email to the IPLS's new executive director, Steve Boers.

Boers was a registered agent of a (now inactive) Iowa company called T3 Techologies Inc., a certified installer of products produced by a company called Leviton. On its website, Leviton listed the Great Ape Trust as a technology transfer partner. There was a picture of Kanzi working with a touchscreen lexigram in a case study describing the use of its products.

Boers thanked me for reaching out. I told him by return email that I needed to check facts about the Trust, and its related organizations. I asked a raft of questions and asked for copies of the organizations' recent reports to the IRS, the names of board members, etc. He responded by asking for more about me and my project. When I explained I was working on a book about intelligence, and why I needed to see documents, he replied that the organization was going through a "rebranding" exercise and felt the need not to be "part of any periodicals at this time" but "thanks for thinking of us."

And that was it until December, 2013, at which point Sue Savage-Rumbaugh's name came up in a different context entirely. My local newspapers ran stories about the Non human Rights Project's legal initiative to free chimpanzees languishing in various facilities in New York State. This charity, whose board members include Jane Goodall, had hit on a very interesting strategy to define the legal rights of great apes. On December 2, 2013, the Project's lawyers launched the first of three motions in three different courtrooms of New York's Supreme Court. The motions were all for an order to show cause and writ of *habeas corpus*— proof of the right of a ruler to hold a prisoner—along with an order for immediate release of four chimpanzees. Two were locked up in a university lab where they were subjects of motion studies; one was locked in a cage in a shed; another was locked up in a cage on private property, a so-called 'sanctuary'. They were located in Suffolk, Niagara, and Fulton counties respectively. The NhRP demanded that the courts release Kiko, Tommy, Leo, and Hercules on the grounds that they are self-aware, autonomous beings who should be recognized as legal persons and given liberty (of a sort), and not treated as tradeable commodities.[212] They argued that New York State had recognized the personhood of slaves long before slavery was abolished, and had recognized the personhood of animals named as beneficiaries of a trust, not to mention the personhood of corporations, so why not recognize the rights of great apes? The judges soon rejected all three of NhRP's motions but the NhRP was not downcast. They'd sued in New York State because it has an automatic right of appeal in *habeas corpus* petitions. The chimps' cases would be heard at the Appellate level where the outcome might be different.

In support of these motions, the NhRP brought forward affidavits from leading scientists who had studied apes for many years, including one from Sue Savage-Rumbaugh.

In her affidavit of November 22, 2013, Savage-Rumbaugh described herself as "currently" Director Emeritus of the Iowa Primate Learning Sanctuary (ILPS), author of 181 published scientific articles, several books, and a frequent public speaker who had also appeared in several documentaries. She listed her qualifications, such as a doctorate, honorary doctorates, and other marks of public esteem, including being selected by *Time Magazine* as one of the world's 100 most influential "scientists."

She explained that the other scientists had shown that the DNA of humans and chimpanzees is ninety-nine percent the same, which indicates that...

> "11...when behavioral studies of chimpanzees suggest that they are capable of self-aware conscious action, the capacity to reason and think, the ability to acquire symbolic language, that is reason to take these results seriously. It also ensures the following...
>
> e)...That, under natural conditions, chimpanzees will develop and utilize self-awareness, self-agency and intelligence to survive...
>
> f) ...they will also construct a social structure that is rule-based, conscious and successful thereby allowing them to survive as a group, by virtue of culture, not by natural instinct...
>
> 16 ...this capacity to 'step-outside' the self and reflect upon ones (sic) own behavior as one might reflect upon the behavior of another—allows one to become the object of one's own thought. This capacity is the root of human cultural, linguistic and moral systems."

It seemed so odd to me that Savage-Rumbaugh should have argued so eloquently that apes held in a research facility are legal persons entitled to liberty. Less than one year earlier, when she was still in charge of science under the direction of the veterinarian Julie Gilmore at the Iowa Primate Learning Sanctuary (IPLS), she had signed an agreement (plus a side agreement without which the first would not have been signed) to settle the ownership of the apes held there. Questions of who owned which bonobo (or half of a bonobo) had come up in 2009 when the Atlanta Zoo, which claimed ownership of Matata, Kanzi's adoptive mother, decided to move her and her son, Maisha, to another zoo for breeding purposes. A letter also arrived from the Democratic Republic of the Congo asking about four bonobos—including Matata—*loaned* to Yerkes by its predecessor government, Zaire, back in 1975. Twenty years after this loan agreement was signed, Matata had been *donated* to the charity that runs the Atlanta Zoo by Yerkes, as if Yerkes owned her. The Zoo had then lent her to Georgia State, which in turn moved her to the Trust. The Zoo wanted to breed her, but the Congo wanted Matata returned for science purposes. By February, 2010, the whole mess landed in the lap of the US Federal Court for disposition.

And a settlement was reached in January of 2013.

The side agreement to the settlement gave Savage-Rumbaugh, and the Bonobo Hope Initiative, Inc. the sole right to decide where all the Trust's bonobos—except for Matata—would go if the Trust went defunct or ceased to do specified studies. Matata was recognized as the property of the Congo.

At the beginning of 2014, Heidi Lyn responded to an email from me asking for information. Things are going to change at the Trust, she said. She was hopeful.

What? I asked, thinking hah! Kanzi and his labmates will be sent to a sanctuary at last.

Lyn said Sue Savage-Rumbaugh had been removed and some researchers she knew were taking over management. Serious science might soon commence.

And sure enough, on January 30, 2014, just a few weeks after Sue Savage-Rumbaugh gave her affidavit in support of the motions of the NhRP, the *Des Moines Register* reported that she had left the IPLS— in May, 2013—for New Jersey where she was pursuing her interests. Dr. Jared Taglialatela of Kennesaw State University and Dr.William Hopkins of Yerkes/Emory (who taught Heidi Lyn how to do science efficiently), had taken over in September, 2013. They took up their positions as the US Congress voted to permit the National Institutes of Health to pay for the upkeep of apes retired from the service of science. Hopkins became the Science Research Director, Taglialatela became the Scientific Program Manager.

I went to Hopkins' and Taglialatela's websites to see what sort of work they do. Both use biological methods to study apes. Hopkins doesn't immerse apes in human culture (although he had published culturally-inflected studies with Sue Savage-Rumbaugh). He mainly correlates brain and gene function with behavior, specifically communication, but he had also published recently on intelligence as an inherited phenomenon. He does fMRIs and PET scans to map ape brains and hunts for gene polymorphisms in chimps—"animal models" as he calls them—that might lead to therapies "for disorders of the social brain" in humans. Taglialatela, (who had studied with Sue Savage-Rumbaugh) works along similar lines, specifically "the neurological foundations of orofacial motor control" with emphasis on ape communicative behavior and its evolutionary origin. It seemed obvious to me that these two would have no problem getting funded by either the NSF or the NIMH. Their work definitely fit the funders' criteria.

Apparently, Sue Savage-Rumbaugh had brought them in to IPLS herself to get the organization back on an even keel while she was on a medical leave of absence.

But Sue Savage-Rumbaugh's exit from the IPLS was not quite as described in the *Des Moines Register.* Though she had spent much of 2013 in New Jersey recovering from a head injury after a fall in the lab, she had returned to Des Moines in November.

Savage-Rumbaugh's lawyer would later tell the judge overseeing the ownership settlement that Savage-Rumbaugh had been ejected without notice from the IPLS board and denied access to the apes and the site beginning in November, 2013. She left Iowa without a job and was owed about half a million dollars in unpaid wages, workmen's compensation, and unsecured loans. But how IPLS managed to eject her in November is a mystery: the corporation had ceased to exist.

In August, 2013, the State of Iowa had dissolved the corporation. IPLS had previously incurred a tax debt of 75,000 dollars due to management's failure to pay unemployment taxes on its payroll. Savage-Rumbaugh had also failed to fill out a form and to pay the small fee required to keep IPLS's corporate status intact. It could have been reinstated if the tax lien had been paid, but there was no money to do that. Bonobo Hope Intiative, Inc., the corporate parent of Bonobo Hope Sanctuary, had also been dissolved, though it had since been reinstated.

On December 18, 2013, a new nonprofit called Ape Cognition and Communication Institute (Ape Cognition and Conservation Initiative according to its website, or ACCI) was incorporated by William Hopkins. On the very same day, the dissolved IPLS wrote out a bill of sale transferring to ACCI all IPLS's assets (but not its liabilities or responsibilities under the settlement agreements) including the bonobos (but not the Congo's Matata) in exchange for one dollar. The bill of sale was signed by William Hopkins for IPLS (and Meg Fitz), and the same William Hopkins (and Meg Fitz) signed for Ape Cognition and Communication Institute. Sue Savage-Rumbaugh's lawyer was asked by email that very day if the Federal Court must be informed about IPLS's dissolution. Savage-Rumbaugh's lawyer had answered yes. The lawyer acting for the defunct IPLS also proposed that an unnamed new corporation should take over IPLS' assets and responsibilities but not some debt that Ted Townsend had left behind. No mention was made that the transfer of assets had already taken place.

In addition to picking up assets for a song, ACCI developed an interesting business strategy. In a letter written to various researchers, Taglialatela suggested that ACCI would get operating revenues from researchers who would come to work at the site with their grant money in tow, and also from some universities willing to put up money to become partners to gain access. He mentioned too that the NIH might send ACCI twenty or thirty chimpanzees who were about to "retire" as federal government research subjects. As he pointed out in his March, 2014 letter, under the title goals and objectives, "financial support to manage and care for the chimpanzees would be largely provided by NIH and therefore increase the number of available apes for behavioral research at no direct cost to ACCI."

Some retirement.

I wondered how Kanzi would enjoy sticking his arm out for injections and having his brain scanned in an fMRI, though I was fairly certain it wouldn't matter much. Any researcher willing to do a study of the bonobos meant money in the bank for ACCI and William Hopkins and Jared Taglialatela had already assigned three of their own funded studies to ACCI. Another four were pending from Hopkins, and one from Heidi Lyn. In addition, the newly constituted Institutional Animal Care and Use Committee—the ethical review committee—consisted of Julie Gilmore (formerly a director of IPLS and a small animal vet); Allyson Bennett (an assistant professor of psychology at University of Wisconsin interested in the neurobiology of emotion); Tami Watson (a volunteer); Jack J. Price (apparently well versed in Des Moines real estate); and Jared Taglialatela. The latter was unlikely to rule against his own proposed protocols.

Savage-Rumbaugh's lawyer, William C. Zifchak, an expert in employment law and a retired partner in Kaye Scholer LLP, told me he met Sue Savage-Rumbaugh after she had been let go by the Trust for the first time. He was appalled to learn then that though she had given up her tenure at Georgia State University to go to Iowa, an Iowa lawyer had advised her to sign a severance agreement which entitled her to only a year's salary. She had signed it before a mutual friend introduced her to Zifchak. She thought she'd been promised a home for the apes and a salary for herself for life. However, as Zifchak explained, Ted Townsend—"that snake and you can quote me"—had put no such thing in writing. When they first met in February, 2008, Savage-Rumbaugh appeared to Zifchak to be so depressed he thought she was near catatonia. Why? Because those bonobos were her family and she'd been denied access to them.

Zifchak had managed to work out a new employment agreement with the Trust which got her back in the door, though she no longer ran the place. He assisted her in being reinstated in 2011 and again in December, 2012. He acted for her over the questions about who owned which bonobo. He had made it clear that the main settlement agreement was contingent upon the side deal that conferred sole rights to resettle the bonobos on Savage-Rumbaugh and Bonobo Hope Intiative Inc., because by then he didn't trust some of the folks at the Trust.

In the summer of 2014, he went back to the same judge who still had jurisdiction over the settlement, asking him to enforce it, or, alternatively, to assert Savage-Rumbaugh's right to decide the fate of the five bonobos still at ACCI. (By then, Matata had died while in ACCI's custody, so Congo was out of the picture.) He argued that since IPLS was defunct, Savage-Rumbaugh and Bonobo Hope Initiative, Inc. had sole discretion about where the remaining bonobos would live. She wanted them to go to a Missouri facility purpose-built to take them. This would leave ACCI holding a big, empty bag.

He was blunt about how much time and effort he had had put into dealing with these issues, *pro bono,* and he made it clear that the lengthy matter had near exhausted him. But on the other hand, a deal is a deal. "They call *New York* lawyers thieves," he said, but in his view, certain Iowa lawyers give New York lawyers a good name. So the litigation continues.

His strategy?

"Scorched earth," he said.

Epilogue

The author of a book should end it neatly knotting the loose ends together in a lively essay rife with pullable quotes for the PR folk and metaphors that will cling like burrs in readers' minds long after the book is shut. Self, I said, waving a metaphoric finger in the metaphoric mirror: deliver a shiny new definition of intelligence for the dust jacket (and the eblurbs). While you're at it, predict the consequences of making robots as smart as plants, and microbial machines, and neuron-like chips that will slide right in there among your own hard-working cells and make you smarter than Ray Kurzweil. Tell us what will result.

I played endlessly with what-is-it-anyway metaphors. First I tried this: intelligence is an emergent property of any struggle for continuance. (I could almost see Inman Harvey waving his hands in mockery, so I set that aside.) Then: intelligence is the narrative of these struggles. Narratives organize information, the interplay between energy and matter. Agents (characters) entangled in dynamic relationships create an ineffable something—a story—that is more than the sum of their actions. Every narrative has an arc: it starts, grows, diminishes, ends. All narratives are situated within larger, more complex entanglements, with different kinds of stories emerging, unpredictably, along the way. None continues on forever like Daisyworld's daisies. That's how you can distinguish life—and versions of life—from computer programs.

For one happy moment I thought this narrative metaphor was so clever it might even have the virtue of being true. But then the journalist in me hooted at my presumption and reminded me that a metaphor is only that—the linkage of a known thing to something elusive.

Upon reflection it seemed to me that no one, not even the folks building Spaun, will ever isolate one must-have feature shared by all the myriad versions of intelligence spewed by that continuous computation called evolution. Bodies will always change to meet new circumstances, producing ever more surprising ways and means to make decisions. It is hard to compare the intelligence of an orangutan who invents fishing to that of the tiny fish that catches dinner by sensing shifts in voltage. Intelligence is utterly malleable and awe-inspiringly particular. Yet it is also communal and cultural. Think of all the transformations that we, the living, make when we act together, weaving ourselves into ever larger tapestries, begetting ever more complex narratives, and ever larger assemblies.

Speaking of presumption. No sane person would dare to predict the future of intelligent machines or machined life when the present is so utterly overwhelming, when every invention kicks off so many more that the speed of machine evolution just keeps accelerating at an ever-increasing pace. In my brief span of years the rate of change has already gone from swift to blur. Beyond comprehension is right around the corner.

Predictions were possible for earlier generations. I don't think my father's father ever felt as fearful about, as excited by, and yet as alienated from the future as I do. My paternal grandfather was born in the southern part of what is now Ukraine just after Darwin died, back when mainstream scientists (but not Darwin) insisted that only humans are intelligent. My grandfather reveled in human genius. He used to say that man made God, not the other way around. He had little formal education (equivalent to about grade three), but he was very smart—by which I mean he was highly adaptable, or, as Louis Herman would say of his dolphins, capable of very flexible behavior far outside his program. He had apprenticed as an ornamental blacksmith. But when he came as a pioneer to the Canadian prairies (for the adventure, or so he said) he turned himself into a different person altogether. He worked on a railroad, he created his own business. He made ploughs, built hearses, wagons, buggies, invented new farm implements and when cars and trucks replaced horses and oxen, he made his smithy into an auto body repair shop. He had grown up under the thumb of a Czar, but in Canada he wore a top hat when he took Prime Minister Mackenzie King around to meet Saskatoon, Saskatchewan's Jewish community. When he was semi-retired, he built an electric car for fun. He plugged it in at home and drove it to the shop on the sidewalk (it only had three wheels so it was legally a tricycle). It was so remarkable that my cousin and I drove it in annual Canada Day parades. Yet, like the rediscovery of Bose's work on plant electrical signals, his car was back-to-the-future: electric cars had been the rage in Europe when he was young, before gas-fueled cars became common.

One Sunday afternoon we found him lying as usual on the couch with a glass of tea cooling beside him and a hunk of chewing tobacco in the ash tray. The TV was on. He was gob-smacked—thrilled—by the space flight of John Glenn, the first human being to orbit the earth. For him, science and technology always marched toward a glorious future. No surprise that two of his four sons became physicians.

Nu, kinder, he said, isn't it wonderful what people can do?

Well yes, I wanted to say, but it can also be terrible. Having experienced the gifted class, I knew even then that human intelligence just *is*: it has no morality. It can do horrors as easily as wonders. I had been taught to listen for the sirens and to duck and cover if the Big One fell from the sky. For my generation, genius meant, and still means, a searing, mushrooming darkness in the world.

A journalist's job is to report on the thoughts and deeds of others in a narrative that will enlighten and amuse. Normally I am content to share the facts and let readers conclude whatever they like about what they might mean. But I can't do that this time. Like Pierre Berton faced with the facts of Huronia, I feel compelled to say flat out what I think about what I've learned.

I think some very smart human beings are destroying most of the boundaries that once divided living things from each other, and machines from living things. I think we have to act now if we are to have any say at all about the future thundering at us from unregulated private and public laboratories. I think we have to think through what intelligent machines and machined life might mean and decide if we want them or not.

Will they change the *telos* of the human story? Of course they will.

For example: IBM's artificial intelligence called Watson is being taught how to research law and predict outcomes in legal battles. One day soon such innovations will free our society from the need for junior lawyers. They will be replaced by smart legal software. Great, you say: Shakespeare's Butcher would have approved. But here is the problem. It will happen at about the same time as newspaper, magazine and TV news writers are replaced by neuroscience-enhanced software. If we want to exert democratic rein control on the powerful, we cannot let software write the news or we really will find ourselves at the end of history.

Smart machines have already remade our cultures, and our cultures will change, are changing, our bodies. Once we can slide neuromorphic chips into our skulls to augment how we think, we will think differently about everything. Now, we yearn to wrap ourselves in robes of beauty, kindness, and reconciliation. We treasure our capacity to love each other, to create ideas, to make pictures, to compose music, to tell stories, to build spectacular machines and institutions which will still be here long after we are gone. But as we invent new machines, and machine ourselves and many other living things, we will change what we want to do, what we want to be.

Our new means to describe, measure, mimic and use so many different forms of intelligence will generate all kinds of innovations. Yet they will not change us at our core. Our smart societies will still harbor human versions of the alpha chimps, Nikkie and Yeroen. The will to dominate can be seen at work in many branches of the evolutionary bush, even among the amoeba. Armed with ever-smarter machines, the princes among us who love power will be so much more dangerous than those who came before them. They will be able to do so much more harm to those who stand in their way.

There will be dialectical consequences to our mastery of smarts. There have been consequences already.

You have been told.

One summer morning, I stood on my back deck, fascinated by the doings in my garden. I was not the same woman who set out to inquire about the nature of intelligence and who displays it. For one thing, I no longer thought of it as a rare commodity unevenly distributed in people and a few other 'higher' animals. I was watching intelligence at work everywhere—enjoying the soap operas large and small unfolding wherever I looked. I had become reconciled to the fact that I could watch it work and yet be unable to either split or lump my way to a precise definition. In short, I saw the world quite differently.

What the plants breathed out, everybody else breathed in, and vice versa. The late blooming roses blew their signals everywhere (why doncha come up and *see* me some time) so the ants and bees and wasps came to call, nattering in chemical compounds and altruistic gestures. The smaller birds swooped down and plucked up worms while the dragonflies made asymmetrical warfare on the mosquitos. A raccoon, the top predator (except for that hawk circling on a thermal), swaggered across the grass and scared everybody, including me, when it knocked over a garbage bin. Under a mound of trimmed euonymous, a black cat with white paws hunched, motionless, then pounced and killed something, but fled in terror when the neighbor's dog howled its fury at being tied up. The grape ivy and the clematis, twining on the same cedar screen, fought it out over rights to sunlight, each determined to get it all and to prevent the other from getting any.

I could almost see the dandelions' minds at work as they cornered untenanted bits of soil, spreading themselves out like dragons on their gold. The fruiting bodies of damp-happy fungi popped up in the wet lawn, while something (slime mold?) deep in the soil gave off a delicious aroma—cool, yet peppery– because it had rained. The spruce and the elm had transduced that rain into growth: the maple out front had too, having already bundled its seeds in green containers shaped like a cross between aircraft propellers and kayak paddles. The giant poplar two yards over had cast itself upon the future via balls of fluff that hooked onto the dogs and cats and squirrels and raccoons (unless the birds grabbed them for nest material).

Everthing in the garden moved with political intent; considered options; made decisions to enhance or defend; communicated; made common cause or made war; made progeny; would later die.

The poor old juniper, bent double by age and heavy snows, had allowed its genome to be taken over by some clever insect. In a matter of minutes, its previously naked branches had coated themselves with bright orange goopy sleeves: though I whacked them off, the larvae in these jelly-like, quaking galls hatched anyway, and infected the nearby pear tree whose leaves turned orange with black spots, which would preclude fruit this year, and wreak havoc on the wasps. The squirrels rolled back strips of turf and dragged forth tasty things hiding below. I stomped the turf back down and waved them off. As soon as I turned my back, they returned: they might as well have shown me the finger. Along with the raccoons, they had already made a mess of the cherry tree. They had hung upside down from its quivering branches, scrambling all the way to the very tips, eating all the fruit it offered before I could shoo them away. The cherry had been dropping broken branches and yellow leaves ever since. Though I couldn't see what the bacteria and viruses were up to, I knew they were there too, meeting, merging, hitching rides, lugging odd bits of adaptative possibilities around in the form of pilfered DNA.

I had gathered these smart life forms together in what had once been an ugly yard dominated by an old garage tilting like a mean drunk over concrete and coal clinkers from the Edwardian period. A friend had taken the garage apart. After someone else took away the concrete, I planted this and that, encouraged connections here and there until the whole space acquired an identity that pleased me. Nothing in it stayed the same from one moment to the next. Everything grew and divided, grew and mated, hatched or gave birth, died, or faded away, which forced all the neighbors to reconsider their situations. And so the whole garden was a story of its own, a narrative generated out of the whole vibrating, agitating, whirling, disintegrating, self-assembling shebang.

Not surprisingly, part of me had begun to think twice before I crushed an ant racing back to its nest with news of food, or yanked dandelions from the grass, or trimmed a healthy branch to make things prettier. In quiet moments, I could see myself as a part of it all, as a hub of hubs instead of an isolated body behind a skin boundary run by a sovereign homunculus squatting in my skull. I knew myself as a part of many different gardens, all adding up to something utterly enormous and beyond my ken. My human smarts serve my needs but are also at the service of so many differently organized narratives, all entangled through a network of networks, a hyper-complicated, never-ending computation.

Evolution is the name of that narrative. It takes a zillion forms and computes a seemingly infinite number of responses to contingencies. As Turing saw so clearly, evolution is a universal computing machine that employs an almost numberless array of particular algorithms to make intelligent decisions.

Evolution may be slow, but it is very, very smart.

This makes Lovelock's Gaia, the summation of all our interactions, the dynamic and self-regulating system of all the systems, the version of life that is greater than all her living parts, the smartest of us all.

Acknowledgements

There is a myth that books are the products of a solitary writer working alone like a two-armed version of an octopus. It is true that every writer spends long hours with only the sound of her own voice for company, muttering imprecations as she drafts and deletes sentence after sentence for failure to satisfy. But the larger truth is that like everything else human beings do, the creation of a nonfiction book is a group endeavor.

I began this one because of a class reunion, a shared experience, but most of all because Stephen Dewar, my best friend, loving husband, and closest colleague, insisted that there was much to discover in this subject. Thank you, Stephen, for making me write the ***New Wilderness*** chimp shows, and for urging me to stick with ***Smarts*** even after more practical folks warned that I was heading into good-money-after-bad territory. And thanks as well to Sam Hiyate, my agent at The Rights Factory, who said yes, there's a book in this, go for it. His encouragement turned into something much more valuable than a cheer: he sold the proposal to others.

During the research phase, my former classmates responded to many queries and phone calls and even hunted down our former teachers who were also very helpful. Thanks to all of you. Many busy scientists welcomed me into their labs, shared their ideas, showed me their experiments, told me their life stories. Just when they thought they'd heard the last of me, I was back in their email inboxes again asking them to read and correct what I'd said of their work. Needless to say, many corrections were necessary. Their patience and kindness to me has made ***Smarts*** a much smarter book than it would have been without their help. A few were so irritated or so pressed for time that they were unable to respond in this second round of queries. I am grateful for the time they gave me during the interview process and take responsibility for any errors that remain.

This brings me to Nick Garrison and Penguin Canada. This book would not have come to life if Nick hadn't read the proposal on the subway home, decided it was fascinating, and said let's do it. But even as I got on planes and trains to get the job done, the book publishing industry entered a fullblown crisis, thoroughly disrupted by waves of technological innovations, including the rise of print-on-demand and instant delivery publishing brought to you by Amazon and other platform makers, smart phones capable of hosting mobile versions of ebooks, and the demise of most big bookstore chains along with many of the small independents. There were massive mergers and acquisitions as a shrinking industry tried to survive this onslaught. Working in publishers' offices became something akin to playing a game of musical chairs. Jobs disappeared. Talented people had to reinvent themselves. When ***Smarts*** was finally finished, Penguin was a different company and no longer in love with it.

I am a doctor's daughter, so I went out and got second and third opinions. Should ***Smarts*** see the light of day? My friend and long-time business associate Charles Greene read it, made many notes, and said yes. My former editor and forever friend, Dawn Macdonald Deme, said the same. Meanwhile, agent Hiyate sold translation rights.

So this book has been brought to you by the very technologies that have put the old form of the book industry in peril and with the aid of wonderful publishing professionals. Philip Turner is a longtime bookstore owner; editorial executive; publishing consultant; and devoted book blogger, writing at TheGreatGrayBridge.com and HonouraryCanadian.com. His company is Philip Turner Book Productions. When Philip was an in-house editorial executive, he published two other books of mine in the US. Philip's edit of ***Smarts*** flushed out syntactical horrors, contradictions, factual errors, and other infelicities. If any remain, it is my fault. ***Smarts*** was designed and produced by a marvelously talented art director, Carmen Dunkjo of pod10.com. Carmen has visually branded many of Toronto's and Canada's most important cultural institutions, such as the National Ballet of Canada, Soulpepper Theatre Company, and the Elgin Winter Garden Theatres. She has been the creative director for national magazines, such as *Shift* and *Saturday Night,* and she has designed books too numerous to name. There are a lot of prizes on her shelves—more than 450. Thank you both for making ***Smarts*** look good and read well. Thanks as well to Anna Dewar Gully who suggested setting out a visual guide to the important names in this book. That led to the Smarts Mind Map. Without the social media marketing undertaken by Danielle Dewar Smollet, you would never have heard of ***Smarts.*** Thank you, Danielle, for helping me to navigate a brave new world.

As always, although I have had so much help along the way I hereby declare that any errors or omissions are my own.

End Notes

1 It seems odd to have to add identifiers for Lorne Greene but he would understand. He used to describe an actor's career arc in the following way. "Who is Lorne Greene?...Get me Lorne Greene....Get me a Lorne Greene look-alike....Who is Lorne Greene?"

2 Gaia is named after the Greek earth goddess which didn't endear him to academic scientists, but Lovelock had become an independent scientist doing his work for hire and he didn't care.

3 For a spectacular work that recounts the lives of Leakey and his family, read the magisterial *Ancestral Passions,* by Virginia Morell, Simon & Schuster, 1996. Leakey tried to get into the beds of all three angels and was certainly very close to Jane Goodall's mother. His wife, Mary, put up with that and much worse while finding many important fossils herself.

4 George John Romanes, *Animal Intelligence,* published by Kegan Paul, Trench and Company, London, 1882.

5 "The Mysterious Reichmanns: The Untold Story" appeared in *Toronto Life Magazine* in November, 1987.

6 Elaine Dewar, *The Second Tree: of Clones, Chimeras and Quests for Immortality,* Random House of Canada, 2004, p. 123.

7 The same Alfred Wallace who created a theory of evolution by natural selection at the same time as Darwin and whose paper on the subject was published with Darwin's first paper when it was read out at a meeting of the Linnean Society. Wallace was an itinerant traveler and collector of samples of previously unknown living things and came to his theory while ill with a fever in Southeast Asia where he was collecting.

8 Galton wrote to his mother on August 16, 1851 as follows: "I have a long story to tell.... The Guamopo are a charming set of niggers but almost all the other nations I have heard of are brutal and barbarous to an almost incredible degree. The Ovaliona, a very extended nation, attacked a village the other day for fun and after killing all the men and women there, tied the childrens' legs together by the ankles and

shoving them head downwards on a long pole which they set horizontally between two trees, then they [put] plenty of reeds together and put them underneath and lighted them and the children were dying poor wretches, half burnt, half suoffocated and they danced a day around them. I made a fine joke of it. Anderson desires particularly to be remembered to all...." Frank Galton. Galton Archives, University of London.

9 Francis Galton, letter to the Editor, *The Times*, June 5, 1873.

10 Theory one: we surged out of Africa into Europe through the Levant about 45,000 years ago and wiped out the Neanderthals because we could talk, use symbols, and were so very smart that we out-competed them. Of course we had no children with such riffraff. Theory two: various *Homo erectus* descendants developed many variations as they moved around the planet, but they interbred. Out of this gene flow, as its main proponent called it, modern humans emerged. A Neanderthal/ modern human full genome comparison would later demonstrate, to the shock of those who did it, that this second theory is closer to the truth. A fourth *Homo* species, Denisovans, whose bones were recently found in a cave in Siberia, is also part of our modern mix.

11 Animals tend to reduce in size over time when confined to small islands but the reduction follows fairly predictable patterns. Among the bones found with the hobbit were those of a now extinct dinosaur called stegadon which had shrunk down to the size of a small pony. Falk and colleagues found that this skull and body in some ways seemed to derive from an *Australopithicine* and in others, *Homo erectus.*

12 Dean Falk, et al," The Brain of LB1, *Homo floresiensis*," *Science*, Vol.308, 8 April, 2005, pp. 242-245

13 Jens Krause, Graeme D. Ruxton and Stefan Krause, "Swarm intelligence in animals and humans," *Trends in Ecology and Evolution*, Vol. 25, No.1, 6 September 2009 pp. 28-34

14 Samuel R. Laycock, *Gifted Children.* Toronto: The Copp Clark Publishing Co. Ltd., 1957, p. 11

15 *Webster's New World Dictionary Second College Edition* (1980). Cleveland: William Collins Publishers, Inc., p. 732

16 www.wikipedia.org/wiki/Nuremberg_trials#Intelligence_tests_and_psychiatric_assessments

17 Mona Gleason, *Normalizing the Ideal of Psychology, Schooling and the Family in Postwar Canada,* University of Toronto Press, 1999, p. 130

18 See Arthur Jensen, "Charles Spearman: Founder of the London School,"at www.galtoninstitute.org/uk/Newsletters/GINL0003/charles_spearman.html

19 Lewis M. Terman, "The Intelligence Quotient of Francis Galton in Childhood," *The American Journal of Psychology,* Vol. 2, No.2, April 1917, pp.209-215

20 http://www.enwikipedia.org/wiki/Lewis_Terman

21 Ibid, Gleason, p. 46

22 Ibid, Laycock, p. 118

23 Laycock UBC bio

24 Ibid, Laycock, p. 10

25 Ibid, Laycock, p.3

26 Ibid, Laycock, p. 27

27 D.O. Hebb, *The Organization of Behavior,* John Wiley & Sons, New York, 1949, p. 302

28 Birute M. Galdikas, *Reflections of Eden,* Little, Brown & Company, 1995, p. 38

29 Ibid, Galdikas, p. 43

30 Ibid, Galdikas, p. 43

31 Ibid, Galdiakas, p. 42. She describes a moment at a lecture when she first hears of Jane Goodall's work with chimpanzees at Gombe. She said it was like hearing the sound of a bell, a bell she knew meant something for her.

32 This rule has been coming undone since the publicly-funded Human Genome Project researchers, afraid the human genome would be patented by private company researchers, insisted that all findings be published on line as they came off the lab bench, establishing

a new kind of ongoing form of publication in which revisions are constantly made.

33 Konrad Z. Lorenz, "Analogy As A Source Of Knowledge", Nobel lecture, December 12, 1973

34 Ibid, Lorenz

35 Anne E. Russon, Alain Compost, Purwo Kuncoro, Agnes Ferisa "Orangutan fish eating, primate acquatic fauna eating and their implications for the origins of ancestral hominin fish eating," *Journal of Human Evolution*, July 2014, online at DOI: 10.1016/j.hevol.2014.06.007

36 Anne E. Russon, Carel P. van Schaik, Purwo Kuncoro, Agnes Ferisa, Dwi P. Handayani and Maria A. van Noordwijk, "Innovation and intelligence in orangutans," in *Orangutans: Variation in Behavioral Ecology and Conservation*, S. A Wich, S.S. Utami, T. Mitra Setia and C.P. van Schaik (eds.), Oxford University Press, 2009, pp. 279-298

37 Anne E. Russon and Kristin Andrews, "Pantomime in Great Apes,"*Communicative & Integrative Biology*, Vol. 4, Issue 3, May-June, 2011, pp. 1-3

38 Ibid, Russon, Andrews

39 Ibid, Russon, Andrews

40 Toshiyuki Nakagaki, Hiroyasu Yamada, Agota Toth," Maze-solving by an amoeboid organism,"*Nature*, vol 407, 28 September, 2000. p. 470

41 Toshiyuki Nakagaki, Hiroyasu Yamada, Agota Toth," Path-finding by tube morphogenesis in an amoeboid organism,"*Biophysical Chemistry*, Vol.92, 2001, pp.47-52

42 Tetsu Saigusa, Asushi Tero, Toshiyuki Nakagaki and Yoshiki Kuramoto," Amoebae anticipate periodic events," *Physical Review Letters*, Vol. 100, Jan. 3, 2008

43 Toshiyuki Nakagaki, Hiroyasu Yamada, Masahioko Hara," Smart network solutions in an amoeboid organism," *Biophysical Chemistry*, 107 (2004) 1-5

44 Andrew Westoll, *The Chimps of Fauna Sanctuary*, Harper Perennial, 2012

45 National Institutes of Health, News Release December 15, 2011, Statement by Dr. Francis Collins

46 It had lately changed its name to the The Iowa Primate Learning/ Bonobo Hope Sanctuary or IPLS.

47 See: Itai Roffman's euology for Panbanisha, at http://www.panbanisha.org/itai-roffman-euology-for-panbanisha

48 Ernest Barker, Ed. *The Politics of Aristotle*, Oxford University Press, 1962, p.6.

49 And identifiable by chimps who can recognize familial relations by looking at pictures. See: "Visual kin recognition in chimpanzees," Lisa A. Parr, Frans B.M. de Waal, *Nature*, Vol.399, 17 June, 1999, p. 647

50 Frans B.M. de Waal, "The Brutal Elimination of a Rival Among Captive Male Chimpanzees," *Ethology and Sociobiology* 7:237-251 (1986)

51 Frans de Waal, *Chimpanzee Politics: Power and Sex Among Apes*, Johns Hopkins Paperback editions, 2007, p. 214. "The newspapers called it 'suicide' but more likely it was a panic attack with a fatal outcome" is how de Waal described this awful denouement in the epilogue to the 2007 re-release of his 1982 book.

52 Andrew Witten, Victoria Horner, & Frans B.M. de Waal, " Conformity to cultural norms of tool use in chimpanzees," *Nature*, Vol. 437/29 September, 2005, p. 737

53 Amy S. Pollick and Frans B.M. de Waal, " Ape gestures and language evolution," *PNAS*, vol. 104, no. 19, pp.8184-8189

54 Seyfarth, R., Cheney D., Marler P. "Monkey responses to three different alarm calls: evidence of predator classification and semantic communication," *Science*, 210: 801, 1980

55 Zanna Clay, Klaus Zuberbuhler, "Bonobos Extract Meaning from Call Sequences, " *PLoS One*/Vol. 6, Issue 4, April 2011

56 Frans de Waal, "Sex Differences in the Formation of Coalitions Among Chimpanzees," *Ethology and Sociobiology* 5 (239-255) 1984

57 Ibid, de Waal

58 He was right about that, of course. Rob Shumaker at the Indianapolis Zoo had already refused to speak to me about the goings on at the

Great Ape Trust when he was there: talk about orangutans yes, Great Ape Trust talk, no. Similarily, in October, 2013, the Toronto Zoo finally sent its three elephants off in their carry crates for a five day nonstop trip to a sanctuary in California. When the caravan of crates, trucks and cars arrived at the US border, the company organizing the move demanded that the Toronto Zoo keepers accompanying the elephants sign a non disclosure agreement promising never to talk about, or take pictures of, or describe to anyone, what happened after they crossed the US border. If they didn't sign, they had to go back to Toronto.(CBC local radio news, October 17, 2013)

59 Andrew Witen, Victoria Horner & Frans B.M. de Waal, "Conformity to cultural norms of tool use in chimpanzees,"*Nature*, Vol.437, 29 September, 2005, pp. 737-740.doi:101038/nature04047

60 Nicer is very much in doubt: dolphin behavior in and between groups can be violent and war-like. Young dolphin males like gang rape and riot as much as young human males do, though female dolphins babysit each others' kids.

61 John C. Lilly, "Dolphin Human Relation and LSD25" first published in *The Use of LSD in Psychotherapy and Alcoholism*, edited by Harold Al Abramson, M.C. Bobbs-Merrill, 1967, pp. 47-52.

62 Esther Herrmann, Josep Call, Maria Victoria Hernandez-Lloreda, Brian Hare, Michael Tomasello, "Humans Have Evolved Specialized Skills of Social Cognition: The Cultural Intelligence Hypothesis," *Science* Vol 317 7 September, 2007

63 Theory of mind means the animal understands that the experimenter has a particular view of the world that might differ from his own.

64 Jamie L. Russell, Heidi Lyn, Jennifer A. Schaeffer and William D. Hopkins,"The role of socio-communicative rearing conditions in the development of social and physical cognition in apes,"*Developmental Science* 14:6 (2011) 1459-1470

65 Ibid, Russell et al

66 Yasuo Nagasaka, Daniel I. Brooks, Edward A. Wasserman, "Amodal completion in bonobos," *Learning and Motivation* 41 (2010) 174-186

67 http://www.mtholyoke.edu/acad/facultyprofiles/karen_hollis

68 The NSF website stipulates that grant money is available for the study of cognition—but only if the work has a direct impact on basic issues of human perception, action, or cognition. Comparative animal behavior research is also supported through its Behavioral Systems cluster—but only if a study involves the biology of behavior from genes to ecosystems. The Evolutionary Processes cluster seemed to focus on the micro and macro effects of the molecular biology of change.

69 That issue made it to the front pages of the *Toronto Star*. The paper had investigated allegations of improper care of animals at Marineland, a large and very popular theme park in southern Ontario where a killer whale and belugas and other smart animals are kept for the entertainment of the paying public. According to the *Star*, Marineland had been deficient in the management of its water system, putting the marine mammals at risk. Disgruntled staff said the marine mammals had sore eyes, some had infections, and the land animals were also living in overcrowded conditions. The *Star* followed up later with a story about dead animals dumped into a mass unmarked grave. The OSPCA finally launched an investigation. The investigation found Marineland to be in compliance with the general rules against cruelty to animals, but it was pointed out that there are no regulations regarding how zoos house their animals.

70 Alan C. Kamil, "A Synthetic Approach to the Study of Animal Intelligence,"in *Behavioral Mechanisms in Evolutionary Biology*, University of Chicago Press, 1994

71 Cichlids are a large family of several thousand species, many still not described. They tend to be social, take care of their eggs, feed their young, and live in groups. They include species such as Tilapia, and various smaller types often seen in home acquariums.

72 John C. Lilly, M.D. "Dolphin-Human Relation and LSD 25 found online at http://www.psymon.com/psychedelia/articles/lilly.htm taken from article in *The Use of LSD in Psychotherapy and Alcoholism*, Harold A. Abramson, M.D. ed., Bobbs-Merrill, 1967, pp.47-52

73 A few scholars do study animal 'languages' using information theory. Without trying to understand the meaning conveyed, the application of Claude Shannon's information theory permits scholars like Laurance Doyle at SETI in Mountain View, California, and Brenda McCowan at University of California, Davis to analyze the information content of animal signals. See also "Using a Self-Organizing Map (SOM) and the Hyperspace Analogue to Language (HAL) Model…". by Kaufman, Allison B. et al., *International Journal of Comparative Psychology*, Vol. 25, 2012, pp. 237-275.

74 Louis M. Herman, "Cognition and Language Competencies of Bottlenosed Dolphins," in *Dolphin Cognition and Behavior: A Comparative Approach.* Ronald J. Schusterman, Jeannette A. Thomas, Forrest G. Wood, eds., Lawrence Erlbaum Associates Inc. Publishers, 1986, pp. 221-262

75 Louis M. Herman, Douglas Richards, James P. Wolz,"Comprehension of sentences by bottle-nosed dolphins," *Cognition*, Vol. 16, Issue two, March 1984, pp. 129-219

76 Louis M. Herman, "Exploring the Cognitive World of the Bottlenosed Dolphin,"in *The Cognitive Animal: Empirical and Theoretical Perspectives on Animal Cognition,* Mark Beloff, Colin Allen, Gordon Burghardt ets., A Bradford Book, 2002

77 Eduardo Mercado III, Scott O. Murray, Robert K. Uyeama, Adam A. Pack, and Louis M. Herman, "Memory for recent actions in the bottlenosed dolphin (*Tursiops truncates*): Repetition of arbitrary behaviors using an abstract rule,"*Animal Learning and Behavior*, 26 (2) 210-218, 1998

78 In the 1970s, after Terrace concluded that poor Nim Chimpsky had only learned the signs for the foods he wanted, but not any syntax, Terrace wrote papers and went on talk shows to spread the word of his own failure and to argue the impossibility of primates using language. Since he was an alpha male among his colleagues, he was listened to.

79 She had already published on the use of underwater keyboards with visual symbols for the dolphins to press so the dolphin would get the object it wanted.

80 George Dyson, *Turing's Cathedral: The origins of the Digital Universe*, Pantheon, New York, 2012.

81 Andrew Hodges' website about Turing: http://www.turing.org.uk

82 George Dyson, *Turing's Cathedral*, p. 48

83 Ibid, Dyson, p. 51.

84 Andrew Hodges, *Alan Turing: The Enigma*, first published by Burnett Books Ltd., 1983, all citations from the Vintage edition, 1992

85 Ibid, Hodges, p. 62

86 Lord Victor Rothschild was once suspected of having been a member of the Cambrige ring as was Ludwig Wittgenstein, the great logician/mathematician. Wittgenstein came from an extremely wealthy family in Austria, with Jewish forebears, which was forced to hand over a great chunk of its wealth to the Nazis. He taught Turing along with Alistair Watson when Turing returned to Cambridge in 1937/38.

87 Newman, formerly Neumann, had been a scholarship student at Cambridge himself, winning the title wrangler, as a result of placing in the first rank in the second part of Cambridge's three-part mathematics examinations. He was a conscientious objector in World War I. He was a topologist whose dissertation was on the use of symbolic 'machines' in physics. This idea prefigured his work with Turing that led directly to the first digital electronic computer, the Colossus, built at Bletchely Park to decipher Nazi codes during World War II. After the war, Newman created the first lab dedicated to computers at University of Manchester and hired Turing.

88 A.M. Turing,"On Computable Numbers, with an Application to the Entscheidungsproblem," *Proceedings of the London Mathematical Society*, Series 2, Vol. 42, 1937

89 Proof, according the brilliant mathematician and statistician I.J. Good, who assisted Turing during WWII, that originality is more important than intelligence. And what is the difference between intelligence and originality? Aren't they the same? To me, originality lies on the far side of intelligence, somewhere near where genius lurks.

90 George Dyson, *Turing's Cathedral*, p. 103

91 Stephen Wolfram, *A New Kind of Science*, Wolfram Media Inc., 2002, p.1107, and p.1184

92 George Dyson, *Turing's Cathedral*, pp. 86-87

93 George Dyson, *Turing's Cathedral*, p.254

94 In the summer of 1940, according to Andrew Hodges, when it looked as if the British would lose the air war and/or starve, Turing took all of his salary, turned it into 250 pounds sterling worth of silver bars and buried those bars near the town of Shenley, close to Bletchley Park. He may have been in despair because his problem was so terribly difficult, or maybe he thought invasion was imminent.

95 Christopher Andrew and Oleg Gordievsky, *KGB: The Inside Story*, Hodder&Stoughton, 1990, pp. 246-248

96 When the Americans entered the War, their secret operatives were briefed by Kim Philby about MI6's counterespionage section, the structure of the British secret services, and decoding operations at Bletchley Park." See: *A Spy Among Friends*, Ben MacIntyre, p. 69.

97 Christopher Andrew and Oleg Gordievsky, Ibid, p. 249

98 The software was developed under contract by William Hamilton through his company Inslaw for the Department of Justice. The copyright reverted to Hamilton due to a change in copyright law: he made a new version which drew the interest of Rafi Eitan acting for Israel's Mossad. He refused to sell. The government then stopped paying on his contract. Hamilton's company was pushed into bankruptcy. The government tried to force liquidation. There were three trials and two findings against the government by two federal judges who were not reappointed. There were two congressional hearings. The Department of Justice was accused of distributing the software to allies such as Israel, Canada, Singapore, Jordan, etc. There was talk of a backdoor in the software, a Trojan horse. The last trial absolved the Department of any wrongdoing.

99 Barton Gellman, Ashkan Soltani,"NSA tracking 5 billion phone records a day," *Toronto Star*, A 39, Dec.5, 2013 taken from the *Washington Post* service.

100 Delilah, Samson's lover, was offered a lot of silver by the Philistines to find out how Samson stayed strong. Samson first told her three false stories, but eventually confessed that if he cut his hair he would lose his strength. So while he slept, Delilah had him shaved. The Philistines blinded him and bound him and made him a slave.

101 George Dyson, *Turing's Cathedral*, p.79

102 Andrew Hodges, *Alan Turing the Enigma*, p. 415

103 A.M. Turing, "Computing Machinery and Intelligence," *Mind* 59, 433-460, 1950

104 A.M. Turing," The Chemical Basis of Morphogenesis, " *Philosophical Transactions of the Royal Society of London. Series B. Biological Sciences*, 14 August, 1952

105 See: Turing.Org.uk.

106 Canadian diplomat E. Herbert Norman was hounded by American security officials from the early 1950s on the grounds that he had been a member of Communist organizations while at school at Trinity College, Cambridge in the 1930s, and favoured the retention of the Communist party as an active political party in Japan after WWII. He committed suicide when he was Ambassador to Egypt in 1957 after a US Senate Commitee began to ask questions about him again. John Watkins, Canadian Ambassador to the Soviet Union, and a gay man, killed himself while under forcible RCMP interrogation in Montreal in 1964.

107 Christopher Andrew, Oleg Gordievsky, *KGB The Inside Story of Its Foreign Operations from Lenin to Gorbachev*, Hodder&Stoughton, London, Sydney Auckland, Toronto, 1990

108 Peter Wright, *Spycatcher: The Candid Autobiography of a Senior Intelligence Officer*, Stoddart Publishing Company Ltd, 1987, pp.362, 363

109 Ben Macintyre, *A Spy Among Friends: Kim Philby and the Great Betrayal*, Signal, McClelland & Stewart, 2014, pp.180-184

110 See: Daily Mail interview, November 17, 2012, with niece Inagh Payne.

111 Inman Harvey, "Robotics: Philosophy of Mind With A Screwdriver," in *Evolutionary Robotics: From Intelligent Robots to Artificial Life, Vol.111,* T.Gomi (ed), AAI Books: Ontario, Canada, 2000, pp. 207-2003

112 Ibid, Harvey

113 Narrative Networks, Solicitation no. : DARPA –BAA-12-03 at http://www.fbo.gov/index?s=opportunity&mode=form&id=fd625a4022

114 Steven Levy, "Can An Algorithm Write A Better News Story Than A Human Reporter? " *Wired,* April 24, 2012. The name of the company is Narrative Science and it is based in Chicago.

115 See Malcolm McIver lab website at: nxr.northwestern.edu/tech-transfer.

116 Jennifer Mather, "Do Cephalopods Have Pain and Suffering?" paper given at International Symposium organized by La Fondation Droit Animal, ethique et sciences, in collaboration with the International Research Group in Animal Law, Paris, October 12-18, 2012

117 Roland C. Anderson, Jennifer A. Mather, Matheiu Q. Monette, Stephanie R.M. Zimsen, "Octopuses (*Enteroctopus dofleini*) Recognize Individual Humans," *Journal of Applied Animal Welfare Science,* 13: 3,261-272, 2010

118 Jennifer A. Mather, "'Home' " Choice and modification by juvenile *Octopus vulgaris* (*Mollusca Cephalopoda*): specialized intelligence and tool use*?" Journal of Zoology,* London, Vol. 233, p.359-368, 1994

119 Jennifer Mather, Tatiana Leite, Roland C. Anderson and James B. Wood, "Foraging and cognitive competence in octopuses," in *Cephalopod Cognition,* Anne-Sophie Darmaillaq, Ludovic Dickel & Jennifer Mather (eds.), Cambridge University Press, 2014

120 Jennifer Mather, "Cephalopod Skin Displays: From Concealment to Communications,"in *Evolution of Communication Systems,* Kim Orler & Lili Griebel (Eds), MIT Press, 2004

121 Ibid, Roland C.Anderson.et al

122 Ibid, Jennifer Mather

123 Sir Francis Galton, "Vox Populi," *Nature,* March 28, 1907, No. 1949, vol.75, pp. 450-451

124 See: Anthony Trewavas, "Le Calcium, Cést la Vie: Calcium Makes Waves." *Plant Physiology*, Vol. 120, May, 1999, pp. 1-6

125 Anthony Trewavas, "Mindless Mastery," *Nature* Vol.415, 21 Feb.2002, p. 841

126 Anthony Trewavas, "Aspects of Plant Intelligence," *Annals of Botany*,92, pp. 1-20, 2003. He wrote: "To commence use of the term intelligence with regard to plant behavior will lead to a better understanding of the complexity of plant signal transduction and the discrimination and sensitivity with which plants construct images of their environment, and raises critical questions concerning how plants compute responses at the whole plant level...."

127 Anthony Trewavas, " Le Calcium, C'est la Vie: Calcium Makes Waves," *Plant Physiology*, Vol.120, May, 1999,pp.1-6

128 Ibid, Trewavas, 1999

129 Trewavas, "Mindless Mastery," *Nature*, Vol. 415, 21 February 2002, p.841

130 Trewavas, "Aspects of Plant Intelligence," *Annals of Botany*, Volume 92, May, 2003, pp. 1-20

131 Anthony Trewavas, "Plant Intelligence," "*Naturwissenshaften*, (2005) 92, pp. 401-403

132 E. Masi, M. Ciszak, G. Stefano, L.Renna, E. Azzarello, C. Pandolfi, S. Mugnai, F. Baluska, F.T. Arecchi and S. Mancuso, "Spatiotemporal dynamics of the electrical network activity in the root apex," *PNAS*, March 10, 2009, Vol.106, pp. 4048-4053

133 Amadeo Alpi, Nikolaus Amrhein, Adam Bertl, Michael R. Blatt, Eduardo Blumwald, Felice Cervone, Jack Dainty, Maria Ida de Michelis, Emanuel Epstein, Arthur W. Galston, Mary Helen M. Goldsmith, Chris Hawes, Rudiger Hell, Alistair Hetherington, Herman Hofte, Gerd Juergens, Chris J. Leaver, Anna Moroni, Angus Murphy, Karl Oparka, Pierdomenico Perata, Hartmut Quader, Thomas Rausch, Christophe Ritzenthaler, Alberto Rivetta, David G. Robinson, Dale Sanders, Ben Scheres, Karin Schumacher, Herve Sentenac, Clifford L. Slayman, Carlo Soave, Chris Somerville, Linconl Taiz, Gerhard Thiel, and

Richard Wagner. "Plant neurobiology: no brain, no gain?" *Trends in Plant Science,* Vol. 12, No.4, March 2007

134 E.Masi, M. Ciszak, G. Stefano, L. Renna, E. Azzarello, C. Pandolfi, S. Mignai, F. Baluska, F.T. Arecchi and Stefano Mancuso, "Spatiotemporal dynamics of the electrical network activity in the root apex," *PNAS,* March 18, 2009, Vol.106, no. 10, pp.40480-4053

135 Hubert Rehm and Dietrich Gradmann, "Intelligent Plants or Stupid Studies," *Lab Times,* .3, 2010, p. 30

136 See: Frantisek Baluska and Stefano Mancuso, "Deep evolutionary origins of neurobiology," *Communicative and Integrative Biology,* 2: 1, pp.60-65, February 2009.

137 According to J.E.R. Staddon in *Adaptive Dynamics: The Theoretical Analysis of Behavior,* Loeb studied developmental programming in sea urchin eggs, setting them on their development path without sperm simply by chemically modifying the water they were kept in. He was also the teacher of John B. Watson, the 'father'of behaviorism.

138 See: Francis Darwin's speech to the British Association for the Advancement of Science as reproduced in *Science* New Series, Vol.28, No. 716, pp. 353-362.

139 See: Pickard, "Action Potentials in Higher Plants," *The Botanical Rview,* Vol. 39, No.2. April-June 1973.

140 Actually, I found several mentions of excitable cells in plants on the Internet, starting with Wikipedia and going back as far as the 1960s. Interestingly, Dietrich Gradmann, who had complained so bitterly about Mancuso's and his colleagues' use of the word neurobiology in *Lab Times,* had published on such signals in fungi in the 1970s, and in plants in the early 1990s.

141 The paper is called: "Root apices as plant command centres: The unique brain-like status of the root apex transition zone." The first author is Frantisek Baluska, and it appeared in *Biologi (Bratisl.)* Vol. 59, Suppl.13, 2004, pp. 1-13.

142 See:http://www.entecarifirencze.it.

143 Jean Chretien, *My Years as Prime Minister*, Ron Graham Books, Random House of Canada, 2008

144 http://www.mit.edu/bcs/bizzilab/overview/index.html

145 See: Monica Gagliano, Stefano Mancuso and Daniel Robert,"Towards understanding plant bioacoustics," *Trends in Plant Science*,"Vol. 17, Issue 6, June, 2012, pp. 323-325.

146 Marzena Ciszak, Diego Comparini, Barbara Mazzolai, Frantisek Baluska, F. Tito Arecchi, Tamas Vicsek, Stefano Mancuso, "Swarming Behavior in Plant Roots,"*PLoS One*, January 17, 2012

147 AiSung Tseng and Michael Levin, "Cracking the bioelectric code,"*Communicative & Integrative Biology*, "Vol. 6, Issue 1, e22595-1-8, http://dx.doi.org/10.4161/cib.22595, 2012

148 Michael Levin, "The Wisdom of the body: future techniques and approaches to morphogenetic fields in regenerative medicine, developmental biology and cancer," *Regenerative Medicine*, Vol.6, issue 6, pp. 667-673, 2011

149 Julian Jaynes, *The Origin of Consciousness in the Breakdown of the Bicameral Mind,* Houghton Mifflin Company, 1976, pp.6-7

150 Dario Floreano, Laurent Keller " Evolution of Adaptive Behaviour in Robots by Means of Darwinian Selection," *PLoSbiology,* January 26, 2010 http://www.plosbiology.org/article/info%3Adoi%2F10.1371%Fjournal.p

151 Tatiana Giraud, Jes. S. Pedersen and Laurent Keller, "Evolution of Supercolonies: The Argentine ants of southern Europe," *PNAS,* Vol.99, No. 9, pp. 6075-6079, April 30, 2002

152 Ibid, Giraud, et al

153 See: Forward *to The Connection Machine* by W. Daniel Hillis, MIT Press, 1985

154 It is named for the Egyptian god Kheper or Khapir who renews himself every day. Often he is pictured as a scarab, a dung beetle. The dung beetle rolls up a ball of dung, lays its eggs in it, which hatch, eat the dung and emerge to roll balls of dung. The name Khepera has an implication of self invention or self renewal.

155 Inman Harvey, "Evolving Robot Consciousness: The Easy Problems and the Rest," in *Evolving Consciousness*, J.H.Fetzer (ed.), in Consciousness Research Series, John Benjamins, Amsterdam, pp.205-219, 2002

156 Inman Harvey, "Robotics: Philosophy of Mind using a Screwdriver," in *Evolutionary Robotics: From Intelligent Robots to Artificial Life*, Vol.111, 2000, pp. 207-230

157 Andrew J. Watson, James E. Lovelock, "Biological homeostasis of the global environment: the parable of Daisyworld, Tellus B, Vol 35B, Issue 4, pp.284-289, September 1983

158 Timothy Lenton and Marcel van Oijen, "Gaia as a complex adaptive system," *Philosophical Transactions of the Royal Society, London B* (2002), Vol. 357, pp.683-695

159 J. McDonald-Gibson, J.G. Dyke, E.A. Di Paulo, I.R. Harvey, "Environmental Regulation can arise under Minimal Assumptions," *Journal of Theoretical Biology*, Vol.251, Issue 4, 21 April, 2008, pp. 653-666

160 Inman Harvey, "Homeostasis and Rein Control: From Daisyworld to Active Perception," in *Artificial Life IX, Proceedings of the Ninth International Conference on the Simulation and Synthesis of Living Systems*, J. Pollack, M. Bedau, P. Husbands, T. Ikegashi, and R. Watson,(eds.) MIT Press, pp. 309-315, 2004

161 And I did. Geoffrey Hinton was born in Wimbledon and brought up in Bristol. His father, who'd been raisedin the US, studied in the UK before WWII to become a leading entomologist, eventually awarded an FRS. Howard Everest Hinton's British ancestors included the founder of mathematical logic, George Boole. Geoffrey Hinton started in philosophy at Cambridge the same year as Harvey. He switched to psychology in his last year, and then spent a year learning carpentry before studying psychology at Edinburgh from a man who had once been interested in neural networks. Hinton had become interested in neural networks thanks to his friend Inman Harvey. Hinton went on to innovate in the field of machine learning, developing

randomized neural networks capable of learning and 'internal representations,' along with another world leader, Terry Sejnowksi. Hinton would be in the news a few months after I returned to Toronto when he did a deal with Google. Google bought the company he and his postdocs set up whose only assets appeared to be a blank website and a new learning algorithm useful for 'deep learning' which mimics the way networked neurons connect through layers of the neocortex. Uncharacteristically, Google agreed to let Hinton retain his academic position and publication freedom so long as he agreed to spend afternoons at Google's Toronto office and a few months each year hanging out at the Google campus in Mountain View, California.

162 Jan Wong, *Red China Blues*, Doubleday Canada, 1996, p. 16-19

163 Dr. Denis Marrian, a chemist, came from Manchester to Cambridge as a Teaching Fellow in 1959 (while the infamous Kim Philby, still under suspicion by MI-5 but cleared by MI-6, was pretending to be a journalist but actually spying in the Middle East). He became a Tutor in 1963 and a Senior Tutor in 1964. He tutored Prince Charles (at the same time as Harvey) and the son of Lee Kwan Yew, Prime Minister of Singapore, which suggests someone in the upper reaches of the British government thought him worthy of an extremely high degree of trust.

164 R.C. Lewontin, *Biology as Ideology, the Doctrine of DNA*, House of Anansi Press, 1991, p. 13

165 See remarks of Jerry Coyne, at http://www.whyevolutionistrue.wordpress.com/2011/04/12/lynn-margulis-disses

166 James A. Shapiro, "Bacteria as Multicellular Organisms," *Scientific American*, June, 1988, pp. 82-89

167 J.A. Shapiro, "Bacteria are small but not stupid: cognition, natural genetic engineering and socio-bacteriology," *Studies in History and Philosophy of Biological and Biomedical Sciences*, Vol 38, (2007) 807-819

168 Myron L. Smith, Johann N, Bruhan & James B. Anderson, " The fungus *Armillaria bulbosa* is among the largest and oldest living organisms," *Nature* 356, 02 April, 1992, pp. 428-431

169 James A. Shapiro, *Evolution: A View from the 21st Century*, FT Press Science, 2011

170 Example: Francis Crick held that hereditary information flows in one direction only, from a gene to an RNA molecule which commands the creation of a protein which does a job in a cell, never the other way around. He called this the Central Dogma. It was overturned by the late 1970s.

171 To see an animation of a *lac* operon, go to: http://www.biology-animations.blogspot.com/2007/11/lac-operon-animation.html

172 Jacob and Jacques Monod had won the Nobel for discovering that *E. coli* regulates its own gene expression. They found that a repressor involved with lactose metabolism prevents the transcription of a DNA sequence into RNA, thus preventing the assembly of an enzyme that digests lactose until lactose is actually available. When lactose is available, this repressor binds to it, freeing the gene to express so RNA will make more of the digesting enzyme. In effect, this is a dynamic molecular communication system.

173 Barbara McClintock, "The Significance of Responses of the Genome to Challenge," Nobel lecture, December 8, 1983

174 James A. Shapiro, " Bacteria as Multicellular Organisms," *Scientific American,* June 1988, pp.82-89

175 James A. Shapiro," The significance of baterial colony patterns," *BioEssays,* Vol. 17, No.7, 1995, pp. 597-607

176 Cairns, J., Overbaugh J., Miller, S. "The origin of mutants." *Nature,* September 8, 1988, (335(6186), pp. 142-145

177 J.A. Shapiro, "Bacteria are small but not stupid: cognition, natural genetic engineering and socio-bacteriology," *Studies on History and Philosophy of Biological and Biomedical Sciences,* Vol.38, 2007, pp. 807-819

178 Thomas Walkom, "Tory cost cuts could doom 4 species," *Toronto Star,* Feb.19, 2014, p. A9

179 Chris Eliasmith, Terrence C. Stewart, Xuan Choo, Trevor Bekolay, Travis DeWolf, Yichuan Tang, Daniel Rasmussen, "A Large-Scale

Model of the Functioning Brain,"*Science,* Vol. 338, No.6111, 30 November, 2012, pp. 1202-1205

180 Quoc V. Le, Marc Aurelio Ranzato, Rajat Monga, Matthieu Devin, Kao Chen, Greg S. Corrado, Jeffrey Dean, Andrew Y. Ng, "Building High-Level Features Using Large Scale Unsupervised Learning, International Conference on Machine Learning, Edinburgh, Scotland, June 2012

181 Adam Clark Estes,"Google's Robot Army: It's Growing,"*Gizmodo,* Jan. 27,2014

182 Chris Eliasmith and Charles H. Anderson, *Neural Engineering: Computation, Representation, and Dynamics in Neurobiological Systems,* the MIT Press, 2003

183 In a digital representation, information is ciphered, stored, and manipulated as ones and zeros. An analog system is actually a physical presence, a representation of one sort of physical signal as another kind that is easier to store or transmit. For example, a non-digital tape recorder turns sounds into voltages and stores them as a pattern of magnetic particles on a tape. Reverse the process, you reproduce the sound.

184 Ibid, Kurzweil, p. 205

185 Ray Singh and Chris Eliasmith, "Higher-Dimensional Neurons Explain the Tuning and Dynamics of Working Memory Cells," *The Journal of Neuroscience,* 26(14) April 5, 2006, pp. 3667-3678.

186 Chris Eliasmith, Terrence C. Steart, Xuan Choo, Trevor Bekolay, Trvis DeWolf, Yichuan Tang, Daniel Rasmussen, "A Large-Scale Model of the Functioning Brain," *Science,* 30 November, 2012, Vol. 338, no.6111, pp. 1202-1205

187 This paper was finally published in 2014. See: Daniel Rasmussen and Chris Eliasmith, "A spiking neural model applied to the study of human performance and cognitive decline on Raven's Advanced Progressive Matrices," *Intelligence,* Vol. 42, 2014, pp. 53-82

188 Spazbot stands for semantic pointer architecture Z bot.

189 James Randerson, "How many neurons make a human brain? Billions fewer than we thought" *The Guardian.com,* Tuesday Feb. 28, 2012

190 This power and size problem is being resolved quickly. According to Carver Mead's website, as of 2014 "reduction of power dissipation by a factor of 10,000, and of silicon area by a factor of 100 are being demonstrated."

191 Bill Joy, "Why the future doesn't need us: Our most powerful 21st century technologies-robotics, genetic engineering, and nanotech—are threatening to make humans an endangered species," *Wired*, http://www.wired.com/wired/archive/8.04/joy_pr.html

192 The National Security Agency took the biggest hit from the Snowden leaks. The NSA was created in great secrecy (its nickname for years was No Such Agency) after a catastrophic failure to protect American military signals intelligence at the start of the Korean War by its organizational predecessor. Though there was evidence the Chinese had 300,000 troops on the border with Korea, General Douglas MacArthur blandly assumed the Chinese would not get involved in the conflict, and the American signals intelligence group, which had previously been able to decrypt Moscow's communications with China and North Korea, suddenly came up with nothing. The capacity to decrypt Russian signals disappeared with a mole named William Weisband, a Russian translator working in the agency. The East's signals went dark at the pivotal moment and Weisband decamped to Moscow. Before Snowden, NSA called the Weisband affair "perhaps the most significant intelligence loss in US history." Now Snowden tops that pantheon. See: *Legacy of Ashes: The History of the CIA* by TimWeiner, Anchor Books, 2008, pp.57-58.

193 See *Black Code*, Ronald J. Deibert, McClelland &Stewart, 2013, pp. 76-77

194 Ibid, Deibert, pp.24-25

195 Boutilier, Alex," Millions of data requests made per year," *Toronto Star*, July 22, 2014, p.1.

196 Gellman was not a *Post* staffer. He had been on staff for years, but had left the paper and become a contributer to *Time Magazine* and

a Century Foundation Senior Fellow. He apparently published these stories in the *Post* because he believed only a newspaper of the *Post*'s size and stature could protect the source (though the source had made it known early on that he had no intention of being protected). Yet even as he worked on the story on the fifth floor of the *Post* where its legal offices are, the *Post* was on the auction block because it was doing so poorly. It was soon sold to Jeff Bezos, the owner of Amazon, for $250 million. Pierre Omidyar, founder of eBay, had also been invited to kick the *Post's* tires. Omidyar later backed Glenn Greenwald, Poitras, and Jeremy Schahill, to the tune of $250 million, in an online publishing venture called First Look Media. It began to publish Snowden's purloined material through its online magazine *Intercept.*

197 Luke Harding, *The Snowden Files: The Inside Story of the World's Most Wanted Man,* Guardian Books, 2014, pp. 136-138

198 Josh Wingrove, "Canadians are lax on privacy, Senate Committee hears," *The Globe and Mail,* May 28, 2014, http://www.theglobeandmail.com/news/

199 The article was finally published in 2014 in *Oecologia,* Vol.195, Issue 1, pp. 63-72, under the title: "Experience teaches plants to learn faster and forget slower in environments where it matters," You had to get to the last line of the abstract to find the animal comparison: "…long lasting behavioral change as a result of previous experience matches the persistence of habituation effects observed in many animals."

200 Susan J. Murch, Colleen B. Simmons, Praveen K. Saxena, "Melatonin in feverfew and other medicinal plants," *Lancet,* Vol.350, Issue 9091, 29 November, 1997, pp. 1598-1599

201 Paul Alan Cox, Sandra Anne Banack, Susan J. Murch," Biomagnification of cyanobacterial neurotoxins and neurodegenerative disease among the Chamorro people of Guam," *PNAS,* Vol. 110, No. 23, 11 November, 2003, pp. 13380-13383.Susan J. Murch, Paul Alan Cox and Sandra Anne Banack, "A mechanism for slow release of biomagnified cynaobacterial neurotoxins and

neurodegenerative disease in Guam,"*PNAS*, Vol.101, no. 33, 1 August, 2004, pp. 12228-12231

202 C. Furukawa, T. Karaki, T. Kawano, "Micro-particle transporting system using galvantotactically stimulated apo-symbiotic cells of Paramecium bursaria," *Zetischrift fur Naturforschung C*, 2009.

203 Shunsuke Furukawa and Tomonori Kawano, "Enhanced Microsphere Transport in Capillary by Conditioned Cells of Green Paramecia Used as Living Micromachines controlled by Electric Stimulii," *Sensors and Materials*, Vol. 24, No.7, pp. 375-386, 2012

Kohei Otsuka, Sayka Maruta, Atsuko Novinsu, Kohji Nakazawa and Tomonori Kawano, "Single Cell Traffic of Swimming Green Paramecia on Microchips with Micro-flow Channels Fabricated by Micro-casting," *Advanced Materials Research*, Vol. 875-877, pp. 2224-2228, 2014

204 K. Kokawa, T. Kagenishi, T. Kawano, S. Mancuso, F. Baluska, "Illumination of Arabidopsis roots induces immediate burst of ROS production,"*Plant Signaling and Behavior*, Vol. 6, pp. 1457-1461, 2011

205 Denise Grady, "Pathogen Mishaps Rise as Regulators Stay Clear," *New York Times*, page 1, Sunday July 20, 2014, and also: Ryan Calo, "The Case for a Federal Robotics Commission," at http://www.Brookings.edu/ research reports2/2014. See also Wikipedia for nanotechnology regulator and regulations: there are none.

206 I knew Kirk Baert and his colleague Celeste Poltak of the Toronto law firm Koskie Minsky LLP because they had been lead members of a team of lawyers who handled two class action lawsuits I was involved with regarding electronic rights to freelance works published in Canadian newspapers and magazines. These actions were known as Robertson I and Robertson II, after the lead plaintiff, the very brilliant and very tenacious journalist and author Heather Robertson. The legal team won significant, multimillion dollar settlements for freelancers, from which I benefited, in both cases.

207 Kate Rossiter and Analise Clarkson, "Opening Ontario's 'Saddest Chapter:'A Social History of Huronia Regional Centre," *Canadian Journal of Disability Studies,* Vol.2, No.3, 2013

208 Between: Marilyn Dolmage as Litigation Guardian of Marie Slark and Jim Dolmage as Litigation Guardian of Patricia Seth, Plaintiffs, and, Her Majesty the Queen in Right of the Province of Ontario, Defendant, Ontario Superior Court of Justice, Plaintiff's Opening Argument (Common Issues Trial Commencing September 16, 2013), Part I, paragraph 7, Court File No. CV-09-376927CP00

209 Ibid, Rossiter and Clarkson

210 Ibid, Plaintiff, p. 94

211 The court eventually approved all three settlements: the total cash portion of all three settlements came to $67,700,000. Koskie Minsky was awarded $8,117,617.13 in fees (including tax) for the Huronia case. Koskie Minsky earned fees of $4,821,953.20 for Rideau, and $2,814,913.80 for Southwestern (both also including tax), for a total of $15,754,484.13, or about 23% of the total, which is below the maximum for class action remuneration.

212 In this case, liberty entailed the chimps being sent to live out their days in as wild a circumstance as possible in North America, in a sanctuary that is part of the North American animal sanctuary alliance.

Selected Bibliography

Books

Andrew, Christopher, Oleg Gordievsky. *KGB: The Inside Story of its Foreign Operations from Lenin to Gorbachev.* London, Sydney, Auckland, Toronto: Hodder & Stoughton, 1990

Barker, Ernest. *The Politics of Aristotle,* New York: A Galaxy Book, Oxford University Press, 1962

Blum, William. *The CIA: A Forgotten History.* London and New Jersey: Zed Books Ltd., 1986

Brand, Stewart. *The Media Lab: Inventing The Future At MIT.* New York: Viking Penguin, 1987

Cave Brown, Anthony and Charles B. MacDonald. *On a Field of Red: The Communist International and the Coming of World War II.* New York: G.P. Putnam's Sons, 1981

Darwin, Charles. *The Formation of Vegetable Mould Through the Action of Worms With Observtions on Their Habits,* Bibliobazaar, 2007

*The Power of Movement in Plants,*16973217R00167. Lexington, KY, August 2012

Deibert, Ronald J. *Black Code: Surveillance, Privacy, and The Dark Side Of The Internet.* Toronto: Signal, 2013

De Waal, Frans. *Chimpanzee Politics: Power and Sex Among Apes.* Baltimore: The Johns Hopkins University Press, 2007

The Bonobo and the Atheist: In Search of Humanism Among the Primates. New York: W.W.Norton & Company, 2013

Dewar, Elaine. *Bones: Discovering the First Americans.* Toronto: Random House of Canada, 2001

The Second Tree: Of Clones, Chimeras and Quests for Immortality. Toronto: Random House of Canada, 2004

Doidge, Norman. M. D. *The Brain That Changes Itself.*New York: Viking Penguin, 2007

Dyson, George. *Turing's Cathedral: The Origins of the Digital Universe.* New York: Pantheon Books, 2012

Darwin Among The Machines: The Evolution of Global Intelligence. New York: Basic Books, 1997

Flynn, James R. *What Is Intelligence? Beyond the Flynn Effect,* Cambridge: Cambridge University Press, 2009

Are We Getting Smarter? Rising IQ in the Twenty-First Century. Cambridge: Cambridge University Press, 2012

Fossey, Dian. *Gorillas In The Mist,* New York: First Mariner Books, 2000

Galdikas, Birute M.F. *Reflections of Eden: My Years with the Orangutans of Borneo.* Boston: Little, Brown and Company, 1995

Galdikas, Birute M.F., Nancy Erickson Briggs, Lori K. Sheeran, Gary L. Shapiro, Jane Goodall, Eds. *All Apes Great and Small, Vol. 1: African Apes.* New York: Kluwer Academic/Plenum Publishers, 2001

Gleason, Mona. *Normalizing the Ideal: Psychology, Schooling and The Family in Postwar Canada.* Toronto: University of Toronto Press, 1999

Greenwald, Glenn. *No Place to Hide: Edward Snowden, the N.S.A. and the U.S. Surveillance State.* Toronto: Signal, 2014

Goodall, Jane. *My Life With The Chimpanzees,* New York: Aladdin Paperbacks, 2002

Goodall, Jane with the Jane Goodall Institute. *Jane Goodall 50 Years At Gombe: A tribute to Five Decades of Wildlife Research, Education, and Conservation.* New York: Stewart Taboir &Chang, 2010

Goodall, Jane with Phillip Berman. *Reason for Hope: A Spiritual Journey.* New York: Warner Books, 2000.

Grand, Steve. *Creation: Life And How To Make It.* London: Phoenix, 2001

Harding, Luke. *The Snowdon Files: The Inside Story Of The World's Most Wanted Man.* London: Guardian Books, 2014

Hebb, D. O. *The Organization of Behavior: A Neuropsychological Theory.* New York: John Wiley & Sons, 1949.

Hillis, W. Daniel. *The Connection Machine.* Cambridge, Mass: The MIT Press, 1985

Hodges, Andrew. *Alan Turing: The Enigma.* London: Vintage, 1992

Hume, David. *Dialogues Concerning Natural Religion:* (Edited and introduced by Norman Kemp Smith). New York: The Bobbs-Merrill Company Inc., 1947

A Treatise of Human Nature: Mineola, NY: Dover Publications Inc., 2003

An Enquiry concerning Human Understanding: New York: Oxford University Press, 2007

Jaynes, Julian. *The Origin of Consciousness in the Breakdown of the Bicameral Mind.*Boston: Houghton Mifflin Company, 1976

Kelly, Kevin. *Out of Control: The New Biology of Machines, Social Systems, and the Economic World.* New York: Basic Books, 1994

Kurzweil, Ray. *How To Create A Mind: The Secret of Human Thought Revealed.* New York: Viking Penguin, 2012

Laycock, Samuel R. *Gifted Children: A Handbook for the Class-Room Teacher.* Toronto: The Copp Clark Publishing Co. Limited, 1957

Levy, Steven. *In The Plex: How Google Thinks, Works, And Shapes Our Lives.* New York: Simon &Schuster, 2011

Lewontin, R.C. *Biology as Ideology: The Doctrine of DNA.* Toronto: House of Anansi Press, 1991

Lily, John C. *Lilly On Dolphins: Humans of the Sea.* New York: Anchor Books, 1973

Lovelock, James. *Gaia: A New Look At Life on Earth.* Oxford: Oxford University Press, 2009

Homage to Gaia: The Life of an Independent Scientist. Oxford: Oxford University Press, 2011

The Revenge of Gaia: Why the Earth is Fighting Back—and How We Can Still Save Humanity. London: Penguin Books, 2007

MacIntyre, Ben. *A Spy Among Friends: Kim Philby and the Great Betrayal*, McClelland & Stewart, 2014

Margulis, Lynn. *Symbiotic Planet [A New Look at Evolution].* New York: Basic Books, 1998

Morell, Virginia. *Ancestral Passions: The Leakey Family and the Quest for Humankind's Beginnings.* New York: Touchstone, 1996

Romanes, George John. *Animal Intelligence.* Nabu Public Domain reprint, USA, 313131LV00015B/415/P (scan of original publication London: Kegan, Paul, Trench, &Co., 1882.).

Savage-Rumbaugh, Sue, and Roger Lewin. *Kanzi: The Ape at the Brink of the Human Mind.* New York: John Wiley & Sons, Inc., 1994

Shapiro, James A. *Evolution: A View From The 21st Century.*Upper Saddle River, N.J.: FT Press Science, 2011

Singer, P.W. *Wired For War: The Robotics Revolution and Conflict In The 21st Century.* New York: Penguin Books, 2010

Spalding, Linda. *A Dark Place in The Jungle: Science, Orangutans, and Human Nature.* Chapel Hill, N.C.: Algonquin Books of Chapel Hill, 1999

Weiner, Tim. *Legacy of Ashes: The History of the CIA.* New York: Anchor Books, Random House, 2008

Westoll, Andrew. The *Chimps of Fauna Sanctuary: A Canadian Story Of Resilience And Recovery.* Toronto: Harper Perennial, 2012

Wilson, Edward O. *The Social Conquest of Earth.* New York: Liveright Publishing Corporation, 2012.

Wolfram, Stephen. *A New Kind of Science.* Champaign, Ill: Wolfram Media Inc., 2002

Wong, Jan. *Red China Blues: My Long March From Mao to Now.* Toronto, New York, London, Sydney, Auckland: Doubleday/Anchor, 1996

Wright, Peter. *Spy Catcher: The Canadid Autobiograpnhy of a Senior Intelligence Officer.* Toronto: Stoddart Publishing Company Ltd., 1987

Articles and Book Chapters

Adamatzky, Andrew, "Physarum attraction: Why slime mold behaves as cats do," *Communicative and Integrative Biology*, Vol.5, Issue 3, May/June, 2012

Adamatzky, Andrew and Jeff Jones, "Road Planning with slime mould: If *Physarum* built motorways it would route M6/M74 through Newcastle." *Journal of Bifurcation and Chaos*, Vol. 20 Issue 10, October, 2010

Anderson, James B., Caroline Sirjusingh, Ainslie B. Parsons, Charles Boone, Claire Wickens, Leah E. Cowen and Linda M. Kohn, "Mode of Selection and Experiemental Evolution of Antifungal Drug Resistance in *Saccharomyces cerevisiae*," *Genetics*, Vol. 163, April, 2003, pp. 1287-1298

Anderson, James. B., Caroline Sirjusingh and Nicole Ricker, "Haploidy, Diploidy and Evolution of Antifungal Drug Resistance in *Saccharomyces cerevisiae*," *Genetics* Vol. 168, December, 2004

Anderson, Roland C., Jennifer Mather, Mathieu Q. Monette, and Stephanie R.M. Zimsen, "Octopuses (*Enteroctupus dofleini*) Recognize Individual Humans," *Journal of Applied Animal Welfare Science*,"Vol. 13, 2010, pp. 261-272

Anderson, Roland C., Jennifer Mather, "The packaging problem: Bivalve prey selection and prey entry tenchiques of the octopus Enteroctopus dofleini," *Journal of Comparative Psychology*, Vol.121, No.3, 2007, pp. 300-305

Baluska, Frantisek, " Recent surprising similarities between plant cells and neurons," *Plant Signaling & Behavior*, Vol. 5, No.2, February, 2010, pp. 1-3

Baluska, Frantisek, Simcha Lev-Yadun, Stefano Mancuso, "Swarm intelligence in plant roots," *Trends in Ecology and Evolution*,"Vol. 25, No.12, December, 2010, pp. 682-683

Baluska, Frantisek, Stefano Mancuso, "Deep evolutionary origins of neurobiology: Turning the essence of 'neural'upside down,"*Communication and Integrative Biology*, Vol.2, no. 1, January/February 2009, pp.60-65

Baluska, Frantisek, Stefano Mancuso, "Plant neurobiology: from sensory biology, via plant communication, to social plant behavior," *Cognitive Process*, Vol.10, (Suppl.1), 2009, pp. S3-S7

Baluska, Frantisek, Stefano Mancuso, "Plants and Animals: Convergent Evolution in Action?" in *Plant-Environment Interactions, Signaling and Communication in Plants*, Frantisek Baluska, ed., Springer-Verlag, Berlin, 2009, pp. 285-301

Baluska, Frantisek, Dieter Volkmann, Andrej Hlavacka, Stefano Mancuso, Peter W. Barlow, "Neurobiological View of Plants and Their Body

Plan," in *Communication in Plants*, F. Baluska, S. Mancuso, D. Volkmann, eds., Springer-Verlag, Berlin, 2006

Bender, Courtney E., Denise L. Herzing, David F. Bjorklund, "Evidence of teaching in atlantic spotted dolphis (*Stenella frontalis)* by mother dolphins foraging in the presence of their calves," *Animal Cognition*, 2008

Bermudes, D. G. Hinkle, L. Margulis, "Do prokaryotes contain microtubules?"*Microbiological Reviews*, Vol. 58, No.3, 1994, pp. 387-400

Bigge, Bill, Inman R. Harvey, "Programmable Springs: Developing Actuators with Programmable Compliance for Autonomous Robots," *Robotics and Autonomous Systems*, Vol. 55, Issue 9, pp. 728-734, September 2007

Blumberg, Mark S. and Edward A. Wasserman, "Animal Mind and the Argument From Design," *American Psychologist*, March 1995, pp. 133-144

Brown, P., T. Sutikna, M.J. Morwood, R.P. Soejono, Jatmiko, E. Wayhu Saptomo, Rokus Awe Due, "A new small-bodied hominin from the Late Pleistocene of Flores, Indonesia," *Nature*, Vol.431, 28 October 2004, pp. 1055-1061

Byrne, Richard W., Philip J. Barnard, Iain Davidson, Vincent M. Janik, William C. McGrew, Adam Mikiosi and Polly Wiessner, "Understanding culture across species," *Trends in Cognitive Sciences*, Vol. 8, No.8, August 2004

Choudhary, Swadesh, Steven Sloan, Sam Fok, Alexander Neckar, Eric Trautmann, Peiran Gao, Terry Stewart, Chris Eliasmith, Kwabena Boahen, "Silicon Neurons That Compute," *Artificial Neural Networks and Machine Learning—ICANN 2012. Lecture Notes in Computer Science* 7552, 2012, pp. 121-128

Ciszack, Marzena, Diego Comparini, Barbara Mazzolai, Frantisek Baluska, F. Tito Arecchi, Tamas Vicsek, Stefano Mancuso, "Swarming Behavior in Plant Roots," *PLoS One*, Vol. 7, Issue 1, January, 2012, http://www.plosone.org/article/infor%3Adoi%2F10.1371%2Fjournal.pone.0029759

Clay, Zanna, Klaus Zuberbuhler, "Bonobos Extract Meaning from Call Sequences," *PLoS One*, Vol.6, Issue 4, April 2011. http://www.plosone.org

Clark, Travis A., James B. Anderson, "Dikaryons of the Basidiomycete Fungus *Schizophyllum communes*: Evolution in Long-Term Culture," *Genetics* Vol. 167, August 2004, pp. 1663-1675

Cleermans, Axel, "The radical plasticity thesis: how the brain learns to be conscious," *Frontiers in Psychology*, Vol.2, article 86, May 2011, 12 pages

Coqueugniot, H., J.-J.Hublin, F. Veillon, F.Houet, T. Jacob," Early brain growth in *Homo erectus* and implications for cognitive ability," *Nature*, Vol.431, 16 September 2004, pp. 299-301

Couzin, Iain, "Collective Minds," *Nature*, Vol.445 15 February, 2007

Cox, Paul Alan, Sandra Anne Banack, Susan J. Murch,"Biomagnification of cyanobacterial neurotoxins and neurodegenerative disease among the Chamorro people of Guam," *PNAS*, Vol.100, No.23, 11 November 2003, pp.13380-13383

De Luccia, Thiago Paes de Barros,"*Mimosa pudica, Dionaea muscipula* and anesthetics," *Plant Signaling and Behavior*, Vol. 7, No.9, September, 2012, pp. 1163-1167

Dettman, Jeremy R., Caroline Sirjusingh, Linda M. Kohn & James B. Anderson," Incipient speciation by divergent adaptation and antagonistic epistasis in yeast," *Nature*, Vol.447, No. 31, May 2007, pp. 585-588

De Waal, Frans B.M.," A century of getting to know the chimpanzee," *Nature*, Vol. 437, 1 September 2005, pp. 56-59

De Waal, Frans B.M., "The Brutal Elimination of a Rival Among Captive Male Chimpanzees," *Ethology and Sociobiology* 7, 1986, pp.237-251

De Waal, Frans B.M., "Evolutionary Psychology: The Wheat and the Chaff," *Current Directions in Psychological Science*, Vol 11, Number 6, December 2002

De Wall, Frans B.M., "Sex Differences in the Formation of Coalitions Among Chimpanzees," *Ethology and Sociobiology* 5: 1984, pp. 239-255

De Wall, Frans B.M., "Primates—A Natural Heritage of Conflict Resolution," *Science*, Vol. 289, 28 July 2000

De Wall, Frans B.M., "The Antiquity of Empathy,"*Science*, Vol.336, 18 May, 2012, pp. 874-876

De Wall, Frans B.M., "Darwin's Last Laugh,"*Nature*, Vol.460.No.9, July 2009, p. 175

De Wal, Frans B.M., Pier Francesco Ferrari, "Towards a bottom-up perspective on animal and human cognition," *Trends in Cognitive Sciences*, Vol.14, No.5, April 2010, pp. 201-207

Dicke, Ursula, Gerrard Roth, "Animal Intelligence and the Evolution of the Human Mind," *Scientific American Mind*, Vol. 19, pp. September/ August 2008, pp. 70-77

Doyle, Laurance R. "Quantification of Information in a One-Way Plant-to-Animal Communication System," *Entropy*, Vol 11, 2009, pp. 431-442

Eliasmith, Chris, Terrence C. Stewart, Trevor Bekolay, Travis DeWolf, Yichuan Tang, Daniel Rasmussen, "A Large-Scale Model of the Functioning Brain," *Science*, Vol. 338, No.6111, 30 November, 2012, pp. 1202-1205

Fammartino, Alessandro, Francesca Dardinale, Cornelia Gobel, Laurent Mene-Saffrane, Joelle Fournier, Ivo Feussner, Marie-Therese Esquerre-Tugaye, "Characterization of a Divinyl Ether Biosynthetic Pathway Specifically Associated with Pathogenesis in Tobacco," *Plant Physiology*, Vol. 143, January, 2007, pp. 378-388

Flack, Jessica C. and Frans B.M. de Waal, "'Any Animal Whatever': Darwinian Building Blocks of Morality in Monkeys and Apes," *Journal of Consciousness Studies*, 7, no. 1-2, 2000, pp. 1-29

Floreano, Dario, Laurent Keller, " Evolution of Adaptive Behaviour in Robots by Means of Darwinian Natural Selection," *PLoS Biology*, Vol.8, No.1, 2010, http://www.plosbiology.org/article/ info%3Adoi%2F10.1371%2F

Floreano, Dario, Sara Mitri, Stephane Magnenat, Laurent Keller, "Evolutionary conditions for the emergence of communication in robots,"*Current Biology*, Vol.17, 20 March, 2007, pp. 1-6

Flynn, James R. "Searching for Justice: the Discovery of IQ Gains Over Time,"*American Psychologist*, Vo. 54, No.1, pp. 5-20

Furukawa, Shunsuke and Tomonori Kawano, "Enchanced Microsphere Transport in Capillary by Conditioned Cells of Green Paramecia Used as Living Micromachines Controlled by Electric Stimulii," *Sensors and Materials*, Vol. 24, No.7, pp. 375-386, 2012

Gillespie-Lynch, Kristen, Patricia M. Greenfield, Heidi Lyn, Sue Savage-Rumbaugh, "The role of dialogue in the ontogeny and phylogeny of early symbol combinations: A cross-species comparison of bonob, chimpanzee, and human learners," *First Language*, Vol.31, No.4, 2011, pp.442-460

Ginsburg, Simona, Eva Jablonka, "The evolution of associative learning: A factor in the Cambrian explosion," *Journal of Theoretical Biology*, Vol.266, 2010, pp. 11-20

Giraud, Tatiana, Jez. S. Pedersen, Laurent Keller, "Evolution of supercolonies: The Argentine ants of southern Europe," *PNAS*, Vol.99, No.9, 16 April 2002, pp. 6075-6079

Gould, S.J. and R. C. Lewontin, "The spandrels of San Marco and the Panglossian paradigm: a critique of the adaptationist programme,"*Proceedings of the Royal Society London, B.*, Vol. 205, 1979, pp. 581-598

Harvey, Inman,"Robotics: Philosophy of Mind using a Screwdriver," at http://www.nmit-schlag.com

Harvey, Inman. "Evolving Robot Consciousness: The Easy Problems and the Rest,"in *Consciousness Evolving*, James H. Fetzer (ed.), Amsterdam: John Benjamins Publishing Co., 2002, pp. 205-211

Harvey, Inman, Ezequiel Di Paulo, Elio Tuci, Rachel Wood and Matt Quinn, " Evolutionary Robotics: a new scientific tool for studying cognition," *Artificial Life*, Vol 11, issues 1-2, pp. 79-98, January 2005

Harvey, Inman. "Homeostasis and Rein Control: From Daisyworld to Active Perception," in *Artificial Life IX: Proceedings of the Ninth International Conference on the Simulation and Sythesis of Living Systems*, J. Pollack, M. Bedau, P. Husbands, T. Ikegami, R.Watson (eds.) MIT Press, 2004

Harvey, I., P. Husbands, D. Cliff, A. Thompson and N. Jakobi, "Evolutionary Robotics: The Sussex Approach," *Robotics and Autonomous Systems*, Vol. 20, Issues 2-4, pp. 205-224, June 1997

Herman, Louis M., "Exploring the Cognitive World of the Bottlenosed Dolphin," in *The Cognitive Animal: Empirical and Theoretical Perspectives on Animal Cognition*, Mark Bekoff, Colin Allen, Gordon Burghardt eds.A Bradford Book, 2002

Herman, Louis M. "Cogntion and Language Competencies of Bottlenosed Dolphins," in *Dolphin Cognition and Behavior: A Comparative Approach,* Ronald .J. Schusterman, Jeanette A. Thomas, Forrest G. Wood, eds., Lawrence Erlbaum Associates Inc., Publishers, 1986, pp. 221-252

Herrmann, Esther, Josep Call, Maria Victoria Hernandez-Lloreda, Brian Hare, Michael Tomasello, "Humans Have Evolved Specialized Skills of Social Cognition: The Cultural Intelligence Hypothesis," *Science,* Vol.317, pp.1360-1366, 7 September 2007

Herrmann, Esther, Brian Hare, Josep Call and Michael Tomasello, "Differences in the Cognitive Skills of Bonobos and Chimpanzees," *PLoS One,* 5 (8) August 27, 2010. http://www.ncibi.nlm.nih.gov/pmc/articles/PMC2929188

Hinton, Geoffrey E., "How Neural Networks Learn from Experience," *Scientific American,* September 1992, pp. 145-151

Hooper, Stacie, Diana Reiss, Melissa Carter, Brenda McCowan, "Importance of Contextual Saliency on Vocal Imitation by Bottlenose Dolphins,"*International Journal of Comparative Psychology,* Vol. 19, 2006, pp. 116-128

Janik, Vincent M. "Whistle Matching in Wild Bottlenose Dolphins *(Tursiops truncates)*, *Science,* Vol. 289, 25 August 2006, pp. 1355-1357

Janik, V.M., L. S. Sayigh and R.S. Wells, "Signature whistle shape conveys identity information to bottlenose dolphins," *PNAS,* vol. 103, no. 21, 23 May 2006, pp. 8293-8297

Johnstone, George A., "Advantages of Ehylene-Oxygen As a General Anesthetic," *California and Western Medicine,* Vol.27, No.2, August 1927, pp. 216-218

Jones, Jeff. " The Emergence and Dynamical Evolution of Complex Transport Networks from Simple Low-Level Behaviours,"*International Journal of Unconventional Computing,* Vol. 6, pp.125-144, 2008

Jones, Jeff. "Characteristics of Pattern Formation and Evolution in Approximations of Physarum Transport Networks," *Artificial Life* 16: 1-27 (2010)

Jones, Jeff, Andrew Adamatzky, " Emergence of self-organized amoeboid movement in a multi-agent approximation of *Physarum polycephalum*," *Bioinspiration and Biomimetics,* Vol. 7, 2012

Jones, Jeff, Andrew Adamatzky, "Towards *Physarum* binary adders," *Biosystems,* Vol. 101, 2010, pp. 51-58

Jones, Thony B., Alan Kamil, "Tool-Making and Tool-Using in the Northern Blue Jay," *Science,* New Series Vol. 180, no. 4090, June 8 1973, pp. 1076-1078

Kamil, Alan, "A Synthetic Approach to the Study of Animal Intelligence," in *Nebraska Symposium on Motivation* 1987, Vol.35: Comparative Perspectives in Modern Psychology, Richard A.Dienstbeir & Daniel W. Leger, Eds., pp. 257 -308

Kamil, Alan, "Sociality and the Evolution of Intelligence," *Trends in Cognitive Sciences,* Vol.8, no.5, May 2004, pp. 195-197

Kamil, Alan, "On the Proper Definition of Cognitive Ethology," in *Animal Cognition in Nature: The Convergence of Psychology and Biology in Laboratory and Field,* Russell P. Balda, Irene M. Pepperberg and Alan C. Kamil, eds. Academic Press, 1998, pp. 1-28

Katsikopoulos, Konstantionos V., Andrew J. King, "Swarm Intelligence in Animal Groups: When Can a Collective Out-Perform an Expert?" *PLoS ONE* 5 (11):e15505. 24 November 2010 doi:10.1371/journal.pone.0015505

Kaufman, Allison B., Sean R. Green, Aaron R. Seitz, Curt Burgess, "Using a Self-Organizing Map (SOM) and the Hyperspace Analog to Language (HAL) Model to Identify Patterns of Syntax and Structure in the Songs of Humpback Whales," *International Journal of Comparative Psychology,* Vol.25, 2012, pp. 237-275

Krause, Jens, Graeme D. Ruxton and Stefan Krause, "Swarm intelligence in animals and humans," *Trends in Ecology and Evolution,* Vol.25, No.1, 6 September 2009. doi: 10.1016/j.tree.2009.06.016

Krieger, Michael J.B., Jean-Bernard Billeter, Laurent Keller, "Ant-like task allocation and recruitment in cooperative robots," *Nature,* Vol.406, 31 August, 2000, pp. 992-995

Kuba, Michael J., Ruth A. Byrne, Daniela V.Meisel, Jennifer Mather,"Exploration and Habituation in Intact Free Moving *Octopus vulgaris*," *International Journal of Comparative Psychology*, Vol 19, 2006, pp. 426-438

Laland, Kevin N., Vincent M. Janik, "The animal culture debate," *Trends in Ecology and Evolution*, Vol.21, No. 10, June 2006

Lenton, Timothy M. and Marcel van Oijen,"Gaia as a complex adaptive system,"*Philosophical Transactions of the Royal Society, London B*, Vol. 357, 2002, pp.683-695

Levin, Michael, "The wisdom of the body: future techniques and approaches to morphogenetic fields in regenerative medicine, developmental biology and cancer,"*Regenerative Medicine*, Vol. 6, No.6, 2011, pp. 667-673

Lily, John C. "Dolphin-Human Relation and LSD 25," in *The Use of LSD in Psychotherapy and Alcoholism*, Harold A. Abramson, M.D. ed., Bobbs-Merrill, 1967, pp. 47-52

Lyn, Heidi, E. Sue Savage-Rumbuagh, "Observational word learning in two bonobos (*Pan paniscus*): ostensive and non-ostensive contexts," *Language & Communication*, Vol. 20, 2000, pp. 255-273

Lyn, Heidi, Becca Franks, E. Sue Savage-Rumbaugh, "Precursors of morality in the use of the symbols 'good' and 'bad' in two bonobos (*Pan paniscus*) and a chimpanzee (*Pan troglodytes*)," *Language & Communication*, Vol. 28, 2008, pp. 213-224

Lyn, Heidi, Patricia Greenfield, Sue Savage-Rumbaugh, "The development of representational play in chimpanzees and bonobos: Evolutionary implications, pretense, and the role of interspecies communications," *Cognitive Development*, Vol. 21, 2006, pp. 199-213

Lyn, Heidi, Patricia M. Greenfeild, Sue Savage-Rumbaugh, Kristen Gillespie-Lynch, William D. Hopkins, "Nonhuman primates do declare!A comparison of declarative symbol and gesture use in two children, two bonobos, and a chimpanzee," *Language & Communication*, Vol. 31, 2011, pp. 63-74

Lyn, Heidi, Peter Pierre, Allyson J. Bennett, Scott Fears, Roger Woods, William D. Hopkins, "Planum temporale grey matter asymmetries in chimpanzees (*Pan troglodytes*), vervet (*Chlorocebus aethiops sabaeus*), rhesus (*Macaca mulatto*) and bonnet (*Macaca radiata*) monkeys," *Neuropsychologia,* Vol.49, 2011, pp. 2004-2012

Mancuso, Stefano,"Federico Delpino and the foundation of plant biology,"*Plant Signaling and Behavior,*"Vol.5, No.9, September 2010, pp. 1057-1071

Margulis, Lynn, "The Origin of Plant and Animal Cells,"*American Scientist,* Vol. 59, No.2, March-April 1971, pp. 230-232

Margulis, Lynn, Michael F. Dolan, Ricardo Guerrero, "The chimeric eukaryote: Origin of the nucleus from the Karyomastigont in amitochondriate protists," *PNAS,* Vol. 97, no. 13, June 20, 2000, pp. 6954-6959

Margulis, Lynn, "Gaia is a Tough Bitch," in *The Third Culture: Beyond the Scientific Revolution,* John Brockman, ed., New York: *Touchstone,* 1995. http://www.edge.org/documents/ThirdCulture/n-Ch.7

Marino, Lori, Richard C. Connor, R. Ewan Fordyce, Louis M. Herman, Patrick R. Hof, Louis Lefebvre, David Lusseau, Brenda McCowan, Esther A. Nimchinsky, Adam A. Pack, Luke Rendell, Joy S. Reidenberg, Diana Reiss, Mar D. Uhen, Estel Van der Gucht, Hal Whitehead, "Cetaceans Have Complex Brains for Complex Cognition," *PLoS Biology,* 15 May 2007

Marsh, Heidi L., Suzanne E. MacDonald, "Information seeking by orangutans: a generalized search strategy?" *Animal Cognition,* Vol. 15, 2012, pp. 293-304

Marsh, Heidi L., Laura Adams, Cahterine Floyd, and Suzanne E. MacDonald, " Feature Versus Spatial Strategies by Orangutans (Pongo abelii) and Human Children (Homo sapiens) in a Cross-Dimensional Task," *Journal of Comparative Psychology,* 2012

Masi, E., M. Ciszak, G. Stefano, L. Renna, E. Azzarello, C. Pandolfi, S. Mugnai, F. Baluska, F.T. Arecchi, S. Mancuso, "Spatiotemporal dynamics of the electrical netowork activity in the root apex,"*PNAS,* Vol. 106, No.10, 10 March, 2009, pp. 4048-4053

Mather, Jennifer A. "'Home' choice and modification by juvenile *Octopus vulgaris* (Mollusca: Cephalopoda): specialized intelligence and tool use?" *Journal of The Zoological Society of London*, Vol. 233, 1994, pp.359-368

Mather, Jennifer A. "Cephalopod consciousness: Behavioural evidence," *Consciousness and Cognition*, Vol. 17, 2008, pp. 37-48

Mather, Jennifer A. "Cephalopod Skin Display: From Concealment to Communication," in *Evolution of Communication Systems*, Kim Orler and Lili Greibel, Eds. *The MIT Press*, 2004, pp. 193-213

Mather, Jennifer A. "Sand Digging in Sepia offi
cinalis: Assessment of a Cephalopod Mollusc's 'Fixed' Behavior Pattern," *Journal of Comparative Psychology*, Vol. 100, No. 3, 1986, pp. 315-320

Mather, Jennifer A. "Foraging and the development of intelligence in octopuses," in *Cephalopod Cognition*, Darmaillaq, Dickel & Mather eds., Cambridge University Press, 2014

McCowan, Brenda, Sean F. Hanser, Laurance R. Doyle, "Quantitative tools for comparing animal communication systems: information theory applied to bottlenose dolphin whistle repertoires," *Animal Behavior*, Vol.57, 1999, pp. 409-419

McDonald-Gibson, J. J.G.Dyke, E.A. Di Paolo, I.R. Harvey, "Environmental Regulation can arise under Minimal Assumptions," *Journal of Theoretical Biology*, Vol.251, Issue 4, pp. 653-666 21 April, 2008

Mercado III, Eduardo, Scott O. Murray, Robert K. Ukeyama, Adam A. Pack and Louis M. Herman, "Memory for rection actions in the bottlenosed dolphin (Tursipos truncatus): Reptition or arbitrary behaviors using an abstract rule," *Animal Learning & Behavior*, Vol. 26, No.2, 1998, pp. 210-218

Mirolli, Marco, Domenico Parisi, "How producer biases can favor the evolution of communication: An analysis of evolutionary dynamics,"*Adaptive Behavior*, Vol. 16, No.1, February 2008, pp. 27-52

Morwood, M.J., R.P. Soejono, R.G. Roberts, T. Sutikna, C.S.M. Turney, K.E. Westaway, W. J. ERink, J.-X. Zaho, G. D. van den Bergh, Rokus Awe Due, D. R. Hobbs, M. W. Moore, M.I. Bird, L.K Fifield, "Archaeology and age of a new hominin from Flores in eastern Indonesia," *Nature*, Vol. 431, 28 October 2004, pp. 1087-1091

Murch, Susan J., Paul Alan Cox, Sandra Anne Banack," A mechanism for slow release of biomagnified cyanobacterial neurotoxins and neurodegenerative disease in Guam,"*PNAS* Vol.101, No.33, 17 August 2004, pp.12228-12231

Nakagaki, Toshiyuki, Hiroyasu Yamada, Agota Toth, "Path finding by tube morphogenesis in an amoeboid organism," *Biophysical Chemistry* 92 (2001) 47-52

Nakagaki, Toshiyuki, Hiroyasu Yamada, Masahiko Hara, "Smart network solutions in an amoeboid organism," *Biophysical Chemistry* Vol.107, 2007, pp. 1-5

Nakagaki, Toshiyuki, Hirosan Yamada, Agota Toth, "Maze-solving by an amoeboid organism,"*Nature,* Vol. 407, 28 September 2000

Nakagaki, Toshiyuki, R. Kobayahsi, Y. Nishiura and T. Ueda, "Obtaining multiple separate food sources: behavioural intelligence in the Physarum plasmodium," *Proceedings of the Royal Society London Biology,* Vol. 271, 2004, pp. 2305-2310

Otsuka, Kohei, Sayka Maruta, Atsuko Novinsu, Kohji Nakazawa and Tomonori Kawano, "Single Cell Traffic of Swimming Green Paramecia on Microchips with Micro-flow Channels Fabriacted by micro-casting," *Advanced Materials Research,* Vol.875, No.871, pp. 2224-2228, 2014

Pack, Adam A., Lewis M. Herman, "Bottlenosed Dolphins (*Tursiops truncatus*) Comprehend the Referent of Both Static and Dynamic Human Gazing and Pointing in an Object-Choice Task,"*Journal of Comparative Psychology,* Vol.118, No.2, 2004, pp. 166-171

Pan, Zhiqiang, Bilal Camara, Harold W. Gardner, Ralph A. Backhaus, "Aspirin Inhibition and Aceylation of the Plant Cytochorom P450, Allene Oxide Synthase, Resembles that of Animal Prostaglandin Endoperoxide H. Synthase," *The Journal of Biological Chemistry,* Vol.273, No. 29, July 17, 1998, pp. 18139-18145

Parr, Lisa A., Frans B.M. de Waal, "Visual kin recognition in chimpanzees," *Nature,* Vol. 399, 17 June 1999, p.647

Parreiras, Lucas S., Linda M. Kohn, and James B. Anderson, "Cellular Effects and Epistasis among Three Determinants of Adaptation in Experimental Populations of *Saccharomyces cervisiae, Eukaryotic Cell,* October 2011, pp. 1348-1356

Pavlovic, Anrej, L'udmila Slovakova, Camilla Pandolfi, Stefano Mancuso, "On the mechanism underlying photosynthetic limitation upon trigger hair irritation in the carnivorous plant Venus flytrap (*Dionaea muscipula* Ellis)" *Journal of Experimental Biology,* February 2, 2011. http://www.jxb.oxfordjournals.org

Pollick, Amy S., Frans B.M. de Waal, Ape Gestures and language evolution," *PNAS,* Vol.104, no. 19, 8 May 2007, pp.8184-8189

Preuschoft, Signe, Xiu Wang, Filippo Aueril, and Frans B.M. de Waal, "Reconciliation in Captive Chimpanzees: A Reevalutaion with Controlled Methods," *International Journal of Primatology,* Vol 23, No.1, Fberuary 2002

Quick, Nicola J. and Vincent M. Janik, "Whistle Rates of Wild Bottlenose Dolphins (*Tursiops truncatus*): Influences of Group Size and Behavior," *Journal of Comparative Psychology,* Vol.122, No.3, 2008, pp. 305-311

Rushton, J. Philippe, Arthur R. Jensen, "Thirty Years of Research on Race Differences in Cognitive Ability," *Psychology, Public Policy and Law,* Vol. 11, No. 2, 2005, pp. 235-294

Rasmussen, Daniel, Chris Eliasmith, " A spiking nueral model applied to the study of human performance and cognitive delicne on Raven's Advanced Progressive Matrices," *Intelligence,* Vol. 42, 2014, pp. 53-82

Rossiter, Kate, Annalise Clarkson, "Opening Ontario's 'Saddest Chapter': A Social History of Huronia Regional Centre," *Canadian Journal of Disability Studies,* Vol. 2, No.3, 2013

Russell, Jamie L., Heidi Lyn, Jennifer A. Schaeffer, William Hopkins,"The role of socio-communicative rearing environmens in the development of social and physical cognition in apes," *Developmental Science,* Vol. 14, No.6, 2011, pp. 1459-1470

Russon, Anne E., Birute M. Galdikas, "Constraints on Great Apes' Imitation: Model and Action Selectivity in Rehabilitant Orangutan

(*Pongo pygmaeus*) Imitation," *Journal of Comparative Psychology*, Vol. 109, No. 1, 1995, pp. 5-17

Russon, Anne E., "Evolutionary reconstructions of great ape intelligence," in *The Evolution of Great Ape Intelligence*, A.E. Russon and D.R. Begun (Eds.) Cambridge University Press, 2004, pp. 1-14

Russon, Anne E., "Great ape cognitive systems,"in *The Evolution of Great Ape Intelligence*, A.E. Russon and D.R.Begun (Eds.), Cambridge University Press, 2004, pp.76-100

Russon, Anne E., "Pretending in Free-Ranging Rehabilitant Orangutans," in *Pretending in animals and humans*, R.W. Mitchell (Ed.), Cambridge University Press, 2002, pp. 229-240

Russon, Anne E., "Life History: The Energy-Efficient Orangutan," *Current Biology*, Vol. 20, No. 22, October 2010

Russon, Anne E. and David R. Begun, "Evolutionary origins of great ape intelligence: an integrated view," in *The Evolution of Great Ape Intelligence*, A.E.Russon and D.R.Begun,(Eds.), Cambridge University Press, 2004,, pp.353-368

Russon, Anne E. and Kristin Andrews, "Orangutan pantomime: elaborating the message,"Biology Letters, published online 11 August 2010. http://www.doi.10.1098/rsbl.2010.0564

Russon, Anne E. and Kristin Andrews, "Pantomime in Great Apes," *Communicative and Integrative Biology*, Vol. 4, Issue 3, May/June 2011

Russon, Anne E., Alan Compost, Purwo Kuncoro & Agnes Ferisa, "Orangutan fish eating, primate acquatic fauna eating and their implications for the origins of ancestral hominin fish eating," *Journal of Human Evolution*, Vol.77, December, 2014, pp.50-63

Russon, Anne E., Agnes Ferisa, Purow Kuncoro, Dwi Putri Handayani, "How Orangutans (Pongo pygmaeus) Innovate for Water," *Journal of Comparative Psychology*, Vol. 124, No.1, 2010, pp.14-28

Russon, Anne E., Carel P. van Schaik, Purwo Kuncoro, Agnes Ferisa, Dwi P. Handayani and Maria a. van Noordwijk, "Innovation and intelligence in orangutans,"in *Orangutans: Goegraphic Variation in*

Behavioral Ecology and Conservation, S.A. Wich, S.S. Utami, T. Mitra Setia & C.P. van Schaik (eds), Oxford University *Press*, 2009, pp. 270-288

Russon, Anne E., Serge A. Wich, Marc Ancrenaz, Tomoko Kanamori, Cheryl D. Knott, Noko Kuze, Helen C. Morrogh-Bernard, Peter Pratje, Hatta Ramlee, Peter Rodman, Azrie Sawang, Kade Sidiyasa, Ian Singelton and Carel P. van Schaik, "Geographic variation in orga-nutan diets, Ibid,. pp. 135-156

Saigusa, Tetzu, Atsushi Tero, Toshiyuki Nakagaki, Yoshiki Kuramoto, " Amoebae anticipate periodic events,"*Physical Review Letters*, Vol.100, 3 January 2008

Shapiro, James A., "Thinking about bacterial populations as multicellular organisms," *Annual Review of Microbiology*, Vol. 52, 1998, pp. 81-104

Shapiro, James A., "The significance of bacterial colony patterns," *BioEssays*, Vol. 17, No 7, 1995, pp. 597-607

Shapiro, James A., "Bacteria as Multicellular Organisms,"*Scientific American*, June, 1988, pp. 82-89

Shapiro, James A. "Bactera are small but not stupid: cognition, natural genetic engineering and socio-bacteriology," *Studies in History and Philosophy of Biological and Biomedical Sciences*," Vol.38, 2007, pp. 807-819

Shapiro, James A., "Mobile DNA and Evolution in the 21st Century,"*Mobile DNA*, Vol.1, No.4, 2010

Sickler, Jessica, John Fraser, Thomas Webler, Diana Reiss, Paul Boyle, Heidi Lyn, Katherine Lemcke, Sarah Gruber, "Social Narratives Surrounding Dolphins: Q Method Study,"*Society and Animals*, Vol. 4, No.4, 2006, pp. 351-382

Singh, Ray, Chris Eliasmith, " Higher-Dimensional Neurons Explain the Tuning and Dynamics of Working Memory Cells," *Journal of Neuroscience*, Vol.14, April 5, 2006, pp.3667-3678

Snyder, James B. Mark E. Nelson, Joel W. Burdick, Malcolm A. MacIver, "Omnidirectional Sensory and Motor Volumes in Electric Fish,"*PLoS Biology*, Vol.5, Issue 11, November 2007, 12 pages

Stewart, Terrence C., Feng-Xuan Choo, Chris Eliasmith, "Spaun: A Perception-Cognition-Action Model Using Spiking Neurons," Proceedings of the 34th Annual Conference of the Cognitive Science Society, 2012

Stringer, C.B., P. Andrews, "Genetic and Fossil Evidence for the Origin of Modern Humans," *Science*, Vol. 239, 11 March 1988, pp. 1263-1268

Sundstrom, Lisolette, Michel Chapuisat, Laurent Keller, "Conditional Manipulation of Sex Ratios by Ant Workers: A Test of Kin Selection Theory," *Science*, Vol. 274, 8 November 1996, pp.993-995

Terman, Lewis, " The Intelligence Quotient of Francis Galton in Childhood," *The American Journal of Psychology*, Vol. 28, No.2, April 1917, http://www.jstor.org/stable/10.2307/1413721

Trewavas, Anthony, "Mindless Mastery,"*Nature*, Vol.415, 21 February, 2002, p. 841

Trewavas, Anthony, "Le Calcium, C'est la Vie: Calcium Makes Waves,"*Plant Physiology*, Vol. 120, May, 1999, pp.1-6

Trewavas, Anthony, "A Brief History of Systems Biology,"*The Plant Cell*, Vol.18, October, 2006, pp. 2420-2430

Trewavas, Anthony, "Aspects of Plant Intelligence,"*Annals of Botany*, Vol.92, 2003, pp. 1-20

Trewavas, Anthony, "Plant Intelligence,"*Naturwissenschaften*, Vol. 92, 2005, pp. 401-413

Trewavas, Anthony, "Green plants as intelligent organisms,"*Trends in Plant Science*, Vol.10, No.9, September, 2005, pp. 413-419

Trewavas, Anthony, "What is plant behaviour?" *Plant, Cell and Environment*, Vol.32, Issue 6, June 2009, pp. 606-616

Trewavas, Anthony, "Information, Noise and Communication: Thresholds as Controlling Elements in Development," in *Biocommunication of Plants*, G. Tizany and F. Baluska, eds., Springer-Verlag, Berlin, 2012

Trewavas, Anthony, Frantisek Baluska, " The ubiquity of consciousness," *EMBO Reports*, Vol.12, Issue 12, December 2011, pp. 1221-1225

Tseng, AiSun, Michael Levin, "Cracking the bioelectric code: Probing endogenous ionic controls of pattern formation,"*Communicative and Integrative Biology*, Vol.6, Issue 1, 2012

Tsuda, Soichiru, Jeff Jones, Andrew Adamatzky, "Towards Physarum Engines,"*Applied Bionics and Biomechanics*, Vol. 00, no 00, Month 200x 1-23 22 February 2012, http://www.informaworld.com

Turing, A.M., "On Computable Numbers, With an Application to The Entscheidungsproblem," *Proceedings of the London Mathematical Society*, Series 2, Vol.42, 1937, pp.230-265

Turing, A.M., FRS. "The Chemical Basis of Morphogenesis," *Philosophical Transactions of the Royal Society of London*, Vol.237 B, No. 641, 14 August 1952

Turing, A.M,"Computing Machinery and Intelligence," *Mind*, Vol.59, 1950, pp.433-460

Tyack, Peter L. "Convergence Calls as Animals Form Social Bonds, Active Compensation for Noisy Communication Channels, and the Evolution of Vocal Learning in Mammals,"*Journal of Comparative Psychology*, Vol 122, No.3, 2008, pp.319-331.

Urban, B.W., M. Bieckwenn, "Concepts and correlations relevant to general anaesthesia," *British Journal of Anaesthesia*, Vol. 89, No.1, 2002, pp.3-16

Vonk, Jennifer, Suzanne E. MacDonald, "Levels of Abstraction in Orangutan (*Pongo abelii*) Categorization," *Journal of Comparative Psychology*, Vol. 118, no 1, 2004, pp. 3-13

Vonk, Jennifer and Suzanne E. MacDonald, Natural Concepts in a Juvenile Gorilla (*Gorilla gorilla gorilla*) at Three Levels of Abstraction," *Journal of the Experimental Analysis of Behavior*, 78, 2002, pp. 315-332

Ward, Ashley J.W., David J.T. Sumpter, Iain D. Couzin, Paul J.D. Hart, Jens Krause, "Quorum decision-making facilitates information transfer in fish shoals," *PNAS*, Vol.105, No.19, 13 May 2008, pp. 6948-6953

Wasserman, Edward A. and Leyre Castro, "Comparative Cognition," in *Handbook of Psychology, Vol. 3, Behavioral Neuroscience, 2nd edition*, Irving B. Weiner ed. John Wiley & Sons, 2013, pp. 480-508

Wasserman, Edward A. and Thomas R. Zentall, "Introduction to the Oxford Handbook of Comparative Cognition, *in Oxford Handbook of*

Comparative Cognition, Thomas R. Zentall and Edward Wassmeran eds, Oxford Library of Psychology, 2012

Wasserman, Edward A. "Humans, animals and computers: Minding machines?" *Revista do Psicolgia,* Vol. XVIII, No. 2, 2009, pp. 25-41

Whiten, Andrew, Victoria Horner & Frans B.M. de Waal, "Conformity to cultural norms of tool use in chimpanzees," *Nature,* Vol 427, 29 September, 2005

Wischmann, Steffen, Dario Floreano, Laurent Keller,"Historical contingency affects signaling strategies and competitive abilities in evolving populations of simulated robots," *PNAS,* Vol.109, no.3, 17 January, 2012, pp.864-868

Yokawa, Ken, Tomoko Kagenishi, Tomonori Kawano, Stefano Mancuso, Frantisek Balusk, "Illumination of Arabidopsis roots induces immediate burst of ROS production," *Plant Signaling & Behavior,* Vol.6, No.10, October, 2011, pp.1460-1464

Yorzinski, Jessica, Gail Patricelli, "Birds adjust acoustic directionality to beam their antipredator calls to predators and conspecifics," *Proceedings of the Royal Society B,* Vol. 277, 2010, pp. 923-932

Invited Papers

Harvey, Inman, "Opening Stable Doors: Complexity and Stability in Nonlinear Systems," *European Conference on Artificial Life,* MIT Press, 2011

Harvey, Inman, "Homeostasis and Rein Control: From Daisyworld to Active Perception," *Artificial Life IX, Proceedings of the Ninth International Conference on the Simulation of Living Systems,* eds. Jordan Pollack, Mark Bedau, Phil Husbands, Takashi Ikegami, and Richard A. Watson, the *MIT Press,* pp. 309-315, 2004

Harvey, Inman, "Misrepresentations," *Artificial Life XI, Proceedings of the 11th International Conference on the Simulation and Synthesis of Living Systems,*

Seth Bulloch, Jason Noble, Richard Watson and Mark Bedau, eds., the MIT Press, pp.227-234, 2008

Lorenz, Konrad Z., "Analogy as a Source of Knowledge," Nobel Lecture, Physiology or Medicine, December 12, 1973 http://www.nobelprize.org/nobel_prizes/medicine/laureates/1973

Lorenz, Konrad Z., autobiographical essay. http://www.nobelprize.org/nobel_prizes/medicine/laureates/1973, Lorenz-bio

Mather, Jennifer A, "Do Cephalopods Have Pain and Suffering?" International Symposium of La Fondation Droit Animal, ethique et sciences in collaboration with the International Research Group in Animal Law, Paris, October 18-19, 2012

McClintock, Barbara, "The Significance of Responses of the Genome to Challenge,"Nobel Lecture, Physiology or Medicine, 8 December, 1983

Tinbergen, Nikolaas, autobiographical essay. http://www.nobelprize.org/nobel_prizes/medicine/laureates/1073/tinbergen

Von Frisch, Karl, authobiographical essay. http://www.nobelprize.org/nobel_prizes/medicine/laureates/1973/frisch-bio

Dissertations

Shapiro, Gary L. "Factors Influencing Variance in Sign Learning Performance by Four Juvenile Orangutans," Dissertation: University of Oklahoma, August, 1985

Letters

Alpi, Amedeo, et al., "Plant neurobiology: no brain, no gain?" *Trends in Plant Science,* Vol.12, No 4, March, 2007, pp. 15-136

Brenner, Eric D., Rainer Stahlberg, Stefano Mancuso, Frantisek Baluska, Elizabeth Van Volkenburgh, "Response to Alpi et al.: Plant neurobiology: the gain is more than the name," *Trends in Plant Science,* Vol.12, No.7, 25 June 2007, pp.285-286

Trewavas, Anthony, "Response to Alpi et al.: Plant neurobiology—all metaphors have value," Trends in Plant Science, Vol. 12. No.6, 2007, pp. 231-233

Trewavas, Anthony, "Plants are intelligent too," EMBO Reports, Vol.13, Issue 9, September, 2012, pp. 772-773

Harvey, Inman. "Tool-makers versus Tool-users:Division of Labour," *Adaptive Behavior,* Vol.17, No 4, 2009

Abstracts

Banerjee, Sumana, Lynn Margulis, "Mitotic arrest by melatonin," *Experimental Cell Research,* Vol.73, Issue 2, April 1973, pp. 314-316

Baluska, Frantisek, "Cell-Cell Channels, Viruses and Evolution," *Annals of the New York Academy of Sciences,* Vol. 1178, *Natural Genetic Engineering and Natural Genome Editing,* October 2009, pp. 1060119

Bull, Larry, Adam Budd, Chrisopher Stone, Ivan Uroukov, Ben de Lacy Costello, Andrew Adamatzky, "Towards Unconventional Computing through Simulated Evolution: Control of Nonlinear Media by a Learning Classifier System," *Artificial Life,* Vol. 14, No.2, 2008, pp. 203-222

Chevalier-Skolnikoff, Suzanne, Birute M.F. Galdikas, Alan Z. Skolnikoff, "The adaptive significance of higher intelligence in wild orang-utans: a preliminary report," *Journal of Human Evolution,* Vol. 11, Issue 7, November 1982, pp. 639-652

Crawford, Charles, Birute M.F. Galdikas, "Rape in non human animals: an evolutionary perspective," *Canadian Psychology,* Vol.27, No.3, July 1986, pp. 215-230

Descovich, Kristin A., Birute M.F. Galdikas, Andrew Tribe, Allan Lisie, Clive J. Phillips, "Fostering Appropriate Behavior in Rehabilitant

Orangutans (*Pongo pygmaeus*), "*International Journal of Primatology*, Vol.32, No.3, 2011, pp. 616-633

Downing, Keith and Peter Zvirinsky, "The Simulated Evolution of Biochemical Guilds: Reconciling Gaia Theory and Natural Selection," *Artificial Life*, Vol.5, no 5, Fall, 1999, pp. 291-318

Duarte, Ana, Ido Pen, Laurent Keller, Franz J. Weissing, " Evolution of self-organized division of labor in a response threshold model," *Behavioral Ecology and Sociobiology*, Vol. 66, No 6, 2012, pp. 947-957

Dyer, John R. G., Anders Johansson, Dirk Helbing, Iain D. Couzin, Jens Krause, "Leadership, consensus decision making and collective behaviour in humans," *Philosophical Transactions of the Royal Society B.*, Vol. 364, No. 1518, 27 March 2009, pp. 781-789

Falk, Dean et al, "The Brain of LB1, Homo floresiensis," *Science*, Vol. 380, 8 April 2005, pp. 242-245

Furukawa, S., C. Karaki, T. Kawano, "Micro-particle transporting system using galvanotactically stimulated apo-symbiotic cells of *Paramecium bursaria*," *Zeitschrift fur Naturforschung C*, Vol. 64, No 5, p.421, 2009.

Galdikas, Birute M.F. "The orangutan long call and snag crashing at Tanjung Puting Reserve," *Primates*, Vol. 24, No 3, 1983, pp. 371-384

Galdikas, Birute M.F., "Orangutan sociality at Tanjung Puting,"*American Journal of Primatology*, Vol.9, Issue 2, 1985, pp. 101-119

Galdikas, Birute M.F., " Adult Male Sociality and Reproductive Tactics among Orangutans at Tanjung Puting," *Folia Primatologica*, Vol. 45, No.1, 1985, pp. 9-24

Galdikas, Birute M.F. "Subadult male orangutan sociality and reproductive behavior at Tanjung Puting, "*American Journal of Primatology*, Vol. 8, no 2, 1985, pp. 87-99

Galdikas, Birute M.F., "Orangutan diet, range, and activity at Tanjung Putting, Central Borneo," *International Journal of Primatology*, Vol. 9, no 1, 1988, pp.1-35

Galdikas, Birute M.F, Geza Teleki, "Variations in Subsistence Activities of Female and Male Pongids: New Perspectives on the Origins of

Hominid Labor Division," *Current Anthropology*, Vol. 22, No.3, June 1981, pp. 241-256.

Galdikas, Birtue M.F., James W. Wood, "Birth spacing patterns in humans and apes,"*American Journal of Physical Anthropology*, Vol. 83, Issue 2, October 1990, pp. 184-191

Green, R. E., et al., "A Draft Sequence of the Neandertal Genome," *Science*, Vol.328, No 5779, May 7 2010, pp. 710-722

Guerrero, Ricardo, Carlos Pedros-Alio, Isabel Esteve, Jordi Mas, David Chase, Lynn Margulis, "Predatory prokayotes: Predation and primary consumption evolved in bacteria," *PNAS*, Vol.83. No 7, April 1, 1986, pp. 2138-2142

Halloy, J., G. Sempo, G. Carari, C. Rivault, M. Asadpour, F. Tache, I.Said, V. Durier, S.Canonge, J.M.Ame, C.Detrain, N. Correll, A. Martinoli, F. Mondada, R. Siegwart and J.L. Deneubourg, "Social Integration of Robots into Groups of Cockroaches to control Sel-Organized Choices," *Science*, Vol. 318, No. 5853, 16 November 2007, pp. 1155-1158

Hamilton, Ruth A., Birute M.F. Galdikas, "A preliminary study of food selection by the orangutan in relation to plant quality," *Primates*, Vol. 35, No 3, 1994, pp.255-263

Harvey, Inman, Ezequiel Di Paolo, Rachel Wood, Matt Quinn, Elio Tuci, Elio Tuci Iridia, "Evolutionary Robotics: A New Scientific Tool for Studying Cognition," *Artificial Life*, Vol 11 Issue 1-2, January 2005, pp.79-98

Harvey, Inman, "The Microbial Genetic Algorithm," Advances in Artificial Life: Darwin Meets Von Neumann, Lecture Notes in Computer Science, 2011, Vol.5778/2011 pp. 126-133

Higham, Tom, Tim Compton, Chris Stringer, Roger Jacobi, Beth Shapiro, Erik Trinkaus, Barry Chandler, Flora Groning, Chris Collins, Simon Hillson, Paul O'Higgins, Charles FitzGerald & Micahel Fagan, "The earliest evidence for anatomically modern humans in northwestern Europe," *Nature*, Vol. 479, 24 November 2011, pp. 521-524

Iglesias, T.L., R. McElreath, G.I. Patricelli, "Western scrub-jay funerals: cacaphonous aggregations in response to dead conspecifics," *Animal*

Behavior, August 27, 2012 http://www.sciencedirect.com/science/article/pii/S0003347212003569

Jaeggi, Adrian, Lynda P. Dunkel, Maria A. Van Noordwijk, Serge A. Wich, Agnes A.L Sura, Carel P. Van Schaik, "Social learning of diet and foraging skills by wild immature Bornean orangutans: implications for culture," *American Journal of Primatology*, Vol. 72, Issue 1, January 2010, pp. 62-71

Jaeggi, Adrean, Maria A. van Noordwijk, Carel P. van Schaik, "Begging for information: mother-offspring food sharing among wild Bornean orangutans,"*American Journal of Primatology*, Vol.70, Issue 6, June 2008, pp. 533-541

Kawano, T., T. Karano, T. Kosaka, H. Hosayu, "Green paramecia as an evolutionary winner of oxidative symbiosis,: a hypothesis and supporting data," *Zeitschrift fur Naturforschung, C.*, Vol. 59, Nos. 7 and 8, pp. 538-542, 2004

Kokawa, K., T. Kagenishi, T.Kawano, S. Mancuso, F. Baluska, "Illumination of Arabidopsis roots induces immediate burst of ROS production," *Plant Signaling & Behavior*, Vol .6, pp. 1457-1461, 2011

Kuze, Noko, David Dollatore, Graham L. Banes, Peter Pratje, Tomoyuki Tajima, and Anne E. Russon, "Factors affecting reproduction in rehabilitant female orangutans: young age at first birth and short interbirth interval," *Primates*, Vol. 53, No. 2, 2012, pp. 181-192

Lovelock, J.E., "Electron Absorption Detectors and Technique for Use in Quantitative and Qualitative Analysis by Gas Chromatography," *Analytical Chemistry*, Vol. 35, 4, 1963, pp. 474-481.

Lovelock, James E., "Geophysiology, the science of Gaia,"*Reviews of Geophysics*, Vol. 27, No.2, 1989, pp. 215- 222

Lovelock, J.E., R. J. Maggs, R. Rasmussen, "Atmospheric Dimethyl Sulphide and the Natural Sulphur Cycle," *Nature*, Vol, 237, 23 June 1972, pp.452-453

Lovelock, James E., Lynn Margulis, "Atmospheric homeostasis by and for the biosphere: the Gaia hypothesis," *Tellus*, Vol. 26, Issue 1-2, 1974, pp. 2-10

Lyn, Heidi, Patricia M. Greenfield, E. Sue Savage-Rumbaugh, "Semiotic combinations in *Pan*: A comparison of communication in a chimpanzee and two bonobos," *First Language*, Vol. 31, no.3, August 2011, pp.300-325

Lyn, Heidi, Jamie L. Russell and William D. Hopkins, "The impact of Environment on the Comprehension of Declarative Communication in Apes," *Psychological Science*, Vol. 21, no 3, March 2010, pp.360-365

Margulis, Lynn, "Symbiotic theory of the origin of eukaryotic organelles: criteria for proof,"*Symposia of the Society for Experimental Biology*, Vol. 29, 1975, pp. 21-38

Margulis, Lynn, "Genetic and evolutionary consequences of symbiosis, "*Experimental Parasitology*, Vol. 39, Issue 2, April 1976, pp. 277-349

Margulis, Lynn, "Origins of species: acquired genomes and individuality," *Biosystems*, Vol. 31, Issues 2-3, 1993, pp.121-125

Margulis, Lynn, J.E. Lovelock, "Biological modulation of the Earth's atmosphere," *Icarus*, Vol. 21, Issue 4, 1974, pp. 471-489

Margulis L., Bermudes D., "Symbiosis as a mechanism of evolution: status of cell symbiosis theory,"*Symbiosis*, Vol. 1, 1985, pp. 101-124

Mather, Jennifer A. "To boldly go where no mollusk has gone before: Personality, play, thinking, and conciousness in cephalopods,"*American Malacological Bulletin*, March, 2008, pp. 51-58

Mather, Jennifer A., Roland C. Anderson, "Personalities of octopuses," *Journal of Comparative Psychology*, Vol. 107, No.3, September 1993, pp. 336-340

Mather, Jennifer A., Roland C. Anderson, "Ethics and invertebrates: a cephaolopod perspective," *Diseases of Acquatic Organisms*, Vol. 75, No. 2, May 4, 2007, pp. 119-129

Mathger, Lydia M., Nada Shashar and Roger T. Hanlon," Do cephalopods communicate using polarized light reflections from their skin?" *Journal of Experimental Biology*, Vol. 212, July 15, 2009, pp. 2133-2140

MacDonald, Suzanne E., Maria M. Agnes, "Orangutan (Pongo abelii) spatial memory and behavior in a foraging task,"*Journal of Comparative Psychology*, Vol. 113(2) June 1999, 213-217

Margulis, Lynn, " The Conscious Cell," Annals of the New York Academy of Sciences, Vol. 929, Cajal and Conciousness: Scientific Approaches to Consciousness, April 2001, pp. 55-70

McDanile, Michael A. "Big-brained people are smarter,"*Intelligence*, Vol.33, Issue 4, July-August 2005, pp.337-346

Murch, Susan J., Colleen B. Simmons, Praveen K. Saxena, " Melatonin in feverfew and other medicinal plants," *Lancet*, Vol. 350, Issue 9091, 29 November 1997, pp. 1598-1599

Murch, Susan J., Ali R. Alan, Jin Cao, Praveen K. Saxena, "Melatonin and serotonin in flowers and fruits of *Datura metel* L.," *Journal of Pineal Research*, Vol.47, Issue 3, October 2009, pp.277-283

Murch, Susan J., Skye S.B. Campbell, Praveen Saxena, "The role of serotonin and melatonin in plant morphogeneisis: Regulation of auxin-induced root organogenesis in *in vitro* cultured explants of st. John's Wort (*Hypericum perforatum* L.)" *In Vitro Cellular and Developmental Biology*—Plant, Vol.37, Issue 6, November-December 2001, pp. 786-793

Murch, Susan J., Praveen Saxena, "Melatonin: A potential regulator of plant growth and development?" *In Vitro, Cellular and Developmental Biology—Plant*, Vol. 38, Issue 6, November-December 2002, pp.531-536

Murch, Susan J., Barbara A. Hall, Cuong H. Le, Praveen K. Saxena, "Changes in the levels of indoleamine phytochemicals during veraison and ripening of wine grapes,"*Journal of Pineal Research*, Vol. 49, Issue 1, 1 June 2010, pp. 95-100

Pennisi, Elizabeth, "The Man Who Bottled Evolution," *Science*, Vol. 342, no 6160, 15 November 2013, pp. 790-793

Pentland, Alex, " On the Collective Nature of Human Intelligence,"*Adaptive Behavior*, Vol. 15, No.2, June 2007, pp. 189-198.

Prufer, Kay et al,"The Complete Genome Sequence of a Neandertal from the Altai Mountains," *Nature*, Vol.505, No. 7481, 02 January 2014, pp.43-49

Russon, Anne E ., "The nature and evolution of intelligence in organutans," *Primates*, Vol. 39, No.4, 1998, pp. 485-503

Sagan, Lynn, "On the origin of mitosing cells," *Journal of Theoretical Biology*, Vol. 14, Issue 3, March, 1967, pp. 225-274

Smith, Myron L., Johann N. Bruhn, James B. Anderson, "The fungus *Armillaria bulbosa* is among the largest and oldest living organisms," *Nature*, Vol. 356, 2 April, 1992, pp. 428-431

Tennie, Claudio, Josep Call and Michael Tomasello, "Ratcheting up the ratchet: on the evolution of cumulative culture,"*Philsophical Transactions fot eh Royal Society B: Biological Sciences*, vol. 364 no. 1528 27 August 2009, pp. 240502415

Tero, Atsushi, Seiji Takagi, Tetsu Saigusa, Kentaro Ito, Dan P. Bebber, Mark D. Fricker, Kenji Yumiki, Ryo Kobayashi, Toshiyuki Nakagaki, "Rules for Biologically Inspried Adaptive Network Design," *Science*, Vol. 327 No.5954, 22 January 2010. pp. 439-442

Turi, C.E., Susan J. Murch, "Spiritual and Ceremonial Plants in North America: An Assessment of Moerman's Ethnobotanical Database Comparing Residual, Binomial, Bayesian and Imprecise Dirichlet Model (IDM) Analysis,"*Journal of Ethnopharmacology*, Vol.148, No.2, 9 July 2013, pp. 386-94

Van Noordwijk, Maria A., Carel P. van Schaik, "Development of ecological competence in Sumatran orangutans," *American Journal of Physical Anthropology*, Vol. 127, Issue 1, May, 2005, pp.79-94

Van Noordwijk, Maria A., Carel P. van Schaik, "Intersexual food transfer among orangutans: do females test males for coercive tendency?"*Behavioral Ecology and Sociobiology*, Vol. 63, no 6, 2009, pp. 868-891

Van Schaik, Carel P., Maria A. van Noordwijk, Serge A. Wich," Innovation in wild Bornean orangutans *(Pongo pygmaeus wurmbii)*," *Behavior*, Vol. 143, No. 7, 2006, pp.839-876

Van Schaik, Carel P., Marc Ancrenaz, Gwendolyn Borgen, Birute Galdikas, Cheryl D. Knott, Ian Singleton, Akira Suzuki, Sri Suci Utami, Michelle Merrill, "Orangutan Cultures and the Evolution of Material Culture," *Science*, Vol. 299, No. 5603, 3 January 2003, pp. 102-105

Watson, Andrew J., James E. Lovelock, "Biological homeostasis of the global environment: the parable of Daisyworld," *Tellus B*, Vol. 35B, Issue 4, September 1983, pp. 284-289

Yerkes, Robert M., "The Intelligence of earthworms," *Journal of Animal Behavior*, Vol. 2 no 5, September-October 1912, 332-352

Younger, K.B., S. Banerjee, J.K. Kelleher, M. Winston and Lynn Margulis, "Evidence that the Synchornized Production of New Basal Bodies is not Associated with Dna Synthesis in Stentor Coeruleus," *Journal of Cell Science*, Vol. 11, September 1972, pp. 621-637

Documents

ECAgents: Emodied and Communicating Agents, January 1, 2004, ***http://ecagents.istc.cnr.it/index.php?***

Amended Statement of Claim, Dec. 8, 2010, in Ontario Superior Court of Justice: Between: Marilyn Dolmage as Litigation Guardian of Marie Slakr and Jim Dolmage as Litigation Guardian of Patricia Seth, Plaintiff, and her Majesty the Queen in Right of the Province of Ontario, Defendant. Court File No. CV-09-376927CP00

Plaintiffs' Opening Argument in Ontario Superior Court of Justice: Between: Marilyn Dolmage as Litigation Guardian of Marie Slark and Jim Dolmage as Litigation Guardian of Patricia Seth, Plaintiff, and Her Majesty the Queen in Right of the Province of Ontario, Defendant, Court File No.CV-09-376927CP00

Summary of Key Settlement Terms in Ontario Superior Court of Justice: Between: Marilyn Dolmage as Litigation Guardian of Marie Slark and Jim Dolmage as Litigation Guardian of Patricia Seth, Plaintiff, and Her Majesty the Queen in Right of the Province of Ontario, Defendant, Court File No.CV-09-376927CP00

Tradecraft Developer, SCEC-Network Analysis Centre: "IP Profiling Analytics & Mission Impacts" Top Secret Powerpoint presentation, May 10, 2012

Videos

Jeff Jones, International Centre for Unconventional Computing, "Sonification of self-organized oscillatory amoeboid movement in particle model of *Physarum polycephalum* slime mould," http://www.uncomp.uwe.ac.uk/jeff/

Popular and Citizen Press, Websites, Blogs, Reports

Alexander, Victoria N., "Beyond Darwin: Rebel Scientist Lynn Margulis Honored in Amherst," March 25, 2012. http://www.digitaljournal.com/article 321533

Andrews, Suzanna, Bryan Burroughs, Sarah Ellison, "The Snowden Saga: A Shadowland of Secrets and Light," *Vanity Fair*, May 2014

Arnold, Carrie, "The hologenome: A new view of evolution," *New Scientist*, Issue 2899, 17 January 2013, http://www.newscientist.com/article/mg2178992.000-the-hologenome-a-new-view

Ash, Stacey, Tobi Day-Hamilton,"Waterloo researchers draw crowds at AAAS in Boston,"*Waterloo Daily Bulletin*, February 20, 2013, http://www.bulletin.uwaterloo.ca/2013/feb-20we

"Archaeologist joins in astounding discovery, 10 March 2005, http://www.tinyurl.com/6m6e

Beeman, Perry, "Board Chairman: Des Moines ape sanctuary passed USDA inspection Wednesday," 13 September, 2012, http://www.desmoines-register.com/article/20120913/NEWS/120913012/

Borghino, Dario, "Scientists build the most accurate computer simulation of the brain yet," *Gizmag*, Dec.6, 2012. http://www.gizmag.com/brain-computer-simulation/25349/?utm_source= Gizmag+Subscribers

Boutilier, Alex,"Millions of data requests made per year," *Toronto Star*, July 22, 2014, p.1

Browne, Malcolm W., "Is That a Grimace? Ask a Pigeon," *The New York Times*, May 2, 1989

Cartmill, Matt, "The Status of the Race Concept in Physical Anthropology," *American Anthropologist* 100 (3) 1999, pp. 651-660

Chimp Trainer's Daughter, 24 August, 2012, http://www.bonbo-hope-needs-vision

Chimp Trainer's Daughter, 10 September 2012, http://www.chimptrainers-daughter.blogspot.ca/2012/09/former-employees-others

Chimp Trainer's Daughter, February 2013, http://www.chimptrainersdaughter.blogspot.ca/p/the-latest-crisis-in-ongoing30year

Ciochon, Russell, Roy Larick, "Early Homo erectus Tools in China,"*Archaeology*, Vol. 53, No.1, January-February 2000

Coyne, Jerry. "Lynn Margulis disses evolution in *Discover* magazine, embarrasses both herself and the field," http://www.whyevolutionistrue.wordpress.com/2011/04/12/lynn-margulis-disses

Culotta, Elizabeth, "Battle Erupts Over the 'Hobbit'Bones," *Science*, Vol.307, p. 1179. 2 February, 2005, http://www.sciencemag.org.cgi/content/full/307/5713/1179

Dalby, Beth, "Behind the Curtain at Great Ape Trust," *West Des Moines Patch*, 21 September, 2012, http://www.westdesmoines.patch.com/articles/an-insider-s-look-behind-the-curtain

Dayton, Leigh, "Hobbit skeleton 'damaged'," *The Courier-Mail*, 11 March 2005, http://www.thecouriermail.news.com/au/commonstory_page/0,5936,12509550%255E401,00.html

Dayton, Leigh, "'Bones of Contention'," *The Australian*, 11 March 2005, http://www.theaustralian.news.com/au/common/story_page/0,5744,12505745%255E30417,00.html

Des Moines Register, "Iowa Primate Learning Sanctuary announces new lead scientists," Blogs.desmoinesregister.com. http://www.blogs.desmoinesregister.com/drm/index.php/2014/01/30/iowa-primate-learning-sanctuary-announces-new-lead-scientists/article?nclick_check=1

Dreifus, Claudia, "A Conversation With: Martin Wells; He Studied Squid and Octopus, Then He Ate Them," *New York Times*, 1998, http://www.nytimes.com/1998/12/08/science/a-conversation-with-martin

El Akad, Omar, " Google taps U of T professor to teach context to computers," *The Globe and Mail*, March 11, 2013, http://www.theglobeandmail.com/technology/tech-news/google-taps-u-of-t-professor-to

Eliasmith, Chris, "Building a Behaving Brain." Centre for Theoretical Neuroscience, University of Waterloo

Empson, Rip, "Google Scoops Up Neural Networks Startup, DNN Research To Boost Its Voice And Image Search Tech," *techcrunch.com*, 2013. http://www.techcrunch.com/2013/03/12/google-scoopes-up-neural-networks-start

Findlay, Alan and John Messenger, "Martin Wells: Distinguished biologist inspired by the hidden wonders of marine life," *The Guardian*, 2009, http://www.guardian.co.uk/science/2009/feb/25/obituary-martin-wells

Gibbons, Ann, "Modern Humans' First European Tour," *ScienceNOW*, November, 2011. http://www.news.sciencemag.org/sciencenow/2011/11/modern-humans-first-eur

Grady, Denise, "Pathogen Mishaps Rise as Regulators Stay Clear," *New York Times*, Sunday July 20, p. 1

Great Ape Trust, "Dr. Sue Savage-Rumbaugh, Executive Director," scientists biographies, downloaded 22 November 2012, http://greatapetrust.org/science/scientists-biographies-Sue-Savage

Great Ape Trust, "Founder, Ted Townsend," downloaded 23 November, 2012, http://www.greatapetrust.org/scientists/biographies

Great Ape Trust, "History of Ape Language," http://www.greatapetrust.org/science/history-of-ape-language

Great Ape Trust, "Additional Studies into Ape Language and Primate Intelligence," downloaded 22 November 2012. http://www.greatapetrust.org/science/history-of-ape-language/additional-studies

Greene, Kate, "TR10: Reality Mining: Sandy Pentland is using data gathered by cell phones to learn about human behavior," *Technology Review*

(MIT), March/April 2008, http://www.technologyreview.com/read_article.aspx?ch=specialsection&s

Harding, Louette, "Life story: Why code-breaker Alan Turing was cast aside by postwar Britain," Mail Online, 17 November, 2013, http://www.dailymail.co.uk/home/you/article-2507393?Life-story-Why-co

Hayes, J.D., "Pentagon researching 'narrative networks'as way to hijack the brain with false stories," *Natural News*, May 15, 2012 http://www.naturalnews.com/035872_Pentagon_narrative_networks_propaganda

Hickman, Leo. "James Lovelock: The UK should be going mad for fracking," *The Guardian*, 15 June 2012, http://www.guardian.co.uk/environment/2012/June/15/james-lovelock-interview

Iowa Primate Learning Sanctuary, "Information about Panbanisha's Illness and Treatment," 12 November, 2012, downloaded 23 November, 2012, http://www.Panbanisha.org/statemnt-about-panbanishas-illness-and-treatment

Jenkins Jr., Holman W., "Will Google's Ray Kurzweil Live Forever?" The Weekend Interview, *The Wall Street Journal*, 12 April, 2013, http://www.online.wsj.com/article/SB10001424127887324504704578412581 38

Joy, Bill, "Why the future doesn't need us," *Wired Magazine*, April, 2000, http://www.wired.com/wired/archive/8.04/joy_pr.html

Kaufman, Scott Barry, "Men, Women, and IQ: Setting the Record Straight," *Psychology Today*, July 20, 2012, http://www.psychologytoday.com/blog/beautiful-minds/201207/me-women

Kaufman, Scott Barry, "The Truth about the 'Termites'," *Psychology Today*, September 9, 2009. http://www.psychologytoday.com/blog/beautiful-minds/200909/the-truth-about-the-termites

Kennedy, Pagan, "The Cyborg in Us All," *New York Times Magazine*, 14 September 2011. http://www.nytimes.com/2011/09/18/magazine/the-cyborg-in-us-all.html

Lemley, Brad, "Machines that Think," *Discover Magazine*, 1 January 2001 http://www.discovermagazine.com/2001/jan/featmachines?searchterm

Leslie, Mitchell, "The Vexing Legacy of Lewis Terman,"*Stanford Magazine*, July/August 2000, http://www.alumni.stanford.edu/get/page/magazine/article/?article_id=40678

Levy, Steven, "How Ray Kurzweil Will Help Google Make the Ultimate AI Brain,"*Wired,* April 25, 2013, http://www.wired.com/business/2013/04/kurzweil-google-ai

Lovelock, James, "Nuclear Power is the only green solution," *The Independent,* 24 May, 2004 http://www.independent.co.uk/voices/commentators/james-lovelock-nuclear

Lunau, Kate, "Orangutans and *gorillas go bananas for tablet computers,"* *Maclean's* Magazine, 23 February 2012, http://www2.macleans.ca/2012/02/23/theres-an-ape-for-that

Marcum, Alicia Heather, "George John Romanes," *PSYography,* downloaded 18/10/2012. http://www.faculty.frostburg.edu/psyography/romanes.html

Markoff, John,"Google Cars Drive Themselves, in Traffic," *The New York Times,* http://www.nytimes.com/2010/10/10/science/10google

Matthews, Robert, "Believe it or not, they're all the same species,"*Telegraph,* December 26, 2004. http://www.telegraph.co.u/news/main.html

Miller, Peter, "The Genius of Swarms," *National Geographic Magazine,* July, 2007. http://www.nationalgeographic.com/print/2007/07/swarms/miller-text

Nadia, Steve, "Look who's talking," *New Scientist,* 12 July 2013. http://www.peterrussel.com/Dolphin/DolphinLang.php

Nippert, Matt, "Eureka!" *Commentary,* Issue 3517, October 6, 2012

Nolan, Tanya, "Experts conclude 'hobbit'belongs to new species// Academics debate implications f 'hobbit'discovery,"" 4 March 2005, *abc.net.* http://www.abc.net.au/pm/content/2005/s1316541.htm

Norvig, Peter, "On Chomsky and the Two Cultures of Statistical Learning," http://www.norvig.com/chomsky.html

Nuzzo, Regina, "Profile of Frans B.M. de Waal,"*PNAS,* Vol.102 no.32, 8 August 2005, pp.11137-11139

Pennisi, Elizabeth, "Iowa Bobono Sanctuary Mired in Controversy," *ScienceInsider,* 18 September, 2012, http://www.news.sciencemag.org/sceinceinsider/2012/09/bonobo-hope-sanctuary

Petitto, Laura Ann, "Impish chimp changed science," *McGill Reporter,* 2013. http://www.mcgill.ca/reporter/32/13/chimpsky

Rehm, Hubert, Dietrich Gradmann, "Plant neurobiology: Intelligent Plants or Stupid Studies,"*Lab Times*, 3, 2010, pp.30-32

Roffman, Itai, "Itai Roffman's eulogy for Panbanisha," downloaded 23 November, 2012. http://panbanisha.org/itai-roffman-eulogy-for-panbanisha

Rumbuagh, Duane and William Fields, "Use of Human Language By Captive Great Apes," Great Ape Trust, downloaded 22 November 2012, http://www.greatapetrust.org/science/historyh-of-ape-language/use

Shapiro, James A., "In the Details... What?" *National Review*, September 16, 1996, pp. 62-65

Shapiro, James A.,"A Third Way," *Boston Review*, 10/5/2003, http://www.bostonreview.net/br22.1/shapiro

Science Daily, "Some Features of Human Face Perception Are Not Uniquely Human, Pigeon Study Shows." http://www.sciencedaily.com/releases/2011/04/110411171847.html

Science Daily, "Predatory Bacterial Swarm Uses Rippling Motion to Reach Prey," 15 November 2008. http://www.sciencedaily.com/releases/2008/10/081029121820html

Smith, Deborah, "The people that time forgot," *The Age*, 12 December, 2004, http://www.theage.com.au/news/Science/The-people-that-time-forgot/2004/12/12/1102786949312.html

Smith, Deborah, "Hobbits triumph tempered by tragedy," *Sydney Morning Herald*, 5 March 2005

Somers, James, "The Man Who Would Teach Machines to Think," *The Atlantic*, November, 2013 http://www.theatlantic.com/magazine/archive/2013/11/the-man-who-would-teach

Stewart, Terrence C,"A Technical Overview of the Neural Engineering Framework,"Centre for Theoretical Neuroscience technical report, Oct. 29, 2012

Taubes, Gary, "Evolving a Conscious Machine," *Discover Magazine*, June 1, 1998. http://www.discovermagazine.com/1998/jun/evolvingaconcsio1453

Teresi, Dick, "Discover Interview Lynn Margulis Says She's Not Controversial, She's right," *Discover Magazine*, April, 2011

Thacker, Eugene, "Networks, Swarms, Multitudes Part Two," *c.theory.net*, May 18, 2004, http://www.ctheory.ent/articles.aspx?id=423

The Daily Galaxy—Great Discoveres Channel: Sci, Space, Tech, "The Google Brain—Are Humans Entering a New Epoch of Evolution?" http://www.dailygalaxy.com/my_weblog/2013/05/the-google-brain-are-humans-entering

The Tutte Institute, "Biography of William T. Tutte,"Communications Security Establishment Canada, 10/31/2011, http://www.cse-cst.gc.ca/tutte/tutte-bio-eng.html

Vergano, Dan, "Fresh scandal over old bones," *USA Today*, 22 March 2005

Walkom, Thomas, "Authorities Absent in Marineland Saga," *Toronto Star*, 19 September, 2012, p.A6

Weston, Greg, Glenn Greenwald, Ryan Gallagher, "CSEC used airport Wi-Fi to track Canadian travelers: Edward Snowden documents," *CBC News*, Posted 30 January 2014, http://www.cbc.ca/news/politics/csec-used-airport-wi-fi-to-track-canadian-travellers

Wikipedia, "Boltzmann machine," downloaded 14/04/2013, http://www.wikipedia.org/wiki/Boltzmann_machine

Wikipedia, "Physarum polychephalum," downloaded at 25/09/2012. http://www.wikipedia.org/wiki/Physarum_polycephalum

Wikipedia, "Robert Yerkes," downloaded 22/08/2012. http://www.wikipedia.org/wiki/Robert_Yerkes

Wikipedia, "Francis Galton," downloaded 28/10/2012. http://www.wikipedia.org/wiki/Francis_Galton

Wikipedia, "Armillaria gallica", downloaded 31/12/2012 http://www.wikipedia.org/wiki/Amrillaria_gallica

Wikipedia, "Alan Turing,"downloaded 20/11/2013 http://www.wikipedia.org/wiki/Alan_Turing

Wikipedia, "Stevan Harnad,"downloaded 15/11/2013 http://www.wikipedia.org/wiki/Steven_Harnad

Wikipedia, "Emergence," downloaded 10/19/2012 http://www.wikipedia.org/wiki/Emergence

Wikipedia, "Neuromorphic engineering," downloaded 11/04/2013, http://www.wikipedia.org/wiki/Neuromorphic_engineering

Wikipedia, "Tommy Flowers," downloaded 17/04/2013. http://www.wikipedia.org/wiki/Tommy_Flowers

Wired, "The Man Behind the Google Brain: Andrew Ng and the Quest for the New AI," http://www.wired.com/wiredenterprise/2013/05/neuro-artificial-intelligence

Wong, Kate, "The Littlest Human," *Scientific American*, February 2005. http://www.www.sciam.com

Yahya, Harun, "Important Developments Regarding Flores Man, http//www.harunyahya.com/articles/70homo_floresiensis_2.html

Zimmer, Carl. "A Vision of the End," 1 May, 1993, *Discover Magazine*, http://www.discovermagazine.com/1993/may/avisionoftheend222/

Index

A

B

Praise for *Cloak of Green*

Amazing Environmental Expose, July 27 2013.

"This is a great book. Especially the chapters that analyze a few different groups; both corporate and NGO, and exposes them for being all about money, not really about protecting the environment. Some are exposed for just getting rich by advocating to protect the environment, others who publicly, through their PR, proclaim to be a great friend to environmental causes, yet are complete environmental pigs, pillaging the rain forest.... The only downside is, I lent my copy out a few years ago and never got it back, so I came here to buy a new copy and now it is about 2X the price. Obviously, it has become much more popular." — *Love Non-Fiction, Amazon Review*

Praise for *Bones: Discovering the First Americans*

"Awesome journey, well said. Vividly conveys the archeological world of set ways and blasts those concepts that have become entrenched in academia. My Ojibwe people have a creation story and it is considered 'myth'; if so, then the Bible is myth as well."
—*Goodreads review.*

"Pulls together the latest insights into the ancient mysteries clouding the origins of the 'First Americans' in clear and lucid prose [with] valuable original thinking.
—*Tony Hillerman*

"Dewar, a Canadian investigative journalist whose expose *Cloak of Green* probed the dark underbrush of environmental politics, returns here to dust off North American anthropology's skeletons in the closet... The truth is out there, but as Dewar argues, proper research has been thwarted time and again by stiff-necked academic careerism, the 'dirty water of ethnic politics' and just plain carelessness—bones mysteriously 'disappear' from museum storerooms, labs forget to conduct crucial DNA studies and so forth. This is popular rather than hard science, and there are gripping moments... Dewar is a keen observer of place and personality."
— *Publisher's Weekly.*

"With the flair of a mystery writer, Dewar explores the conflicting theories as they are influenced by academic and personal jealousies, government interference, ethnic concerns, mishandled artifacts—all the human and bureaucratic folly that have gotten in the way of the science. A revealing and informative look not only at the archaeology in question but at the convoluted, intricate, and very human difficulties involved in 'doing science,' this book is recommended." — *Ann Forister, Library Journal*

"Canadian journalist Dewar ... develops the provocative idea that the story of the colonization of the New World is as long as that of the Old and that the New and the Old Worlds may not be as separate in human origins as has been believed.... Controversial but not far-fetched, Dewar's narrative, written with the zest of a travel account, will intrigue amateur archaeologists and readers interested in American Indians."
— *Booklist*

"Elaine Dewar's work is aimed always at expanding mental horizons, preserving the sanctity of the national conscience, and telling a fresh and inventive story .... This is a must read."
— *Peter C. Newman*

"Compelling.... The peopling of the Americas is one of the epic chapters in the human story." — *Maclean's*

"Highly provocative.... Dewar's years of skulking around archaeological sites also uncovered a cast of unforgettable characters, including the real-life inspiration for the fictional, swashbuckling hero Indiana Jones."— *Ottawa Citizen*

"Elaine Dewar is a Canadian journalist with an excellent eye for hidden stories."
— *The Globe and Mail*

"The style of writing is lively and immediate and the book is hard to put down, once started...Dewar has done quite an outstanding job in bringing this story together, and it will reward your time spent with it." — *Amazon.com five star review.*

Praise for *The Second Tree: Of Clones, Chimeras and Quests for Immortality* Winner of The Writer's Trust (Hilary Weston) Prize for literary non-fiction

"Canadian journalist Dewar's book is as much a personal journey as an examination of the science, ethics and politics of cutting-edge biology. It's structured around the author's comprehensive set of interviews with leading figures in stem cell research and reproductive cloning technology; with ethicists attempting to come to grips with the complex moral issues these studies raise.... The scientific advances are breathtaking (companies breeding cloned farm animals, scientists growing heart muscle from embryonic stem cells) and the ethical questions perplexing (is manipulating a human egg immoral? when is the potential to become human replaced with the actuality of being human?). Dewar insinuates that many of those involved in this research are looking for personal glory." — *Publisher's Weekly*

"Like John McPhee's books, Dewar's personal quest to make sense of it all is always in the foreground. A former business reporter, she knows her stock options and her IPO's, but she keeps that side of the science in perspective. This book is not a simple prognostication (this is our future) nor is it a gee-whiz puff piece (look at the cool things we may be able to do) but it covers those topics in what seems to be to be a balanced way. Her prose is swift and lean: I was able to read the book in one sitting, and enjoyed it." — *Readersf. Amazon.*

"Alternately funny, terrifying and filled with wonder.... An important book about tough issues— a scary, miraculous story unfolding like the murky clouds behind the tree of life." — *Vancouver Sun*

"Breathtaking, a factual counterpart to Margaret Atwood's *Oryx and Crake*.... Should be on everyone's reading list." — *The Gazette (Montreal)*

"Dewar both entertains and clearly explains complex scientific concepts and the associated ethical and moral dilemmas."— *Winnipeg Free Press*

"Throughout this book, the reader will come into contact with science, history and historical science, mini-biographies of key people, business and conflicts ofinterest, patents, the private versus the public spheres of society, bioethics, eugenics and, of course, politics. Dewar (who admits she's a non-scientific type) frequently injects herself into the main narrative giving her impressions, observations, thoughts, emotions, and feelings. Some readers may see this as a negative attribute of this book but I, however, see it as a positive."
— *Stephen Pletko, Amazon.com Customer Review*

About the Author

Elaine Dewar has been propelled since childhood by boundless curiosity and the joy of storytelling. Her first book, *Cloak of Green,* delved into the dark side of environmental politics and became an underground classic. *Bones: Discovering the First Americans* was a national bestseller. *The Second Tree: of Clones, Chimeras, and Quests for Immortality* won Canada's premier literary nonfiction prize from the Writers' Trust. Dewar has been called "one of Canada's best muckrakers" and aspires to be a happy warrior for the public good. She lives with her husband in Toronto.

Made in United States
Orlando, FL
04 November 2026

90207802R00381